CDMA for Wireless Personal Communications

The Artech House Mobile Communications Series

John Walker, Series Editor

Advanced Technology for Road Transport: IVHS and ATT, Ian Catling, editor

An Introduction to GSM, Siegmund M. Redl, Matthias K. Weber, Malcolm W. Oliphant

CDMA for Wireless Personal Communications, Ramjee Prasad

Cellular Digital Packet Data, Muthuthamby Sreetharan and Rajiv Kumar

Cellular Mobile Systems Engineering, Saleh Faruque

Cellular Radio: Analog and Digital Systems, Asha Mehrotra

Cellular Radio Systems, D. M. Balston, R. C. V. Macario, editors

Cellular Radio: Performance Engineering, Asha Mehrotra

Digital Beamforming in Wireless Communications, John Litva, Titus Kwok-Yeung Lo

GSM System Engineering, Asha Mehrotra

Introduction to Radio Propagation for Fixed and Mobile Communications, John Doble

Land-Mobile Radio System Engineering, Garry C. Hess

Mobile Communications in the U.S. and Europe: Regulation, Technology, and Markets, Michael Paetsch

Mobile Antenna Systems Handbook, K. Fujimoto, J. R. James

Mobile Data Communications Systems, Peter Wong, David Britland

Mobile Information Systems, John Walker, editor

Personal Communications Networks, Alan David Hadden

RF and Microwave Circuit Design for Wireless Communications, Lawrence E. Larson, editor

Smart Highways, Smart Cars, Richard Whelan

Spread Spectrum CDMA Systems for Wireless Communications, Savo G. Glisic and Branka Vucetic

Transport in Europe, Christian Gerondeau

Understanding GPS: Principles and Applications, Elliott D. Kaplan, editor

Vehicle Location and Navigation Systems, Yilin Zhao

Wireless Communications in Developing Countries: Cellular and Satellite Systems, Rachael E. Schwartz

Wireless Communications for Intelligent Transportation Systems, Scott D. Elliott, Daniel J. Dailey

Wireless Data Networking, Nathan J. Muller

Wireless: The Revolution in Personal Telecommunications, Ira Brodsky

For a complete listing of *The Artech House Telecommunications Library*, turn to the back of this book.

CDMA for Wireless Personal Communications

Ramjee Prasad

Artech House
Boston • London

Library of Congress Cataloging-in-Publication Data
Prasad, Ramjee.
CDMA for wireless personal communications / Ramjee Prasad.
p. cm.
Includes bibliographical references and index.
ISBN 0-89006-571-3 (alk. paper)
1. Code division multiple access. 2. Wireless personal communications.
I. Title.
TK5103.45.P73 1996
621.3845—dc20 95-53774
CIP

British Library Cataloguing in Publication Data
Prasad, Ramjee
CDMA for wireless personal communications
1. Code division multiple access 2. Wireless communication systems
I. Title
621.3'82

ISBN 0-89006-571-3

International Standard Book Number: 0-89006-571-3
Library of Congress Catalog Card Number: 96-53774

10 9 8 7 6 5 4 3

To
my parents
Chandrakala and Sabita Nath,
who inspired me in my childhood in making
deductions of geometry that taught me to think of
the horizon of current knowledge of science and technology.

Later on:

To my parents-in-law
Kamal and Aditya Narain,
who motivated me to fly beyond the horizon that taught me to write the thoughts.

I dedicate this book to these four souls to rest with peace in heaven.

Contents

Preface

सहजं कर्म कौन्तेय सदोषमपि न त्यजेत् ।
सर्वारम्भा हि दोषेण धूमेनाग्निरिवावृताः ॥

saha-jaṁ karma kaunteya
sa-doṣam api na tyajet
sarvārambhā hi doṣeṇa
dhūmenāgnir ivāvṛtāḥ

Every endeavour is covered by some fault, just as fire is covered by smoke. Therefore one should not give up the work born of his nature, even if such work is full of fault.

-The Bhagvad-Gita (18.48)

Code division multiple access (CDMA) and hybrid CDMA are discussed in this book as they apply to wireless personal communications. Hybrid CDMA is defined as a combination of any two types of CDMA (e.g., direct sequence DS and frequency hopping FH) or CDMA with any other contention (e.g., ALOHA) or contentionless (e.g., time division multiple access, TDMA) multiple access protocols or CDMA with any other techniques (e.g., orthogonal frequency division multiplexing OFDM). Techniques (e.g., interference cancellation IC, joint detection JD) are introduced to enhance the performance of a wireless personal communication system. CDMA is treated here from the point of view of multiple access protocol. Therefore a separate chapter is devoted to multiple access protocols. This is the first book to give a broad treatment to CDMA as a multiple access protocol for a wireless channel.

This book is the outcome of research contributions by the master, doctoral, and postdoctoral candidates of the Telecommunications and Traffic-Control Systems Group of Delft University of Technology (DUT), The Netherlands, under my supervision. The idea of converting our research contributions into a book came to me in 1992 when I gave a course on "Code Division Multiple Access (CDMA) for Wireless Personal Communications" for the Advanced Studies in Electrical Engineering (ASEE) programme. I realized a shortage of teaching material for this course. The teaching material, which I prepared in 1992 with the help of Casper van den Broek and Michel G. Jansen, was revised during the lecture preparations in 1993, 1994, and 1995. Until now no such book was available on the market. An attempt has been made to fill this gap in the literature.

The research program in the field of CDMA for wireless communications started in DUT in 1989. My first master student for the CDMA studies was Howard Sewberath Misser. He studied direct-sequence spread-spectrum multiple access for indoor wireless communications. Later on Casper A.J. Wijffels and several other students investigated throughput and delay of a CDMA network. The effect of imperfect power control has been studied by Michel G. Jansen. Richard D.J. van Nee and some other students applied the investigation of direct-sequence spread-spectrum to land mobile satellite communications.

In the beginning the study of the direct-sequence spread-spectrum was a very high priority; the research activities in hybrid direct-sequence (DS)/slow frequency (SFH) CDMA had been carried out at a low profile by some master students. Luc Vandendorpe from the Catholique University of Louvain-la-Neuve, Belgium, who worked with me as a postdoctoral research scholar, made interesting contributions in hybrid DS/SFH CDMA. Later on René G.A. Rooimans continued this research activity. He is still active in the investigation of hybrid DS/SFH CDMA with Omar Fatah.

Before the CDMA investigations at DUT, significant research activities had been carried out in the field of random multiple access protocols with the capture effect for wireless communications which was originally developed by Jens C. Arnbak. Jens and I investigated the inhibit sense multiple access (ISMA) protocol and concluded that ISMA is a good random multiple access scheme for indoor wireless communications. Later we also concluded that direct-sequence CDMA is a good access scheme for indoor wireless communications. Therefore we thought that a combination of CDMA and ISMA could be a very good hybrid CDMA protocol for indoor wireless computer communications. This subject was deeply studied by Jos A.M. Nijhof, Huub R.R. van Roosmalen, and master students. Random multiple access protocols in the wireless environment are still one of the key research topics at DUT. Casper van den Broek is currently studying several multiple access protocols for his doctoral research.

Bas W. 't Hart studied code synchronization. Interference cancellation was investigated by Roy K. Sukdeo with Zeke Bar-Ness from the New Jersey Institute of Technology, USA, who was on sabbatical at DUT during 1993-1994. Gerard J.M.

Janssen and Frank van der Wijk developed the concept of groupwise successive interference cancellation. Coexistence of CDMA and TDMA was investigated by Michael B.K. Widjaja. Joint detection CDMA has been investigated in great depth at the University of Kaiserslautern, Germany, by the radio research group headed by Walter Baier. This particular topic was contributed for the sake of completeness of the book by Peter Jung. Research activity in the field of multicarrier CDMA was initiated by Jean-Paul M.G. Linnartz. Later, this subject was investigated in great detail by Shinsuke Hara from the University of Osaka, Japan, who is a visiting scientist at DUT, doing postdoctoral research with me.

Completing this book gives me the same pleasure as a gardener feels upon seeing his garden full of flowers. Since this book has been completed with the help of several research contributions, I have tried my best to make each chapter quite complete in itself. This book provides an overview of CDMA for the benefit of young research students, engineers, and scientists in the field of wireless personal communications. This book will help generate many new research problems and solutions for future wireless personal communications. I cannot claim that this book is errorless. Any remarks to improve the text and correct the errors would be highly appreciated.

Acknowledgments

I would like to express my heartfelt gratitude to colleagues and students without whom this book would have never been completed. Jens Arnbak, with whom I had started this book, gave me full encouragement to finish it. Adriaan Kegel, Jos Nijhof, Dirk Sparreboom, Han Reijmers, Jean-Paul Linnartz, and Gerard Janssen supported me in supervising many graduate students whose results have been used in this book. Nel Kay helped me plan the book and gave her full support while I finished it. Eefje Ooms gave the final shape of the book. During the preparation of the text, Eefje Ooms, Antoinette Steinman, Sophia Chlimintzas, Claudia Hoogervorst, and Esmeralda van Dijke prepared the typescript of the book. Jane Zaat improved the English language.

I had interesting discussions with Zeke Bar-Ness from the New Jersey Institute of Technology, USA.Walter Baier and Peter Jung from the University of Kaiserslautern, Germany, are deeply acknowledged for their valuable contributions. The postdoctoral research scholars Luc Vandendorpe from the Catholique University of Louvain-la-Neuve, Belgium, and Shinsuke Hara from the University of Osaka, Japan, are acknowledged for their contributions. Richard van Nee, Homayoun Nikookar, Casper van den Broek, and Mqhele Dlodlo are thanked for use of their doctoral research results in the book. Michel G. Jansen, Frank van der Wijk, and Bas 't Hart, predoctoral students, are especially thanked for their valuable contributions.

Last but not least, the following master students are greatly acknowledged in generating several new results which have been used throughout in this book: H.S. Misser, C.A.J. Wijffels, E. Walther, R.R.J. Ponson, M.G. Jansen, H.R.R. van Roosmalen, A. Daryanto, R.G.A. Rooimans, P.H. Sinaga, H. Satyanegara, M.B.K. Widjaja, S. Srivastava, A. Soelaksono, R.K. Sukdeo, D.Y. Zhang, F. Çakmak, R.N. van Wolfswinkel, O. Fatah, H.L.A. Le, T.H. Lee, S. Grujev, J. Wigard, Z.Q. Lie, A.R. Prasad, N.R.Prasad, H. Çakil, A. Nollen, and F. Harmsze.

Chapter 1
Introduction

With the current rapid growth of technology, it can now emphatically be said that the objective of today's communication engineers to achieve a future wireless personal communication (FWPC) system, which was yesterday's myth (before 1970), will be tomorrow's reality (beyond 2000). FWPC systems will convert the already shrinking world into a global village. A future wireless personal communication system, defined as being the ultimate goal of today's communication engineers, will provide communication services from any person to any person in any place at any time without any delay in any form through any medium by using one pocket-sized unit at minimum cost with acceptable quality and security through using a personal telecommunication reference number.

1.1 FWPC EVOLUTION

A family tree for the FWPC system is shown in Figure 1.1 [1]. The objectives of the research and development of FWPC systems are focused in three technological platforms, namely, universal mobile telecommunication systems (UMTS), mobile broadband systems (MBS), and wireless customer premises networks (WCPN) [2]. UMTS is a multi-function, multi-service, multi-application digital system, evolving from currently operational second-generation systems, and several other ongoing second-generation systems.

The first generation was introduced in 1980 in analog form to provide local mobile speech services and it was then further extended to nationwide coverage. Various standard systems were developed worldwide: AMPS (advanced mobile phone service) in the United States, NTT (Nippon Telephone and Telegraph) systems in Japan, TACS (total access communications system) in the United Kingdom, NMT (Nordic Mobile Telephones) in European countries, and so on.

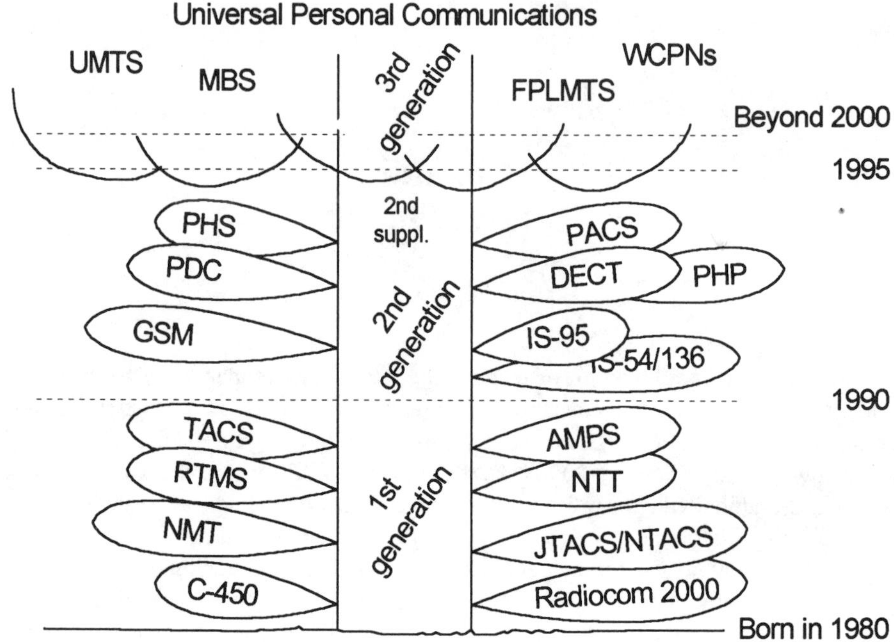

Figure 1.1 Family tree of wireless personal communications systems.

Fast user growth was observed. It penetrated up to 10% of the calls in North America, Western Europe and Japan. The access technique used was frequency division multiple access (FDMA). Capacity and quality were the major problems in the first-generation systems. In addition, systems were not compatible.

The advancement in digital technology gave birth to Pan-European digital cellular mobile (DCM) GSM (Groupe Spécial Mobile) systems in Europe, PDC (personal digital cellular) systems in Japan, and IS-54/136 and IS-95 in North America, which are the second-generation systems. Time division multiple access (TDMA) is used as the access technique, except for IS-95, which is based on CDMA (code division multiple access). The second-generation systems provide digital speech and short message services. These services are expected to penetrate to more than 20% of the call population. GSM has become deeply rooted in Europe and in several other countries worldwide [3]. Now GSM stands for "global systems for mobile communications." The development of new digital cordless technologies gave birth to the second supplement generation systems, namely, PHS (personal handyphone systems, formerly PHP) in Japan, DECT (digital European cordless telephone) in Europe, and PACS (personal access communication services) in North America. It may increase the call penetration depth up to 30% and introduce many new services. Although the second generation and its supplement will cover local, national, and international services, it will still have one major drawback in terms of a universal service facility.

The third-generation is expected to be deployed by the year 2000 via universal personal communications systems (UPCS), which will provide universal speech services and local multimedia services. It is expected that the third-generation system will penetrate up to 50% of the telecommunication services population. The third-generation personal communication systems are in the process of development worldwide by the ITU (International Telecommunications Union) within the framework of the FPLMTS (future public land mobile telecommunications systems) activities. In Europe this is supported by the UMTS program within the European community. Both FPLMTS and UMTS programs are tightly related and expected to lead to consistent and compatible systems. Figure 1.2 shows the possible configuration for UMTS subnetworks and fixed networks.

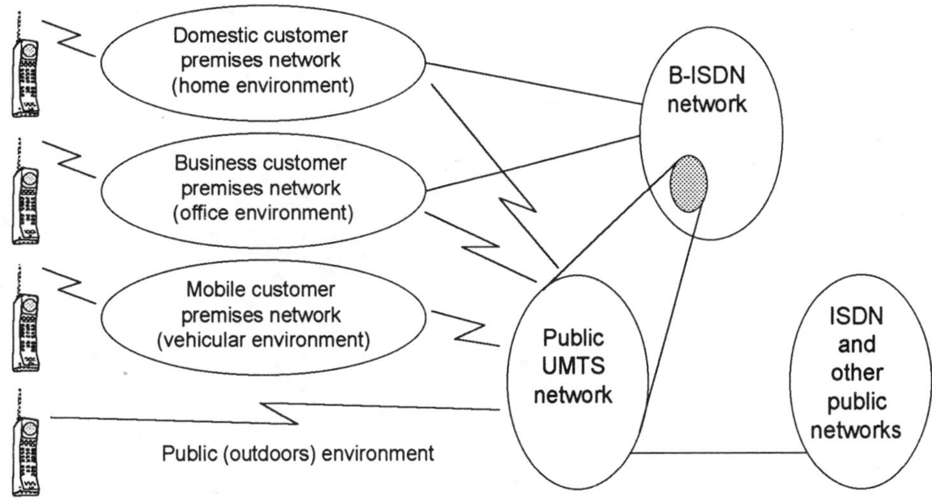

Figure 1.2 UMTS possible service configuration.

UMTS is intended to provide a wide range of mobile services to the users via a range of mobile terminals that enable the use of the pocket telephone in almost any location, indoor or outdoor, in home, office, or street. North America and Japan are also equally engaged in developing the third-generation systems in the UMTS direction. Looking at the research trends going on in Europe, Japan, and North America, it appears that the future access scheme will be either CDMA or one of the hybrid multiple access schemes based mainly on the combination of CDMA and TDMA/FDMA.

The only drawback will be lack of rural (modern) communication coverage particularly in developing countries. Hopefully, this will be solved at the beginning of the 21st century. Land-mobile satellite communications could solve this problem.

Figure 1.3 depicts a view of a global village integrating terrestrial pico-, micro-, and macrocellular systems and satellite networks. An integrated satellite-terrestrial cellular system is the concept of FWPC; that is, UPCS.

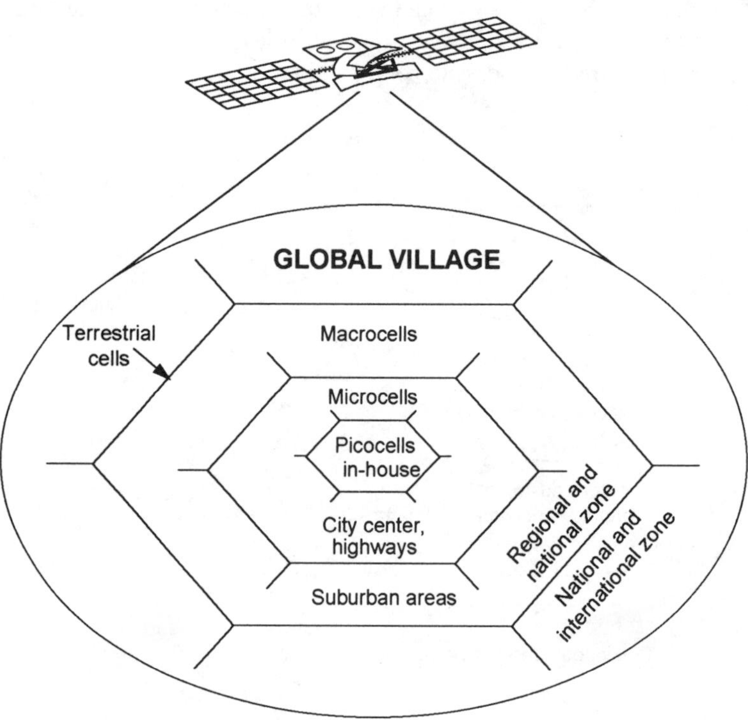

Figure 1.3 View of a future global village.

1.2 CDMA AS A PROTOCOL

The development of an appropriate multiple access technique is going to be one of the major challenges. In fact one of the most important aspects in achieving the ultimate goal of FWPC systems is the proper choice of a multiple access technique. The multiple access protocols are classified into three groups: (1) contentionless protocols, (2) contention protocols, and (3) CDMA protocols [4]. The CDMA protocols are rightly placed between the contentionless (TDMA, FDMA) and contention (ALOHA, ISMA) protocols [5]. The necessary reasons for such placement of CDMA are as follows: (1) it can achieve synchronous communications, (2) the frequency band for each terminal does not need partitioning, (3) simultaneously transmitting two or more terminals can be captured successfully, and (4) the probability of packet success decreases with the number of simultaneously transmitting terminals. Propagation measurements have shown that the personal wireless channels, which include macrocells, microcells, picocells, satellite channels, and overlay channels, are highly hostile in nature and subject to multipath fading, shadowing, near-far effect, etc. A CDMA system may overcome some of these problems depending on the type of CDMA used. For example, a DS (direct-sequence) CDMA system is highly resistant to multipath propagation, a FH (frequency hopping) CDMA system combats the near-far effects, and a hybrid DS/FH CDMA system can combine the effectiveness of direct-sequence systems against multipath fading with the avoidance property of frequency hopping systems to reduce the near-far problem.

In a star connected CDMA network, the base station consists of a bank of spread-spectrum transmitters/receivers, one for each active user. Each user is assigned a unique spread-spectrum code sequence for DS CDMA or a unique hopping pattern for FH CDMA, or a unique spread-spectrum code sequence and a hopping pattern for hybrid DS/SFH CDMA. This means that CDMA offers simultaneous transmission by multiple terminals with successful reception of packets. It may be noted that in the case of ALOHA there is collision if there are simultaneous transmissions. Thus ALOHA is a special limiting case of CDMA systems with only one packet transmission at a time with successful reception.

However in the case of CDMA there is also a threshold value for simultaneous active users. When the number of simultaneous users exceeds the threshold value, all packets are lost due to excessive errors. The lost packets are rescheduled according to a retransmission delay distribution.

1.3 FUTURE CDMA

Recently extensive investigations have been carried out into the application of a CDMA system as an access method for the FWPC systems [4-82] because it appears that CDMA is a strong candidate for FWPC. Some positive points of the CDMA systems are (1) soft capacity (i.e., there is no hard limit on the system capacity), (2) soft handover,

(3) inherent diversity, (4) enhanced spectrum efficiency, (5) unity cluster size, and (6) simplified frequency planning and system deployment. Therefore in the United States, an alternative to digital AMPS (IS-95) is being considered using a CDMA system. However, it is still a big question as to whether a CDMA system is the right choice for the global FPLMTS. Several future CDMA system concepts have been developed for third-generation personal communications, such as, hybrid CDMA/ISMA [34], broadband-CDMA [67], CDMA/OFDM (orthogonal frequency division multiplexing) [69], multirate CDMA [73], and JD (joint detection)-CDMA [78]. A combination of direct-sequence CDMA and p-persistent ISMA is proposed for indoor wireless communications. The combination of these two protocols is called the hybrid CDMA/ISMA protocol. Broadband code division multiple access (B-CDMA) is a technique which allows PCS operation in the cellular frequency band in conjunction with existing cellular service, as well as in the PCS band (1850-1990 MHz in USA) [68]. The combination of CDMA and OFDM (orthogonal frequency division multiplexing is an approach that transforms a highly selective wideband channel into a large number of nonselective narrowband channels which are frequently multiplexed) allows for optimal detection performance, use of the available spectrum in an efficient way, retention of many advantages of a CDMA system, and exploitation frequency diversity [70]. In addition, it allows simple cell separation by using frequency hopping, and a simple hardware realization. The CDMA/OFDM technique is also known as multitone CDMA [72] and multicarrier CDMA [71]. JD-CDMA has evolved from GSM systems by introducing a CDMA component [78]. JD-CDMA applies joint detection techniques at the receiver and provides a flexible and adaptive allocation of radio resources. In Europe, a multinational research project named CODIT (UMTS code division test bed) had been set up within the European RACE (research and development in advanced communication technologies in Europe) program. A multirate DS-CDMA has been proposed [73-75]. The key feature of the proposed system is the coexistence of different chip rates or signal bandwidths on the physical radio channel.

One of the major problems with introducing a new system is that the new must coexist with the old. This means that the performance of CDMA systems in the presence of existing TDMA carrier-based systems is vitally important to the way in which the potential benefits of the CDMA approach can be taken advantage of.

1.4 PLAN OF THE BOOK

The book consists of 11 chapters. It concentrates mainly on the study of a CDMA system from the multiple access point of view. System comparisons are made between wired (cable) and wireless networks. General problems of wireless indoor and outdoor communications, including mobile satellites, are addressed as they relate to the CDMA protocol. Basic propagation characteristics are summarized.

An overview of multiple access techniques is presented in Chapter 2. With the proper classification of various access methods, the behavior of CDMA may be easily recognized. TDMA, FDMA, well-known random multiple access protocols such as ALOHA, CSMA (carrier sense multiple access), ISMA, PRMA (packet reservation multiple access), and stack algorithm, and CDMA are addressed with their specific features. Some interesting results of slotted ALOHA and ISMA in a radio hostile environment with capture effects are discussed.

The CDMA system concept is conceived in Chapter 3. Direct-sequence (DS), frequency hopping (FH), time hopping (TH), and hybrid CDMA are explained using block diagrams. Their advantages and disadvantages are discussed.

Power control problems are addressed in Chapter 4. First, the problem is examined analytically by using a generalized traffic distribution that power control is essential for a CDMA system. Next the effect of imperfection in the transmitted power control (TPC) on performance is investigated. Computational results are obtained for both single and multiple cells in a reverse link. Then the performance of a forward link is analyzed. Two types of power control schemes are also introduced to study the performance of the forward link.

Chapter 5 starts with the application of the CDMA systems. First the application of a CDMA system is discussed for indoor wireless communications. At the beginning, the propagation characteristics of an indoor channel are reviewed. An analytical model for evaluating the bit error probability for DS CDMA systems using DPSK modulation is developed. The effect of FEC (forward error correction) coding and diversity techniques on the system performance is investigated. Finally, using a computer simulation, the analytical results are verified. Throughput is also evaluated.

The outdoor cellular CDMA systems are analyzed in Chapter 6 by using DS CDMA with BPSK and DPSK modulation considering FEC coding, selection diversity, and maximal ratio combining for both macro- and microcells. Bit error probability, throughput, and delay are evaluated and discussed.

Similar to indoor and outdoor CDMA systems, mobile satellite CDMA systems are introduced in Chapter 7 considering direct-sequence spread-spectrum, BPSK modulation, selection diversity, and maximal ratio combining for the log-normal shadowed Rician fading channel. The performance is investigated by evaluating bit error probability, throughput, and delay. The performance of narrowband slotted ALOHA and direct-sequence spread-spectrum CDMA is compared.

Chapter 8 presents the analysis of a hybrid DS/SFH CDMA system for indoor radio communications. Three types of modulations are considered, namely, BPSK, QPSK, and DPSK. The effects of FEC coding and diversity techniques on the performance parameters of bit error probability, throughput, and delay are also studied.

Throughput, delay, and stability of a slotted DS CDMA protocol are analyzed using a discrete time Markov chain model in Chapter 9.

Combining random multiple access protocol and a CDMA scheme may result in enhanced performance and reduced complexities. Such a combination is presented in Chapter 10 by introducing hybrid CDMA/ISMA protocols. First, the performance is analytically obtained by using a Markov chain model for DPSK modulation. Then, the analytical results are verified using computer simulation. The hybrid system is described through considering indoor wireless computer communications.

Even if a CDMA system is accepted for future radio communications, such a system can be implemented only in phases. It is therefore important to study the coexistence of CDMA and TDMA systems. The coexistence of these two systems is introduced in Chapter 11. In addition, important issues related to CDMA systems, such as code synchronization and interference cancellation, are also discussed.

Finally, MC-CDMA and JD-CDMA are briefly explained in the light of future multiple access schemes for the FWPC systems.

REFERENCES

[1] R. Prasad, "Research challenges in future wireless personal communications: microwave perspective," *Proc. 25th European Microwave Conference* (Keynote opening address), Bologna, Italy, pp. 4-11, 4-7 September 1995.

[2] Advanced Communications Technologies and Services (ACTS): *Workplan, European Commission Directorat General XIII: Telecommunications, Information Market and Exploitation of Research,* August 1997.

[3] J.C. Arnbak, "The European (R)evolution of wireless digital networks," *IEEE Comm. Mag.*, Vol. 31, pp. 74-82, September 1993.

[4] R. Prasad, "An overview of code division multiple access systems for universal personal communication networks," *Proc. Fourteenth Symposium on Information Theory in the Benelux*, Veldhoven, The Netherlands, pp. 206-212, May 1993.

[5] R. Prasad, "Throughput analysis of slotted code division multiple access for indoor radio channels," *Proc. IEEE International Symposium on Spread-Spectrum Techniques and Applications*, King's College, London, pp. 12-17, September 1990.

[6] R. Prasad, H.S. Misser and A. Kegel, "Indoor radio communication in a Rician channel using direct-sequence spread-spectrum multiple access with selection diversity," *Proc. IEEE International Symposium on Spread-Spectrum Techniques and Applications*, King's College, London, pp. 6-11, September 1990.

[7] R. Prasad, H.S. Misser and A. Kegel, "Performance analysis of direct-sequence spread-spectrum multiple-access communications in an indoor Rician-fading channel with DPSK modulation," *Electron. Lett.*, Vol. 26, pp. 1366-1367, August 1990.

[8] K.L.A. Sastry and R. Prasad, "Throughput analysis of CDMA with DPSK and BPSK modulation and diversity in indoor radio channels," *Proc. Third IEE Conference on Telecommunications*, Edinburgh, U.K., pp. 90-94, March 1991.

[9] H.S. Misser, C.A.F.J. Wijffels and R. Prasad, "Throughput analysis of CDMA with DPSK modulation and diversity in indoor Rician fading radio channels," *Electron. Lett.*, Vol. 27, pp. 601-602, March 1991.

[10] H.S. Misser and R. Prasad, "Spectrum efficiency of a mobile cellular radio system using direct sequence spread-spectrum," *Proc. Vol. III, Computer and Communication, IEEE Tencon '91*, New Delhi, India, pp. 91-94, August 1991.

[11] R. Prasad, "Capacity of a slotted code division multiple access system for wireless inhouse network communications," *Proc. Vol. III, Computer and Communication, IEEE Tencon '91*, New Delhi, India, pp. 86-90, August 1991.

[12] R.D.J. van Nee, H.S. Misser and R. Prasad, "Bit error probability for direct sequence spread-spectrum in a shadowed Rician fading land mobile satellite channel with diversity," *Proc. 1991 IEEE International Symposium on Personal, Indoor and Mobile Radio Communications*, King's College, London, pp. 125-130, September 1991.

[13] R. Prasad and C.A.F.J. Wijffels, "On performance comparison of slotted CDMA in macro-, micro-, and picocellular environment," European Cooperation in the Field of Scientific and Technical Research, Leidschendam, COST 231 TD (91)-72, 25-27 August 1991.

[14] R. Prasad and C.F.J. Wijffels, "Performance analysis of a slotted CDMA system for indoor wireless communication using a Markov chain model," *Proc. IEEE Global Telecommunications Conference*, Phoenix, Arizona, pp. 1953-1957, December 1991.

[15] R. Prasad, E. Walther and R. Ponson, "Hybrid SFH/DS CDMA for macro-, micro-, and picocellular systems," European Cooperation in the Field of Scientific and Technical Research, Vienna, COST 231 TD (92) 002, January 1992.

[16] R.D.J. van Nee, H.S. Misser and R. Prasad, "Direct-sequence spread-spectrum in a shadowed Rician fading land-mobile satellite channel," *IEEE J. Selected Areas Comm.*, Vol. 10, pp. 350-357, February 1992.

[17] R. Prasad, M. Jansen and A. Kegel, "Cellular DS CDMA systems with imperfect power control, part 1: reverse link," Leeds, COST 231 TD (92) 48, April 1992.

[18] R. Prasad, A. Kegel and M.G. Jansen, "Effect of imperfect power control on cellular code division multiple access system," *Electron. Lett.*, Vol. 28, pp. 848-849, April 1992.

[19] H.S. Misser and R. Prasad, "Bit error probability evaluation of a microcellular spread-spectrum multiple access system in a shadowed Rician fading channel," *Proc. IEEE 42nd VTS Conference*, Denver, Colorado, pp. 439-442, May 1992.

[20] R. Prasad, M. Jansen and A. Kegel, "Power control in cellular DS CDMA systems, part II: forward link," European Cooperation in the Field of Scientific and Technical Research, COST 231 TD (92) 092, Helsinki, 8-11 September 1992.

[21] R. Prasad, E. Walther and R. Ponson, "Performance analysis of hybrid SFH/CDMA networks for personal communication systems," *Proc. Third IEEE International Symposium on Personal, Indoor and Mobile Radio Communications*, Boston, U.S.A., pp. 362-366, October 1992.

[22] R. Prasad, M. Jansen and A. Kegel, "Performance analysis of a direct sequence code division multiple access system considering multiple cells," *Proc. IEEE Symposium on Spread-Spectrum Techniques and Applications (ISSSTA'92)*, Yokohama, Japan, pp. 15-18, November/December 1992.

[23] R. Prasad and E. Walther, "Performance analysis of a hybrid slow frequency hopping/direct sequence CDMA system in an indoor Rician fading channel," *Proc. IEEE Symposium on Spread-Spectrum Techniques and Applications (ISSSTA'92)*, Yokohama, Japan, pp. 119-122, November/December 1992.

[24] R. Prasad, C.A.F.J. Wijffels and K.L.A. Sastry, "Performance analysis of slotted CDMA with DPSK modulation diversity and BCH-coding in indoor radio channels," *Archiv für Elektronik und Übertragungstechnik*, Vol. 46, No. 6, pp. 375-382, November 1992.

[25] H.S. Misser, A. Kegel and R. Prasad, "Monte Carlo simulation of direct sequence spread-spectrum for indoor radio communication in a Rician fading channel," *IEE Proc.-I*, Vol. 139, No. 6, pp. 620-624, December 1992.

[26] L. Vandendorpe and R. Prasad, "Performance analysis of hybrid slow frequency hopping/direct sequence spread-spectrum communication systems with PSK modulation in an indoor wireless environment," Barcelona, Spain, COST 231 TD (93)-05, January 1993.

[27] M.G. Jansen and R. Prasad, "Throughput analysis of slotted CDMA with imperfect power control," IEE Professional Group E8 (Radiocommunications Systems), *Colloquium on Spread-Spectrum Techniques for Radio Communication Systems,* London, U.K., April 1993.

[28] R. Prasad, M.G. Jansen and J.P. van Deursen, "Frequency hopping slotted ALOHA in a shadowed radio environment," *Proc. IEEE Fourteenth Symposium on Information Theory in the Benelux,* Veldhoven, The Netherlands, May 1993.

[29] R. Prasad, H.S. Misser and A. Daryanto, "Throughput and delay analysis of a slotted DS/SS CDMA system in a shadowed Rician channel," *Proc. 43rd IEEE Vehicular Technology Conference,* Secauscus, NJ, U.S.A., pp. 456-459, May 1993.

[30] R. Prasad and M.G. Jansen, "Near-far-effects on performance of DS/SS CDMA systems for personal communication networks," *Proc. 43rd IEEE Vehicular Technology Conference*, Secauscus, NJ, U.S.A., pp. 710-713, May 1993.

[31] C.A.F.J. Wijffels, H.S. Misser and R. Prasad, "A microcellular CDMA -system over slow and fast Rician fading radio channels with forward error correction coding and diversity," *IEEE Trans. Vehicular Technol.,* Vol. 42, pp. 570-580, November 1993.

[32] R. Prasad, M.G. Jansen and A. Kegel, "Capacity analysis of a cellular direct sequence code division multiple access system with imperfect power control," *IEICE Transactions on Communications,* Vol. E76-B, pp. 894-905, August 1993.

[33] R.D.J. van Nee and R. Prasad, "Spread-spectrum path diversity in a shadowed Rician fading land-mobile satellite channel," *IEEE Trans. Vehicular Technol.*, Vol. 42, pp. 131-135, May 1993.

[34] H. van Roosmalen, J.A.M. Nijhof and R. Prasad, "A hybrid/CDMA protocol for computer communications in an indoor radio environment," *Fourth IEEE Int. Symp. on PIMRC 1993,* Yokohama, Japan, pp. 627-631, September 1993.

[35] L. Vandendorpe, R. Prasad and R. Rooimans, "Hybrid slow frequency hopping/direct sequence spread-spectrum communication systems with B- and Q-PSK modulation in an indoor wireless environment," *Fourth IEEE Int. Symp. on PIMRC 1993,* Yokohama, Japan, pp. 498-502, September 1993.

[36] H. van Roosmalen, J. Nijhof and R. Prasad, "Performance analysis of a hybrid CDMA/ISMA protocol for indoor wireless computer communications," *IEEE JSAC on code Division Multiple Access Networks*, Vol. 12, pp. 909-916, June 1994.

[37] R. Prasad, H.S. Misser and A. Kegel, "Performance evaluation of direct-sequence spread-spectrum multiple-access for indoor wireless communication in a Rician fading channel," *IEEE Trans. Comm.*, Vol. 43, pp. 581-592, February/March/April 1995.

[38] M.G. Jansen and R. Prasad, "Capacity, throughput, and delay analysis of a cellular DS CDMA-system with imperfect power control and imperfect detorization," *IEEE Trans. Vehicular Technol.,* Vol. 44, pp. 67-75, February 1995.

[39] W.C.Y. Lee, "Power control in CDMA," *Proc. 41st IEEE Vehicular Technology Conference*, U.S.A., pp. 77-80, May 1991.

[40] W.C.Y. Lee, "Overview of cellular CDMA," *Proc. 41st IEEE Vehicular Technology Conference*, U.S.A., pp. 291-302, May 1991.

[41] K.S. Gilhousen, I.M. Jacobs, R. Padovani, A.J. Viterbi, L.A. Weaver and C.E. Wheatley, "On the capacity of a cellular CDMA-system," *IEEE Trans. Vehicular Technol.*, Vol. 40, pp. 302-312, May 1991.

[42] A. Salmasi and K.S. Gilhousen, "On the system design aspects of code division multiple access (CDMA), applied to digital cellular and personal communications networks," *Proc. 41st IEEE Vehicular Technology Conference*, U.S.A., pp. 57-62, May 1991.

[43] R.L. Pickholtz, L.B. Milstein and D.L. Schilling, "Spread-spectrum for mobile communications," *IEEE Trans. Vehicular Technol.*, Vol. 40, pp. 313-322, May 1991.

[44] D.L. Schilling, L.B. Milstein, R.L. Pickholtz, M. Kullbach and F. Miller, "Spread-spectrum for commercial communications," *IEEE Comm. Mag.*, pp. 66-79, April 1991.

[45] M.A. Beach, A. Hammer, S.A. Allpress, J.P. McGeehan and A. Bateman, "An evaluation of direct sequence CDMA for future mobile communication networks," *Proc. 41st IEEE Vehicular Technology Conference*, U.S.A., pp. 63-70, May 1991.

[46] R. Kohno, H. Imai, M. Hatori and S. Pasupathi, "An adaptive canceller of co-channel interference for spread-spectrum multiple access communication networks in a power line," *IEEE JSAC*, Vol. 8, pp. 691-699, May 1990.

[47] C.L. Weber, G.K. Huth and B.H. Batson, "Performance considerations of code division multiple-access systems," *IEEE Trans. Vehicular Technol.*, Vol. 30, pp. 3-9, February 1981.

[48] M. Kavehrad and P.J. McLane, "Performance of low-complexity channel coding and diversity for spread-spectrum in indoor wireless communications," *AT&T Tech. J.*, Vol. 64, pp. 1927-1965, October 1985.

[49] M. Kavehrad and B. Ramamurthi, "Direct-sequence spread-spectrum with DPSK modulation and diversity for indoor wireless communications," *IEEE Trans. Comm.*, Vol. 35, pp. 224-236, February 1987.

[50] K. Pahlavan and M. Chase, "Spread-spectrum multiple-access performance of orthogonal codes for indoor radio communications," *IEEE Trans. Comm.*, Vol. 38, pp. 574-577, May 1990.

[51] J.M. Musser and J.N. Diagle, "Throughput analysis of an asynchronous code division multiple access (CDMA) system," *Proc. Int. Conf. Commun.*, Philadelphia, pp. 2F.2.2-7, 1982.

[52] D. Raychaudhuri, "Performance analysis of random access packet-switched code division multiple access systems," *IEEE Trans. Comm.*, Vol. 29, pp. 895-901, 1981.

[53] C. Sandeep and S.C. Gupta, "A comparison of CDMA and NPCSMA for an indoor data network," *Proc. IEEE 39th Conf. Vehicular Technology*, San Francisco, Vol. 2, pp. 767-773, 1989.

[54] G.L. Turin, "The effects of multipath and fading on the performance of direct-sequence CDMA-systems," *IEEE JSAC*, Vol. 2, pp. 597-603, 1984.

[55] E.S. Sousa, "Performance of a direct sequence spread-spectrum multiple access frequencies," *IEICE Trans. Comm.*, Vol. E76-B, pp. 906-912, August 1993.

[56] J.T. Miyajima, T. Hasgawa and M. Hameishi, "On the multiuser detection using a neural network in code-division multiple-access communications," *IEICE Trans. Comm.*, Vol. E76-B, pp. 961-968, August 1993.

[57] R. Esmailzadeh and M. Nakagawa, "Pre-rake diversity combination for direct sequence spread-spectrum mobile communications systems," *IEICE Trans. Comm.*, Vol. E76-B, pp. 1008-1015, August 1993.

[58] F. Simpson and J.M. Holtzman, "Direct sequence CDMA power control, interleaving and coding," *IEEE JSAC*, Vol. 11, pp. 1085-1095, September 1993.

[59] A. Higashi and T. Matsumoto, "Combined adaptive rare diversity (ARD) and coding for DPSK DS/CDMA mobile radio," *IEEE JSAC*, Vol. 11, pp. 1076-1084, September 1993.

[60] Y.C. Yoon, R. Kohno and H. Imai, "A spread-spectrum multi access system with co-channel interference cancellation for multipath fading," *IEEE JSAC*, Vol. 11, pp. 1067-1075, September 1993.

[61] A. Klein and P.W. Baier, "Linear unbiased data estimation in mobile radio systems applying CDMA," *IEEE JSAC*, Vol. 11, pp. 1058-1066, September 1993.

[62] A.M. Viterbi and A.J. Viterbi, "Erlang capacity of a power-controlled CDMA-system," *IEEE J. Selected Areas Comm.*, Vol. 11, pp. 892-900, August 1993.

[63] N.D. Wilson, R. Ganesh, K. Joseph and D. Raychandhari, "Packet CDMA versus dynamic TDMA for multiple access in an integrated voice/date PCN," *IEEE J. Selected Areas Comm.*, Vol. 11, pp. 870-884, August 1993.

[64] P. Jung, P.W. Baier and A. Steil, "Advantages of CDMA and spread-spectrum techniques over FDMA and TDMA in cellular mobile radio applications," *IEEE Trans. Vehicular Technol.*, Vol. 42, pp. 357-364, August 1993.

[65] K.I. Kim, "CDMA cellular engineering issues," *IEEE Trans. Vehicular Technol.*, Vol. 42, pp. 345-350, August 1993.

[66] D.L. Schilling and L.B. Milstein, "Broadband CDMA for indoor and outdoor personal communications," *Proc. PIMRC'93*, Yokohama, Japan, pp. 104-108, September 1993.

[67] D.L. Schilling, L.B. Milstein, R.L. Pickholtz, F.B.E. Kanteakis, M. Kullbach, V. Erlang, W. Biederman, D. Fischer and D. Salermo, "Broadband CDMA for personal communication systems," *IEEE Comm. Mag.*, Vol. 29, pp. 86-93, November 1991.

[68] D.L. Schilling, G.R. Romp and S. Garodnick, "Broadband-CDMA overlay," *Proc. PIMRC'93*, Yokohama, Japan, pp. 99-103, September 1993.

[69] K. Fazel and L. Papke, "On the performance of convolutionally coded CDMA/OFDM for mobile communication systems," *Proc. PIMRC'93*, Yokohama, Japan, pp. 468-472, September 1993.

[70] K. Fazel, "Performance of CDMA/OFDM for mobile communication systems," *Proc. ICUPC'93*, Ottawa, Canada, pp. 975-979, October 1993.

[71] N. Yee, J.P. Linnartz and G. Fettweis, "Multicarrier CDMA in indoor wireless radio networks," *Proc. PIMRC '93*, Yokohama, Japan, pp. 109-113, September 1993.

[72] L. Vandendorpe, "Multitone direct sequence CDMA-system in an indoor wireless environment," *Proc. IEEE First Symposium on Communications and Vehicular Technology in the Benelux*, Delft, The Netherlands, pp. 4.1-1 - 4.1-8, October 1993.

[73] A. Baier, "Multi-rate DS-CDMA; a promising access technique for third-generation mobile radio systems," *Proc. PIMRC'93*, Yokohama, Japan, pp. 114-118, September 1993.

[74] A. Baier, “Open multi-rate radio interface architecture based on CDMA,” *Proc. ICUPC'93*, Ottawa, Canada, pp. 985-989, October 1993.

[75] P. Bauer-Trocheris, “CODIT, The project with ambitious tasks,” *Proc. RACE Mobile Telecommunications Workshop*, Metz, France, pp. 71-74, May 1993,

[76] H. Yanikomeroglu and E. Sousa, “CDMA distributed antenna system for indoor wireless communications,” *Proc. ICUPC'93*, Ottawa, Canada, pp. 990-994, October 1993.

[77] C. van Himbeek, I. Deman, M. Dorthey, L. Philips, I. Bolsens, A. Rabaeijs and H. de Man, “Silicon integration of a flexible CDMA/APSK, mobile communication modem,” *Proc. IEEE First Symposium on Communications and Vehicular Technology in the Benelux*, Delft, The Netherlands, pp. 2.4.1 - 2.4.5, October 1993.

[78] J. Blanz, A. Klein, M. Nabhan and A. Steil, “Performance of a cellular hybrid C/TDMA mobile radio system applying joint detection and coherent receiver antenna diversity,” *IEEE JSAC*, Vol. 12, pp. 568-569, 1994.

[79] IEEE JSAC on “Code division multiple access networks,” Vol. 12, May and June 1994.

[80] Proceedings IEEE ISSSTA’94, Oulu, Finland.

[81] J.L. Massey, “Coding and modulation for code-division multiple accessing,” *Proc. Third Int. Workshop on Digital Signal Processing Techniques Applied to Space Communications*, ESTEC, Noordwijk, The Netherlands, pp. 3.1-3.17, 1992.

[82] Z. Siveski, Y. Bar-Ness and D.W. Chen, “Error performance of synchronous multiuser code division multiple access detector with multi-dimensional adaptive canceler,” *European Trans. Telecomm.*, Vol. 5, pp. 719-724, November/December 1994.

Chapter 2
Multiple Access Techniques

2.1 INTRODUCTION

In the last two decades, interest in wireless communication systems has been rapidly growing. The mobile communication market in particular has grown rapidly over the past years and it is expected that this growth will continue in the near future. The number of mobile telephone subscribers in Europe, for example, is expected to more than triple before the year 2000 [1,2]. This wireless boom is mainly caused by three very attractive features of the wireless systems:

- The wiring cost in many "ordinary" communication systems is very high. In telephone systems, for example, these costs can exceed half of the total system investment (see Table 2.1). Wireless systems do require more expensive equipment but obviously no wires, and in a great number of applications the wireless solution will be cheaper.
- Wireless systems allow mobile users and immobile users, for whom connection by wire is either difficult or impractical, to communicate.
- Even for stationary users, a wirelesses communication system offers big advantages in flexibility with respect to changes in the stationary positions (e.g., changing the location of your telephone, fax or terminal becomes a much easier task if you do not have to adapt the wiring).

Wireless systems do however cause new problems that were not present in the wired system. One of these is that we are now faced with numerous users that share a common communication channel. This may lead to conflicts if several users want to transmit at the same time, so there must be some regulation on how the available channel capacity is allocated to the users. These regulations constitute the multiple access protocol rules; each user has to follow in accessing (transmitting on) the common channel.

Table 2.1
Specification of the investment in the public telephone network of the Federal Republic of Germany. As can be seen the total cost of wiring is 53% of the total investment.

Equipment	% of Total Investment
Subscriber lines to local exchanges	46
Trunk connections between exchanges	7
User equipment (about 25 million including Private Branch Exchanges at companies)	7
Radio and international satellite connections	4
Local exchanges	20
Trunk switching centers	9
Test equipment, maintenance, facilities, etc.	7
Total investment (40 million US $)	100

Since the development of the first multiple access protocol in 1970 by the University of Hawaii (the ALOHA protocol [3]), many new protocols have been invented. One of these protocols, the code division multiple access (CDMA) protocol, has received a lot of attention during recent years. This is mainly due to the advantages this protocol offers in channels that experience signal fading as a result of the different paths a transmission can follow from source to destination. An indoor wireless channel is a good example of such a multipath fading channel.

This chapter gives an overview of the multiple access protocols with attention focused on the class of CDMA protocols.

2.1.1 Need for Multiple Access Protocols

Whenever some resource is used and thus accessed by more than one independent user, the need for a multiple access (MA) protocol arises. In the absence of such a protocol, conflicts can occur if more than one user tries to access the resource at the same time. Therefore the multiple access protocol should avoid or at least resolve these conflicts. Of course, one could avoid the need for a multiple access protocol by letting each user have its own resources. But there are two very good reasons to share the resources among the users. First, the resources may be scarce and/or expensive; for example, a mainframe computer is usually time-shared among a number of users for this reason. Second, there may be the need for a user to be able to communicate with all the other users (this is sometimes called the connectivity requirement). A good example of this is the telephone system. The users share common switching centers so a connection can be made between any two users.

The multiple access protocols that are addressed in this section are those used in communication systems in which the resource to be shared is the communication channel. In this case the reason for sharing the resource is mainly the connectivity requirement. In wireless communication systems an added reason is the scarceness of the resource; there is only one ether.

2.1.2 Desired Properties

The design of a protocol is usually accomplished with a specific goal (environment) in mind and the properties of the protocol are mainly determined by the design goal. But if we rule out the environment-specific properties, we can still address a number of properties that any good MA protocol should possess:

- The first and foremost task of the multiple access protocols addressed here is to share the common transmission channel among the users in the system. To do this, the protocol must control the way in which the users access (transmit onto) the channel by requiring that the users conform to certain rules. The protocol controls the allocation of channel capacity to the users.
- The protocol should perform the allocation such that the transmission medium is used efficiently. The efficiency is usually measured in terms of channel throughput and the delay of the transmissions.
- The allocation should be fair toward individual users, that is, not taking into account any priorities that might be assigned to the users, each user should (on the average) receive the same allocated capacity.
- The protocol should be flexible in allowing different types of traffic (e.g., voice and data).
- The protocol should be stable. This means that if the system is in equilibrium, an increase in load should move the system to a new equilibrium point. With an unstable protocol an increase in load will force the system to continue to drift to even higher load and lower throughput.
- The protocol should be robust with respect to equipment failure and changing conditions. If one user does not operate correctly, this should affect the performance of the rest of the system as little as possible.

In this book the wireless mobile environment is the one we are most interested in. In such an environment we can be more specific on some of the protocol properties, especially on the robustness with respect to changing conditions. In the wireless mobile environment the protocol should be able to deal with:

- The hidden terminal problem [two terminals are out-of-range (hidden from) of each other by a hill or by a building or by some physical obstacle opaque to UHF signals but both are within the range of the central or base station].
- The near-far effect (transmissions from distant users are more attenuated than transmissions from users close by).
- The effects of multipath fading and shadowing experienced in radio channels.
- The effects of co-channel interference in cellular wireless systems caused by the use of the same frequency band in different cells.

Many of the protocol properties mentioned above are conflicting and a trade-off has to be made during the protocol design. The trade-off depends on the environment and the specific use for the protocol one has in mind.

2.2 CLASSIFICATION OF MULTIPLE ACCESS PROTOCOLS

Starting in 1970 with the ALOHA protocol (discussed later in this section), a number of multiple access protocols have been developed. Numerous ways have been suggested to divide these protocols into groups [4,6]. For the purpose of this book, the MA protocols are classified into three main groups (Figure 2.1): the contentionless protocols, the contention protocols, and the class of CDMA protocols.

The contentionless (or scheduling) protocols avoid the situation in which two or more users access the channel at the same time by scheduling the transmissions of the users. This can either be done in a fixed fashion where each user is allocated part of the transmission capacity, or in a demand-assigned fashion where the scheduling only takes place between the users that have something to transmit.

With the contention (or random access) protocols, a user cannot be sure that a transmission will not collide because other users may be transmitting (accessing the channel) at the same time. Therefore these protocols need to resolve conflicts if they occur.

The CDMA protocols do not belong to either the contentionless or the contention protocols. As explained in Chapter 3, CDMA falls between the two groups. In principle, it is a contentionless protocol where a number of users are allowed to transmit simultaneously without conflict. However if the number of simultaneously transmitting users rises above a threshold, contention will occur.

The contention protocols can be further subdivided into repeated random protocols and random protocols with reservation. In the latter protocols the initial transmission of a user uses a random access method to get access to the channel. However once the user has accessed the channel, further transmissions of that user are scheduled until the user has nothing more to transmit. Two major types are known as implicit and explicit reservations. Explicit reservation type protocols use a short reservation packet to request transmission at scheduled times. Implicit reservation protocols are designed without the use of any reservation packet.

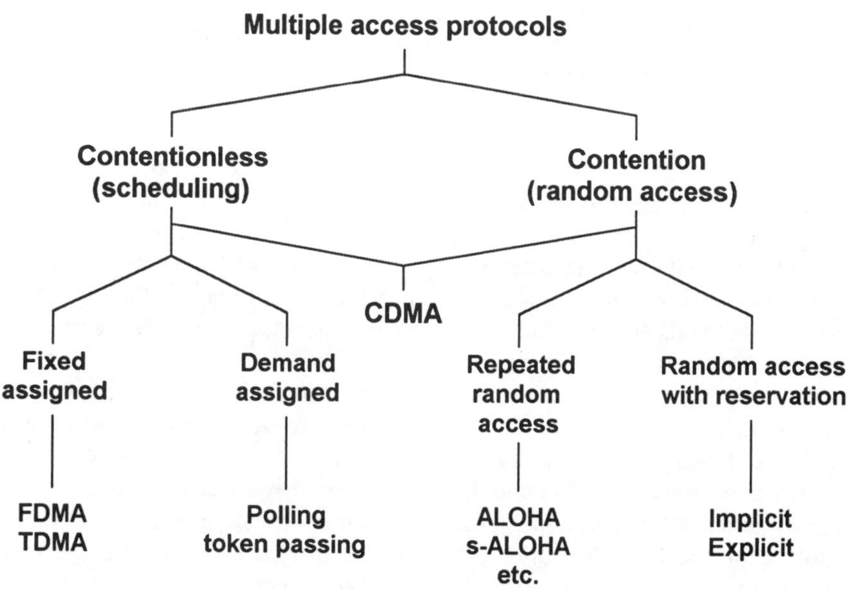

Figure 2.1 Classification of the multiple access protocols.

With CDMA forming a separate group and taking into account the subgroups of the contentionless and contention protocols, we end up with five categories. In the following sections at least two MA protocols for each category are described in some detail. Section 2.3 describes the contentionless MA protocols, first, in subsection 2.3.1, the fixed assignment protocols TDMA (time division multiple access) and FDMA (frequency division multiple access) are described, followed by the demand-assigned polling protocol and the token-bus protocol in Subsection 2.3.2. In Section 2.4 the contention protocols are described, namely, in Subsection 2.4.1 the pure random protocols p(ure)-ALOHA, s(lotted)-ALOHA, CSMA (carrier sense multiple access), ISMA (inhibit sense multiple access), and stack algorithm, and in Subsection 2.4.2 the random protocols with scheduling PRMA (packet reservation multiple access) and r(eservation)-ALOHA. Random access protocols with the capture effect are discussed in Subsection 2.4.3. In Section 2.5 the last class of protocols, the CDMA protocols, are described.

2.3 CONTENTIONLESS (SCHEDULING) MULTIPLE ACCESS PROTOCOLS

The contentionless MA protocols avoid the situation in which multiple users try to access the same channel at the same time by scheduling the transmissions of all users. The users transmit in an orderly scheduled manner so every transmission will be a successful one. The scheduling can take two forms:

1. Fixed assignment scheduling. With these types of protocols, the available channel capacity is divided among the users such that each user is allocated a fixed part of the capacity, independent of its activity. The division can be done in time or frequency. The time division results in the TDMA protocol, where transmission time is divided into frames and each user is assigned a fixed part of each frame, not overlapping with parts assigned to other users. The frequency division results in the FDMA protocol where the channel bandwidth is divided into nonoverlapping frequency bands and each user is assigned a fixed band.
2. Demand assignment scheduling. A user is only allowed to transmit if it is active (if it has something to transmit). Thus the *active* (or ready) users transmit in an orderly scheduled manner. Within the demand assignment scheduling we can distinguish between centralized control and distributed control. With centralized control a single entity schedules the transmissions. An example of such a protocol is the roll-call polling protocol. With distributed control all users are involved in the scheduling process and such a protocol is the token-passing protocol.

2.3.1 Fixed Assignment

With the fixed assignment multiple access protocols, the channel capacity is divided among the users in a static fashion; each user is allocated part of the channel capacity whether it has something to transmit or not. The allocation can take part in time or frequency, which results in the TDMA protocol and the FDMA protocol, respectively.

Time Division Multiple Access

In the *basic* TDMA protocol, the (transmission) time axis is divided into frames of equal duration, and each frame is divided into the same number of time slots. All time slots have equal duration. Each slot position within a frame is allocated to a different user and this allocation stays the same over the sequence of frames. This means that a particular user may transmit during one particular slot in every frame. During this slot it has the whole channel bandwidth at its disposal. Figure 2.2 shows the allocation in a basic TDMA frame with four time slots per frame. The shaded areas in this figure depict the guard times in each time slot in which transmission by a user is prohibited. These guard times are necessary to prevent transmissions of different (spatially distributed) users from overlapping due to transmission delay differences.

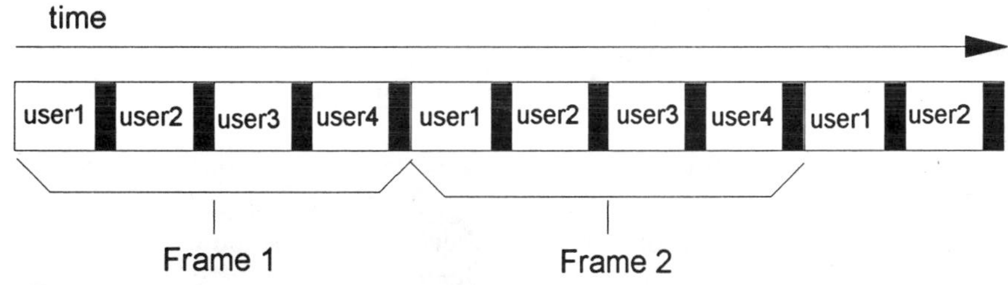

Figure 2.2 Frame and slot structure with basic TDMA.

With the basic TDMA protocol every user is allocated the same capacity, namely, one slot per frame. Thus the amount of traffic that can be transmitted within one slot must be enough to accommodate the user which generates most traffic. But this means that users generating much less traffic waste a lot of capacity. For this reason more generalized TDMA protocols have been developed that allow users to be allocated more than one slot per frame and also allow for the slots within one frame to be of different duration. However capacity is still wasted if a user has nothing to transmit in its allocated time slot.

Despite this capacity-wasting property, TDMA has been and still is used extensively because of its relative simplicity. The only real problem encountered with TDMA is that of achieving the necessary synchronization of all users so that each user knows when and for how long it can transmit.

Frequency Division Multiple Access

With FDMA, the bandwidth of the communication channel is divided into a number of frequency bands with guard bands between them to achieve frequency separation of adjacent bands. Each user is allocated a particular frequency band for its own private use. Thus with FDMA, a user can use part of the transmission channel all the time.

FDMA has the same capacity-wasting properties as TDMA, because if a user has nothing to transmit, its frequency band cannot be used by other users. It also gives a slightly inferior performance with respect to packet delay. FDMA does, however, have the advantage of being even simpler than TDMA because no synchronization of the users is necessary.

2.3.2 Demand Assignment

With the demand-assignment contentionless protocols only users that have something to transmit are allocated part of the channel capacity. The transmissions of these users are scheduled in an orderly fashion so that no contention will occur. No channel capacity is wasted on users that have nothing to transmit. However, the protocols do need some capacity to determine which users require channel access and to schedule the transmissions.

The scheduling can be controlled by a central entity, resulting in the class of centralized demand assignment protocols of which the roll-call polling protocol is an example discussed in this section. The control can also be distributed among all users, resulting in the class of distributed demand assignment protocols of which the token-bus protocol is explained in this section.

Roll-Call Polling

The roll-call polling protocol is probably the most conventional polling protocol around. A central controller polls the users one after another in a predetermined sequence. Each user is polled only once in a sequence. If a user receives a polling message from the central controller, it either transmits a control message stating it has nothing to transmit, or it transmits all messages that have accumulated in its buffer. After the user transmits its last packet, it sends a "ready" message to the central controller. Upon receiving this message, the controller sends a polling message to the next user in the sequence.

The overhead in the roll-call polling protocol consists of the polling messages and the control messages stating that a user has nothing to transmit or has finished its transmission. The overhead grows with the number of users that do not have anything to transmit. More advanced polling protocols have been developed to partially reduce the overhead such as the hub polling protocol described in [7]. Other polling protocols also allow users generating much traffic to be polled more than once during one sequence.

Token-Passing Protocol

One of the examples of a token-passing protocol is the token-bus protocol. In the token-bus protocol a number of users are connected to the same bus. Each user is given a unique address and the users of the bus are ordered in a cyclic fashion (thus forming a logical ring) by letting each user know its successor's address. Transmissions to a certain user take place by appending the address of that user to the information to be transmitted. The user that recognizes its address will know the information is intended for it.

The right of a user to transmit depends on whether or not this user has received a particular piece of control information called a token from its predecessor. If a user has received the token and indeed has something to transmit, it transmits this information. After it is finished, it transmits the token to its successor. If the user has nothing to transmit in the first place, it will immediately transmit the token to its successor. Now the successor holds the token and is permitted to transmit.

A more advanced version of the token-bus protocol has been standardized by the IEEE in the 802.4 standard. A short explanation of this standard can be found in [8].

2.4 CONTENTION (RANDOM) MULTIPLE ACCESS PROTOCOLS

With the contention multiple access protocols there is no scheduling of transmissions. This means that a user getting ready to transmit does not have exact knowledge of when it can transmit without interfering with the transmissions of other users. The user may or may not know of any ongoing transmissions (by sensing the channel), but it has no exact knowledge about other ready users. Thus, if several ready users start their transmissions more or less at the same time, all of the transmissions will fail. This possible transmission failure makes the occurrence of a successful transmission a more or less random process. The random access protocol should resolve the contention that occurs when several users transmit simultaneously .

We can subdivide the contention multiple access protocols into two groups, the repeated random access protocols and the random access protocols with reservation. With the former protocols, every transmission a user makes is as described above. With every transmission there is a possibility of contention. With the latter protocols, only in its first transmission does a user not know how to avoid collisions with other users. However, once a user has successfully completed its first transmission (once the user has access to the channel), future transmissions of that user will be scheduled in an orderly fashion so that no contention can occur. Thus, after a successful transmission, part of the channel capacity is allocated to the user and other users will refrain from using that capacity. The user loses its allocated capacity if, for some time, it had nothing to transmit.

2.4.1 Repeated Random Access Protocols

At the start of each transmission by a user, the user does not know if other users will also begin transmitting. Therefore, contention will occur if two or more users start transmitting at more or less the same time. If the users are also not able to detect an ongoing transmission, then contention will also occur if a new user starts a transmission while another user is already busy. If a user can sense an ongoing transmission, it can defer its own transmission until the channel is free. Contention can then only occur if two or more users start transmitting at the same time.

In this section the following repeated random access protocols are described: pure (p)-ALOHA, slotted (s)-ALOHA, carrier sense multiple access (CSMA), inhibit sense multiple access (ISMA), and stack algorithm. With the CSMA and ISMA protocols, a user senses the channel before transmitting (and defers if necessary). With the other protocols a user has no knowledge of any transmissions in progress.

p-ALOHA

The pure ALOHA protocol (or p-ALOHA) was named after the ALOHA system [3], a packet radio network developed by the University of Hawaii that became operational in 1970. The multiple access protocol used in this system was the first of its kind and was called the ALOHA protocol. Later when variations of the ALOHA protocol were thought of, the original protocol was renamed pure (p)-ALOHA.

The p-ALOHA protocol is a centralized protocol. A number of users can transmit to a base station via an uplink channel and receive from the base station via a downlink channel that is located in a different frequency band. Immediately after a user has generated a packet it will transmit this packet on the uplink channel. If no other users transmit, the base station will receive a correct transmission and send an acknowledgment packet on the down link channel. On reception of the acknowledgment, the user knows its transmission has been successful.

If two or more users transmit simultaneously, a collision will occur. The base station recognizes this occurrence because it receives a garbled transmission and does not transmit an acknowledgment. When a user does not receive an acknowledgment, it assumes its transmission has collided so it will have to retransmit. Simply retransmitting after a fixed time interval will not do, because two users that transmitted at the same time will find out about the collision at about the same time and therefore retransmit at the same time, thus creating another collision. To avoid this deadlock situation, a user experiencing a collision will wait a random amount of time before retransmitting.

Figure 2.3 shows two terminals transmitting on the uplink channel. At $t=t_0$ user 1 generates a packet that it transmits immediately.

The start of the packet arrives at the base station the propagation delay time t_{p1} later. No other transmissions occur so the transmission succeeds. At $t=t_1$ user 2 generates and transmits a packet. The start of this packet will arrive at the base station the propagation delay time t_{p2} later. Normally the times t_{p1} and t_{p2} are different because of the spatial distributions of the users, here $t_{p2} > t_{p1}$.

The transmission of the first packet of user 2 collides with the second packet of user 1. The packet of user 1 is rescheduled a random amount of time t_{r1} later, and the packet of user 2 is rescheduled the random amount of time t_{r2} later. On these occasions both packets are transmitted successfully.

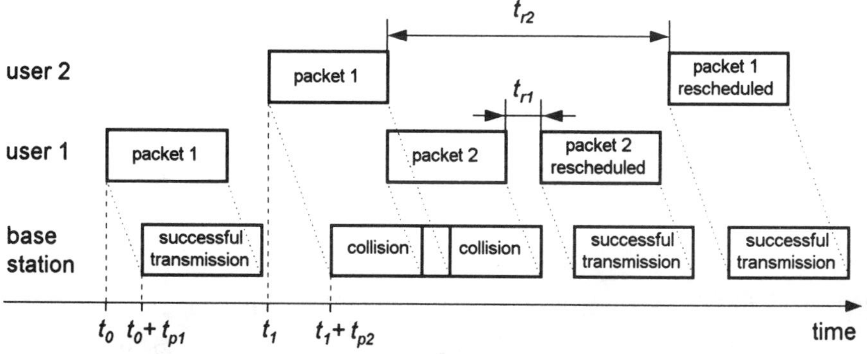

Figure 2.3 Successful transmissions and collisions in p-ALOHA.

In literature the p-ALOHA protocol is sometimes described a little bit differently. Instead of waiting for an explicit acknowledgment from the base station, a transmitting user will "listen" to the channel to hear if there are any other transmissions during the time it takes for its own transmission. If it does not hear other transmissions within this time, its transmission will not have collided. If it does hear another transmission, there will be a collision so a retransmission is needed. This retransmission is again delayed a random amount of time.

This ALOHA-like protocol can be fully distributed because it does not rely on a base station to check whether collisions have occurred.

s-ALOHA

The p-ALOHA protocol is very simple but has a very low channel throughput, only 18%. The main reason for this is that the probability of a user's transmission being interrupted by other transmissions becomes quite large if the traffic on the channel increases. If we look at Figure 2.4, we can see that user 1 starts transmission at $t=t_0$. Assume a transmission takes T seconds, so the transmission of user 1 ends at $t=t_0+T$. As can be seen from the figure, the transmission of a user starting anywhere within the time period between t_0 minus T to t_0+T will collide with the transmission of user 1 (indicated as the shaded area in Figure 2.4).

As a result the transmission of user 1 there is a vulnerable period of $2T$ (2 times the duration of a transmission). Note that we assumed the propagation delay to be negligible compared to the time needed to transmit a packet (T).

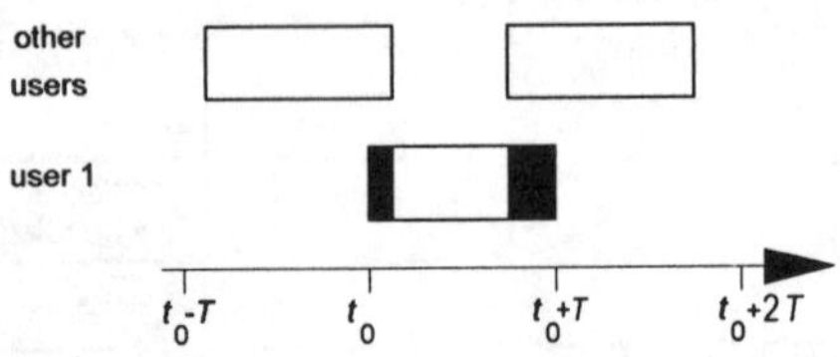

Figure 2.4 Vulnerable period of a transmission with p-ALOHA.

One way to improve the performance of p-ALOHA protocol is to try and make the vulnerable period smaller. This can be done by dividing the transmission time axis into time slots and requiring that a user is only permitted to start its transmission at the start of a time slot.

Figure 2.5 shows what happens if user 1 generates a packet (indicated by a solid vertical black line) between time 0 and T.

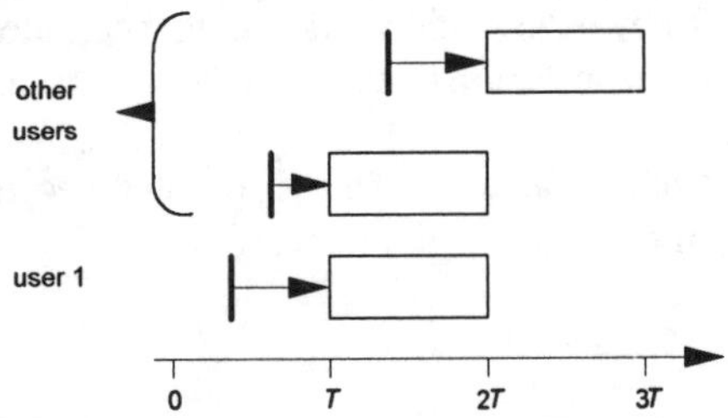

Figure 2.5 Vulnerable period of a transmission in the s-ALOHA protocol.

The transmission of this packet is delayed until time $t=T$ (indicated by an arrow followed by the packet) and only those users that generated a packet between time 0 and T will also transmit at time $t=T$ and collide with the transmission of user 1. Users that generate a packet after time $t=T$ will not start transmission until time $t=2T$ and will therefore not collide with the transmission of user 1.

The vulnerable period of a transmission is now only T so it is halved compared to p-ALOHA. This doubles the maximum channel throughput to 36%. The resulting protocol is called the slotted (s)-ALOHA protocol.

Carrier Sense Multiple Access (CSMA)

CSMA is a class of protocols which we can divide into two subclasses: the nonpersistent CSMA protocols and the p-persistent CSMA protocols.

In the *nonpersistent* CSMA protocols, a user that has generated a packet first "listens" to (senses) the channel for transmissions of other users. If it senses the channel idle, it will transmit; otherwise the user will wait a random time and then try again. At first sight it may seem that this protocol totally avoids collisions however due to the propagation delay between two users, it is possible that a user senses the channel idle and starts its transmission while another transmission is already in progress. Figure 2.6 shows a transmission from user 1 that starts at t_0. With a propagation delay between user 1 and user 2 of t_p, user 2 will sense the channel idle between t_0 and t_0+t_p. Therefore, if user 2 generates a packet within this time a colliding transmission will result.

A user is informed of a collision by the absence of an acknowledgment packet from the receiving station. Upon detecting the collision, the packet is rescheduled for transmission a random time later.

A special case of the *p-persistent* CSMA protocols is the 1-persistent CSMA protocol . The protocol is the same as the nonpersistent CSMA protocol except when a user senses the channel busy. In this case the transmission is not rescheduled a random time later but instead the user keeps sensing the channel until it becomes idle and then immediately transmits its packet.

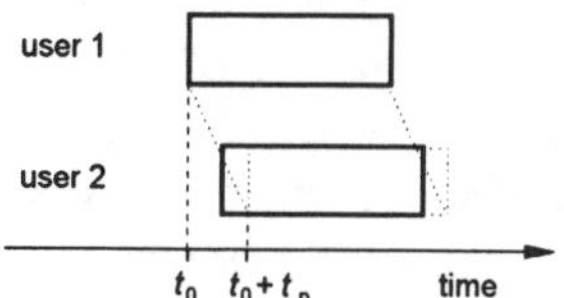

Figure 2.6 Collision in the CSMA protocol due to propagation delay.

As a result of this, all users that become ready during a busy channel will transmit as soon as the channel becomes idle, which leads to a high probability of a collision at the end of a successful transmission.

To avoid the collision of packets accumulated while the channel was busy, the start of the transmission times of the accumulated packets can be randomized. This can be done by letting all users that generate a packet during a busy channel transmit as soon as the channel becomes idle with a probability p. With a probability 1 minus p they will defer their transmission for τ seconds (with τ being the maximum propagation delay between any two users in the system). After the τ seconds the deferred terminal will sense the channel again and apply the same algorithm as before.

With this p-persistent CSMA protocol the probability of more than one ready user transmitting when the channel becomes idle can be made quite small by choosing a small value for p.

With the nonpersistent and p-persistent CSMA protocols, a user will not learn about a collision until after its whole packet has been transmitted. The reason for this is, of course, that an acknowledgment packet will only be sent after the complete packet has been received by the receiving user. Since a collision can only occur within the propagation delay after the start of the transmission, it is a waste of time to transmit more of the packet if a collision has occurred within this period. For this reason the CSMA-CD (carrier sense multiple access with collision detect) protocols have been developed. With these protocols a user keeps monitoring the channel while it is transmitting. If it detects a collision, it aborts its transmission as soon as possible thus saving time. The 1-persistent CSMA-CD protocol is widely used in local area computer networks and has been standardized by the IEEE in the 802.3 standard [9].

Inhibit Sense Multiple Access (ISMA)

With the CSMA protocols each user must be able to detect (to sense) the transmissions of all other users. However, especially in radio channels, this may prove to be very difficult because in such channels it can easily happen that two users are hidden from each other by a building or some other obstacle. As is shown in [10] this hidden terminal problem severely degrades the performance of CSMA. As a solution the ISMA (also called the BTMA, busy tone multiple access) protocol is proposed .

The ISMA protocol is identical to the CSMA protocol except for the way in which the users sense the channel for transmissions of other users. In CSMA the sensing is done by listening to the channel on which the users transmit. In ISMA there is a base station that transmits a busy/idle signal on a separate channel to indicate the presence or absence of a transmission of one of the users. The channel on which the users transmit to the base station is called the inbound channel and the channel on which the base station broadcasts to the users is called the outbound channel. As soon as the base station receives a transmission from a user on the inbound channel, it will generate a busy signal on the outbound channel. If the transmission ends, the base station will transmit an idle signal. Now if two users are hidden from each other but not from the base station they will still be able to determine if the other user is transmitting or not.

Analogous to the CSMA protocols, the ISMA protocols can be divided into the nonpersistent ISMA protocols and the p-persistent ISMA protocols. Since the protocols do not basically differ from the CSMA protocols, no further explanation is given here.

Stack Algorithm

The stack algorithm was developed by Tsybakov [11]. The stack algorithm uses the feedback concept, which states that the feedback of more information from the common receiver to the contending transmitters allows better coordination of their subsequent accesses, thus reducing the risk of repeated collisions and thereby increasing the probability of successful packet transmission. The packets, destroyed in a collision, are divided among two levels of a virtual stack. Packets at one level are transmitted immediately. Packets at the other have to wait until those packets (plus any packets generated intermediately) are successfully transmitted.

The stack algorithm has some very special properties, such as simplicity of implementation, relatively large throughput (about 0.4), and a small delay for small offered traffic intensity. Several feedback algorithms have been developed, e.g., idle/success/failure feedback, conflict/no conflict feedback, idle/transmission feedback. Details of three algorithms can be founu in [11]. The performance of the stack-algorithm in mobile radio channels has been analyzed in [26].

2.4.2 Random Access with Reservation

If a user has a row of packets to transmit, the transmission of the first packet in the row is done in the same manner as every packet in a pure random access protocol is transmitted. When transmitting this first packet, the user will not know for sure that this transmission will succeed because other users may also be transmitting their first packet. Success of the transmission of a first packet is therefore a random process.

The difference between a reservation protocol and a pure random access protocol arises when a user successfully transmits its first packet in a row of packets. Now a fixed part of the channel capacity is allocated to the user for the transmissions of the rest of the packets. The user obtains a reservation. All users are aware of what parts of the channel are allocated to the reserved users. Therefore the transmissions of these users are carried out without contention, and the transmissions are scheduled.

Once a user has transmitted its whole row of packets, it will return the allocated capacity (give up its reservation) so it can be used by other users. If the user wants to transmit a new row of packets, the first packet will again have to contend for the channel.

There are many protocols that fall within the category of random access with reservation. Many of those protocols (probably most) use slotted ALOHA as the random access method to obtain a reservation. These protocols are collectively known as the reservation ALOHA or r-ALOHA protocols. Three of those protocols are described in this section. A fourth protocol described in this section also uses slotted ALOHA for random access, but has a slightly different purpose than the other protocols and is therefore treated separately. This is the packet reservation multiple access (PRMA) protocol.

Reservation ALOHA (r-ALOHA)

Since the reservation ALOHA protocols use the slotted ALOHA protocol to obtain a reservation, the channel is obviously a slotted one. Packet transmission may only start at the beginning of a time slot. The time slots are put into groups or frames, each frame containing the same number of time slots. Users can have one or more slots in each frame allocated to them. Therefore they do not have to contend for the channel. A user that does not have any slots assigned to it has to contend for the free slots by using the slotted ALOHA protocol. Several variations on how the reservations are assigned to the users have been thought of. Three of them are discussed here.

Protocol number ***one*** was devised by Roberts [12] and if one speaks of "the" reservation ALOHA protocol this protocol is often meant. In Roberts' protocol the time axis is divided into frames. However the division of the frames into time slots depends on the state of the system. The system can be in a reserved state or in an ALOHA state. In the reserved state the frame is divided into M+1 time slots. The first M time slots of the frame are for reserved users. Each of the reserved users is allocated one or more of the M slots. The last time slot is divided into V mini-slots. Each mini-slot is big enough for a user to transmit a reservation request packet. The mini-slots within this time slot are accessed by the users using the slotted ALOHA protocol. Thus a user wishing to receive a reservation transmits a request in one of the mini-slots. Upon successful transmission in a mini-slot, a broadcast acknowledgment packet informs this and other users which full width slot is allocated to the user until further notice. Therefore, Roberts' protocol comes in the category of the explicit reservation scheme.

Not all the M time slots for reserved users have to be allocated. It can even happen that none of the slots is allocated because there are no reserved users. If this happens, the system goes to the ALOHA state. In this state all time slots within a frame are divided into V mini-slots to be used to transmit reservation requests. As soon as a user successfully transmits a reservation request in one of the mini-slots, the system goes to the reserved state and a full width slot is allocated to the user.

Figure 2.7 shows the frame in the reserved and the ALOHA state in Roberts' reservation ALOHA.

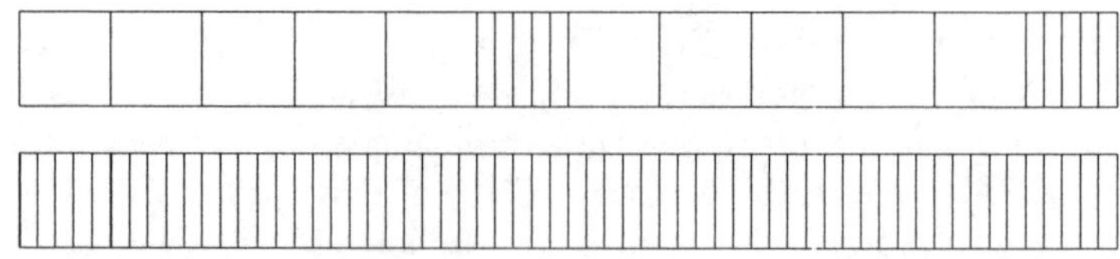

Figure 2.7 Reserved state and ALOHA state in Roberts' r-ALOHA.

Protocol number ***two*** was devised by Crowther [13]. This protocol is an example of an implicit reservation method. In this protocol each frame is divided into the same number of time slots, each time slot being large enough to hold one packet. If a user has a row of packets to transmit, it starts out by transmitting the first packet using the slotted ALOHA protocol. If in a certain time slot this first packet is transmitted successfully, the user gets a reservation for that slot in future frames. The user keeps this reservation until it has transmitted all its packets in the row.

Thus users that have a reservation transmit their packet using time division multiplexing (Figure 2.8). In the third slot of the first frame, users 1 and 2 transmit their first packet so a collision occurs. Using the slotted ALOHA protocol user 1 tries again two slots later which is the first slot of the second frame and succeeds so slot one of the following frames is allocated to user 1. User 2 transmits its first packet again in the third slot of the second frame and now also succeeds, so it gets a reservation for all the third slots.

In the second slot of frame four, user 3 transmits its first packet and receives a reservation for that slot. In frame five user 2 transmitted all its slots so it loses its reservation.

Frame 1				Frame 2				Frame 3				Frame 4				Frame 5				Frame 6				Frame 7			
				1		2		1		2		1	3	2		1	3			1	3			1			

Figure 2.8 Crowther's reservation ALOHA.

The ***third*** and final r-ALOHA protocol that is discussed here is the protocol proposed by Binder [14]. In this protocol there are N equal sized time slots in each frame and $M \leq N$ users in the system. Each user is permanently assigned one slot position (in each frame). If $M < N$, the rest of the slots are not assigned. This protocol is also an example of an implicit reservation scheme.

Of course, it can happen that the owner of a slot has no packet to transmit, leaving the slot empty. An empty slot in a frame (whether it is an unassigned slot or an assigned slot left empty by the owner) is a signal to all users that this slot is available to all users in the next frame (although it still stays assigned to its owner). The empty slots (whether slots not assigned to any user or assigned slots not currently used by their owners) are accessed using the slotted ALOHA protocol.

If the owner of a slot wants its slot back, it simply transmits a packet in its assigned slot. If no other user was transmitting, the owner immediately has its slot back. If another user was also using the slot, a collision will occur. On the occurrence of a collision in an assigned slot, all users that are not the owner of the slot will abstain from transmitting in that slot in the next frame to see if the owner wants the slot back. Thus the owner will always transmit its packet successfully within two slot times.

Packet Reservation Multiple Access (PRMA)

The packet reservation multiple access protocol could also have been assigned to the group of reservation ALOHA multiple access protocols but there is one property that makes this protocol a bit different from the others. PRMA is mainly focused on the transmission of speech and therefore the main issue of the protocol is to put an upper bound on the packet delay to maintain intelligible speech, while occasional loss of packets can be permitted. In the other reservation protocols packet loss is not allowed. The PRMA protocol can be placed in the category of implicit reservation protocol.

With the PRMA protocol transmission time is once more divided into frames and each frame is divided into the same number of time slots. The frame time is chosen such that a user generating speech packets generates exactly one speech packet per frame.

The speech users have a voice activity detector to detect if the user is active or not. Once a user starts generating speech packets, the packets are stored in a first-in/first-out buffer and the user tries to transmit the first packet from the buffer using the slotted ALOHA protocol. If this first packet has not been transmitted within about 40 ms (has been in the buffer for more than 40 ms), the delay of this packet is such that transmission is no longer useful because the speech would become unintelligible. For this reason, the packet is dropped from the buffer and the user tries to transmit the next packet in line (which was generated one frame time later) still using the slotted ALOHA protocol. If this packet has been in the buffer for more than 40 ms, it is also dropped and the user tries to transmit the next packet. This continues until either all the packets have been dropped or a successful transmission has occurred.

Upon a successful transmission the packet is removed from the buffer and the user gets a reservation of the time slot in which it transmitted in future frames. It keeps this reservation until all the packets have been transmitted from the buffer.

A user with allocated reservation generates and transmits exactly one packet per frame so the speech packets have a constant delay, equal to the delay with which the first packet was transmitted. A more detailed description of the PRMA protocol can be found in [15,16,27].

2.4.3 Random Access Protocols with Capture Effect

The development of packet communication protocols has been based mainly on the assumption that when two or more packets collide (overlap in time) all packets are lost. However, a realistic radio receiver is able to be captured by the strongest of the overlapping packets and thus receive this packet correctly. This is called the capture effect [17,25].

The probability that one packet captures the common receiver in the presence of several other (weaker) contenders depends on the power differences with which the packets are received. In a mobile radio channel these power differences, and with these the capture probability, are increased in a natural way as a result of near-far effects [17] and fading [18,19]. The influence of capture effects on the performance of the slotted ALOHA and the unslotted ISMA protocols has been reported in numerous papers, e.g., [17,25]. The studies in these papers show that the channel capacity (maximum throughput) due to an individual effect (e.g. just the near-far effect or just fading) is always less than the capacity in the case of more impairments [23]. From this we can conclude that higher disorder in the environment can lead to higher throughput.

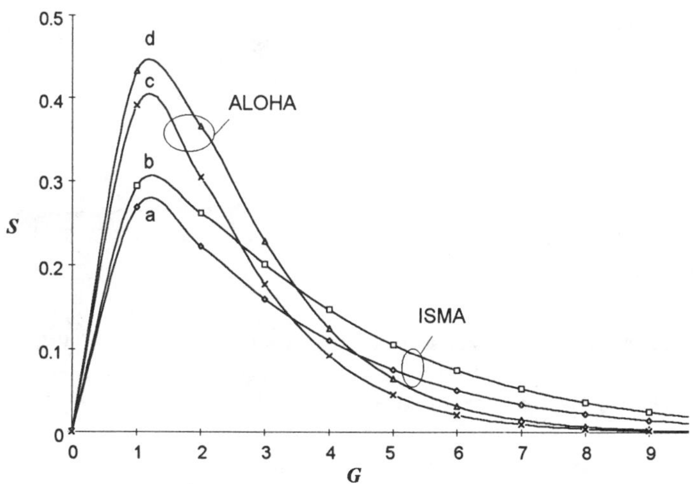

Figure 2.9 Throughput curves for an inhibit delay factor of 0.5 for (a) unslotted ISMA with a Rician parameter of 7 dB (both test signal and interfering signals) and a capture ratio of 3 dB; (b) unslotted ISMA with a Rician parameter of 7 dB (test signal) and 5 dB (interfering signals), and a capture ratio of 5 dB; (c) slotted ALOHA with the same parameters as in *a*; (d) slotted ALOHA with the same parameters as in *b*.

Figures 2.9 and 2.10 compare the throughput performance of the slotted ALOHA and the unslotted nonpersistent ISMA protocols in a Rician fading channel with receiver capture. It can be seen from these figures that the channel capacity of the slotted ALOHA protocol in a Rician fading channel is higher than 0.36 (the channel capacity of ideal slotted ALOHA). Furthermore it can be seen that the channel capacity of ISMA with an inhibit delay factor $d = 0.5$ is lower than the channel capacity of ALOHA while the capacity of ISMA with $d = 0.05$ is higher than that of ALOHA. It is worth mentioning here that the inhibit delay factor is low if the cell size is low. Therefore the unslotted ISMA protocol would be a better choice for indoor wireless communications than the slotted ALOHA protocol.

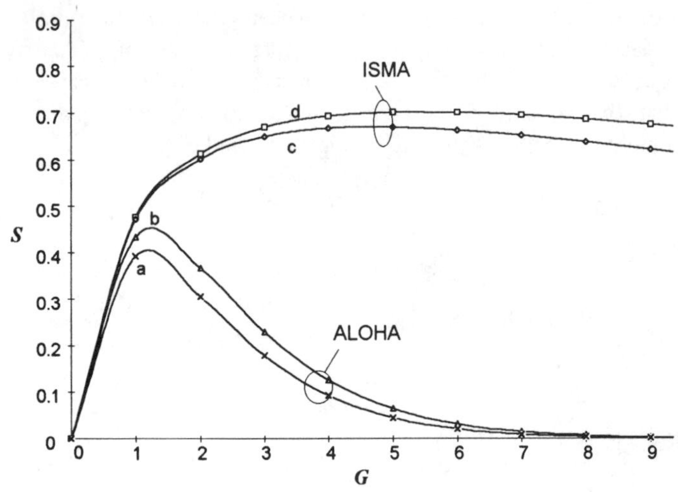

Figure 2.10 Throughput curves for an inhibit delay factor of 0.05 for (a) slotted ALOHA with a Rician parameter of 7 dB (both test signal and interfering signals) and a capture ratio of 3 dB; (b) slotted ALOHA with a Rician parameter of 7 dB (test signal) and 3 dB (interfering signals), and a capture ratio of 5 dB; (c) unslotted ISMA with the same parameters as in *a*; (d) unslotted ISMA with the same parameters as in *b*.

2.5 CODE DIVISION MULTIPLE ACCESS (CDMA) PROTOCOLS

This Section deals with the class of code division multiple access (CDMA) protocols, which rely on coding to achieve their multiple access property. In this Section the basic principles of CDMA protocols are discussed and the placement of the protocols between the contentionless (scheduling) protocols and contention (random access) protocols is explained.

CDMA protocols do not achieve their multiple access property by a division of the transmissions of different users in either time or frequency, but instead make a division by assigning each user a different code. This code is used to transform a user's signal in a wideband signal (spread-spectrum signal). If a receiver receives multiple wideband signals, it will use the code assigned to a particular user to transform the wideband signal received from that user back to the original signal. During this process, the desired signal power is compressed into the original signal bandwidth while the wideband signals of the other users remain wideband signals and appear as noise when compared to the desired signal.

As long as the number of interfering users is not too large, the signal-to-noise ratio will be large enough to extract the desired signal without error. Thus in this case the protocol behaves as a contentionless protocol. However if the number of users rises above a certain limit, the interference becomes too large for the desired signal to be extracted and contention occurs, thus making the protocol interference limited. Therefore, the protocol is basically contentionless unless too many users access the channel at the same time. This is why we place the CDMA protocols between the contentionless and contention protocols.

There are several ways to divide the CDMA protocols into a number of groups. The most common is the division based on the modulation method used to obtain the wideband signal. This division leads to four protocol types: direct-sequence CDMA (DS-CDMA), frequency hopping CDMA (FH-CDMA), time hopping CDMA (TH-CDMA), and hybrid CDMA where the last group of protocols uses a combination of the modulation methods of the other protocols.

In the direct-sequence CDMA protocol the original signal is modulated on a carrier and then multiplied by a binary code sequence with a bandwidth much larger than the original bandwidth. This spreads the spectrum of the original signal, transforming it to a wideband signal. The modulated wideband signal is then transmitted through the channel. During this transmission, the desired modulated signal suffers interference from the modulated signals of other users so at the receiver the desired signal plus interference is received. The receiver correlates the received signal with the code sequence of the original signal. The correlation between the desired signal and the code sequence of the original signal will be substantial, since the desired signal was generated using that code sequence. If the cross-correlations of the code sequences of the interfering signals and the code sequence of the original signal are small, then, after correlation, the power ratio of the desired signal to interfering signals will be large so the desired signal can be demodulated and the original data recovered without error.

In the frequency hopping CDMA protocols, the wideband channel is divided into frequency bands equal to the bandwidth of the original signals of the users. During the transmission of a user's signal, the carrier frequency is changed periodically resulting in a periodic change of the frequency band occupied by the user. The pattern of frequency changes is dictated by the code assigned to the user. At the receiver the demodulator follows the changes of the carrier frequency when demodulating the signal.

We distinguish between two types of frequency hopping protocols, fast frequency hopping and slow frequency hopping protocols. With the *fast frequency hopping* protocols the change of frequency band is chosen so rapidly that a number of changes occur during the transmission of one bit. One bit is thus transmitted in several frequency bands. Because of the different codes assigned to the users, the probability of a user being the only one to transmit in most of the frequency bands during a bit is high, even if multiple users transmit. Thus the received power of the desired bit is higher than the power of interfering bits and therefore the desired bit can be extracted. With the *slow frequency hopping* protocols the hop rate is much slower. During one hop (change of frequency) multiple bits are transmitted. Because of the different codes, the probability is again small that multiple users transmit a number of bits in the same frequency band. Therefore, most of the time the desired signal can be recovered. By interleaving and using error-correcting codes, we are ensured that, if multiple transmissions happen in the same frequency band, data can still be recovered.

In the time hopping CDMA protocols, a user's signal is not transmitted continuously. Instead it is transmitted in short intervals (bursts). The start of each burst is decided by the code assigned to the user. Because the signal is compressed in time, it will need a larger bandwidth during its transmission times. Compared to the frequency hopping CDMA protocol, we see that the TH-CDMA-protocol uses the whole wide-band spectrum for short periods instead of parts of the spectrum all of the time.

The probability of multiple users transmitting at the same time is small because of the different codes assigned to each user. If two users transmit at the same time, interleaving and error-correcting codes ensure the receiver can still recover the signal correctly.

The hybrid CDMA protocols use a combination of the modulation methods of the direct sequence, frequency hopping, or time hopping protocols to obtain the wideband signal. Combining the modulation methods uses the specific advantages that each modulation method offers.

REFERENCES

[1] PT Aktueel, Stam Tijdschriften B.V., Rijswijk, No. 154, pp. 11, 12 April 1989.

[2] J.S. Da silva and B.E. Fernandes, "The European research program for advanced mobile systems," *IEEE Personal Comm.*, Vol. 2, pp. 14-19, February 1995.

[3] N. Abramson, "The ALOHA system — another alternative for computer communications," *Proceedings Fall. Joint Computer Conference (AFIPS)* 37, pp. 281-285.

[4] R. Rom and M. Sidi, *"Multiple Access Protocols Performance and Analysis,"* Springer-Verlag, New York, 1990, pp. 2-4.

[5] J. D. Spragins, J. L. Hammond and K. Pawlikowski, *"Telecommunications Protocols and Design,"* Addison-Wesly Publishing Company, Reading, Mass., 1991, pp. 207.

[6] C. A. Sunshine, *"Computer Network Architectures and Protocols,"* Plenum Press, New York, 1989, pp. 142-143.

[7] M. Schwartz, *"Computer-Communication Network Design and Analysis,"* Prentice-Hall, Englewood Cliffs, N.J., 1977.

[8] A.S. Tanenbaum, *"Computer Networks,"* Prentice-Hall, Englewood Cliffs, N.J., 1989.

[9] ANSI/IEEE standard 802.3-1985, "Carrier-sense multiple access with collision detection," The Institute of Electrical and Electronics Engineers, New York, 1985.

[10] F.A. Tobagi and L. Kleinrock, "Packet switching in radio channels: part II — the hidden terminal problem in carrier sense multiple-access and the busy-tone solution," *IEEE Trans. Comm.*, Vol. 23, pp. 1417-1433, December 1975.

[11] B.S. Tsybakov, "Survey of USSR contributions to random multiple access communications," *IEEE Trans. Inf. Theory,* Vol. IT-31, No. 2, pp. 143-165, March 1985.

[12] L. G. Roberts, "Dynamic allocation of satellite capacity through packet reservation," *Proceedings National Computer Conference (AFIPS),* 42, pp. 711-716, June 1973.

[13] W. Crowther et. al., "A system for broadcast communications: reservation ALOHA," *Proceedings Sixth Hawaii International Conference on System Sciences*, pp. 371-374, January 1973.

[14] R. Binder R. Rettberg, D. Walden, S. Ornstein and F. Heart, "A dynamic packet-switching system for satellite broadcast channels," *Proceedings IEEE International Conference on Communications*, San Francisco, pp. 41.1-41.5, June 1975.

[15] D.J. Goodman and S.X. Wei, "Factors affecting the bandwidth efficiency of packet reservation multiple access," *Proceeding of the 39th IEEE Vehicular Technology Conference*, San Francisco, pp. 292-299, May 1989.

[16] C. van den Broek, D. Sparreboom and R. Prasad, "Performance evaluation of packet reservation multiple access protocol using steady-state analysis of a Markov chain model," *Proceedings COST 227/231 Workshop*, Limerick, Ireland, pp. 454-462, 6-10 September 1993.

[17] J.C. Arnbak and W. van Blitterswijk, "Capacity of slotted ALOHA in Rayleigh-fading channels," *IEEE J. Selected Areas Comm.*, Vol. SAC-5, No. 2, pp. 261-269, February 1987.

[18] J.J. Metzner, "On improving utilization in ALOHA networks," *IEEE Trans. Comm.*, Vol. COM-24, pp. 447-448, April 1976.

[19] F. Kuperus and J.C. Arnbak, "Packet radio in a Rayleigh channel," *Electron. Lett.*, Vol. 18, pp. 506-507, June 1982.

[20] C. Namislo, "Analysis of mobile radio slotted ALOHA network," *IEEE J. Selected Areas Comm.*, Vol. SAC-2, pp. 583-588, July 1984.

[21] R. Prasad and J.C. Arnbak, "Enhanced throughput in packet radio channels with shadowing," *Electron. Lett.*, Vol. 24, No. 16, pp. 986-987, August 1988.

[22] C. van der Plas and J.-P.M.G. Linnartz, "Stability of mobile slotted ALOHA network with Rayleigh fading, shadowing and near-far effect," *IEEE Trans. Vehicular Technol.*, Vol. VT-39, No. 4, pp. 359-366, November 1990.

[23] R. Prasad, "Performance analysis of mobile packet radio networks in real channels with inhibit-sense multiple access," *IEE Proceedings-I*, Vol. 138, No. 5, pp. 458-464, October 1991.

[24] R. Prasad and C.Y. Liu, "Throughput analysis of some mobile packet radio protocols in Rician fading channels," *IEE Proceedings-I*, Vol. 139, No. 3, pp. 297-302, June 1992.

[25] I. Widipangestu, A.J. 't Jong and R. Prasad, "Capture probability and throughput analysis of slotted ALOHA and unslotted np-ISMA in a Rician/Rayleigh environment," *IEEE Trans. Vehicular Technol.*, Vol. 43, pp. 457-465, August 1994.

[26] J.A.M. Nijhof, R.D. Vossenaar and R. Prasad, "Stack algorithm in mobile radio channels," *Proc. IEEE 44th Vehicular Technology Conference,* Stockholm, Sweden, pp. 1193-1197, June 1994.

[27] C. van den Broek and R. Prasad, "Effect of capture on the performance of the PRMA protocol in an indoor radio environment with BPSK modulation," *Proc. IEEE 44th Vehicular Technology Conference,* Stockholm, Sweden, pp. 1223-1227, June 1994.

Chapter 3
CDMA System Concepts

3.1 INTRODUCTION

Code division multiple access (CDMA) protocols constitute a class of protocols in which the multiple access property is primarily achieved by means of coding. Each user is assigned a unique code sequence that it uses to encode its information-bearing signal. The receiver, knowing the code sequences of the user, decodes a received signal after reception and recovers the original data. Since the bandwidth of the code signal is chosen to be much larger than the bandwidth of the information-bearing signal, the encoding process enlarges (spreads) the spectrum of the signal and is therefore also known as spread-spectrum (SS) modulation. The resulting encoded signal is also called a spread-spectrum signal, and the CDMA protocols are often denoted as spread-spectrum multiple access (SSMA) protocols.

It is the spectral spreading of the coded signal that gives the CDMA protocols their multiple access capability. It is therefore important to know the techniques to generate spread-spectrum signals and the properties of these signals.

The precise origin of spread-spectrum communications may be difficult to pinpoint because the modern spread-spectrum communication is the outcome of developments in many directions, such as high-resolution radars, direction finding, guidance, correlation detection, matched filtering, interference rejection, jamming avoidance, information theory and secured communications [1,7].

The spread-spectrum modulation techniques were originally developed for use in military radar and communication systems because of their resistance against jamming signals and a low probability of detection. Only in recent years with new and cheap technologies emerging and a decreasing military market have the manufacturers of spread-spectrum equipment and researchers became interested in the civil applications.

In order to classify as a spread-spectrum modulation technique, two criteria must be fulfilled [8]:

1. The transmission bandwidth must be much larger than the information bandwidth.
2. The resulting radio-frequency bandwidth is determined by a function other than the information being sent (so the bandwidth is independent of the information signal). This excludes modulation techniques like FM and PM.

Therefore, spread-spectrum modulation transforms an information-bearing signal into a transmission signal with a much larger bandwidth. This transformation is achieved by encoding the information signal with a code signal that is independent of the data and has a much larger spectral width than the data signal. This spreads the original signal power over a much broader bandwidth, resulting in a low(er) power density. The ratio of transmitted bandwidth to information bandwidth is called the processing gain G_p of the spread-spectrum system,

$$G_p = \frac{B_t}{B_i} \tag{3.1}$$

where B_t is the transmission bandwidth and B_i is the bandwidth of the information-bearing signal.

The receiver correlates the received signal with a synchronously generated replica of the code signal to recover the original information-bearing signal. This implies that the receiver must know the code signal used to modulate the data.

Because of the coding and the resulting enlarged bandwidth, SS signals have a number of properties that differ from the properties of narrowband signals. The most interesting from the communication systems point of view are discussed below. To have a clear understanding, each property has been briefly explained with the help of illustrations, if necessary, by applying direct-sequence spread-spectrum techniques.

1. Multiple access capability. If multiple users transmit a spread-spectrum signal at the same time, the receiver will still be able to distinguish between the users provided each user has a unique code that has a sufficiently low cross-correlation with the other codes. Correlating the received signal with a code signal from a certain user will then only despread the signal of this user, while the other spread-spectrum signals will remain spread over a large bandwidth. Thus, within the information bandwidth the power of the desired user will be much larger than the interfering power provided there are not too many interferers, and the desired signal can be extracted. The multiple access capability is illustrated in Figure 3.1. In Figure 3.1a, two users generate a spread-spectrum signal from their narrowband data signals. In Figure 3.1b both users transmit their spread-spectrum signals at the same time. At the receiver only the signal of user 1 is "despread" and the data recovered.

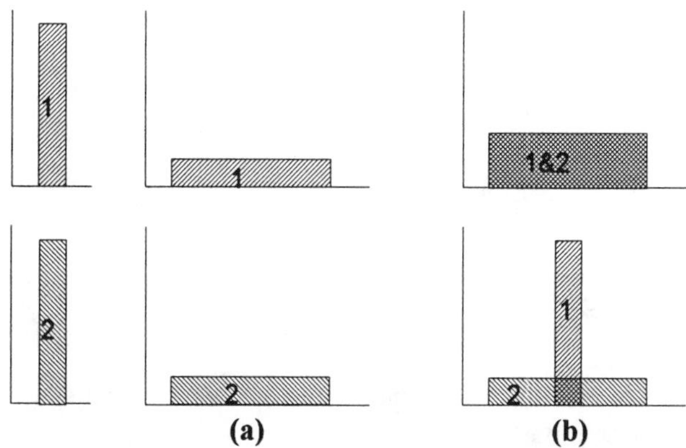

Figure 3.1 Principle of spread-spectrum multiple access.

2. Protection against multipath interference. In a radio channel there is not just one path between a transmitter and receiver. Due to reflections (and refractions) a signal will be received from a number of different paths. The signals of the different paths are all copies of the transmitted signal but with different amplitudes and phases. Adding these signals at the receiver will be constructive at some of the frequencies and destructive at others. In the time domain, this results in a dispersed signal. Spread-spectrum modulation can combat this multipath interference; however, the way in which this is done depends very much on the type of modulation used. In the next section where CDMA protocols based on different modulation methods are discussed, we show for each protocol how multipath interference rejection is obtained.
3. Privacy. The transmitted signal can only be despread and the data recovered if the code is known to the receiver.
4. Interference rejection. Cross-correlating the code signal with a narrowband signal will spread the power of the narrowband signal thereby reducing the interfering power in the information bandwidth. This is illustrated in Figure 3.2. The spread-spectrum signal (s) receives a narrowband interference (i). At the receiver the SS signal is "despread" while the interference signal spreads, making it appear as background noise compared to the despread signal.
5. Anti-jamming capability, especially narrowband jamming. This is more or less the same as interference rejection except the interference is now wilfully inflicted on the system. It is this property together with the next one that makes spread-spectrum modulation attractive for military applications.
6. Low probability of interception (LPI) or covert operation. Because of its low power density, the spread-spectrum signal is difficult to detect.

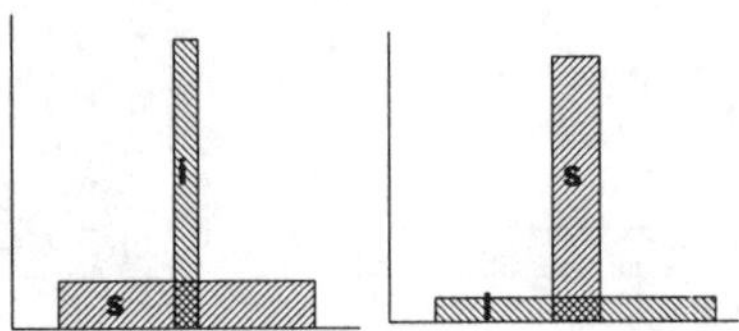

Figure 3.2 Interference rejection.

There are a number of modulation techniques that generate spread-spectrum signals. We briefly discuss the most important ones:

Direct-sequence spread-spectrum. The information-bearing signal is multiplied directly by a fast code signal.

Frequency hopping spread-spectrum. The carrier frequency at which the information-bearing signal is transmitted is rapidly changed according to the code signal

Time hopping spread-spectrum. The information-bearing signal is not transmitted continuously. Instead the signal is transmitted in short bursts where the times of the bursts are decided by the code signal.

Chirp modulation. This kind of spread-spectrum modulation is almost exclusively used in military radars. The radar continuously transmits a low power signal whose frequency is (linearly) varied (swept) over a wide range.

Hybrid modulation. Two or more of the above-mentioned SS modulation techniques can be used together to combine the advantages and, it is hoped, to combat their disadvantages.

In the next section the above-mentioned modulation techniques are used to obtain the multiple access capability that we want for CDMA (SSMA) protocols. However, the remainder of the chapters mainly concentrates on direct-sequence (DS)-CDMA and its related subjects, e.g., hybrid DS/SFH CDMA, hybrid DS CDMA/ISMA protocol, etc.

3.2 SPREAD-SPECTRUM MULTIPLE ACCESS

We can classify the SSMA or CDMA protocols in two different ways: by concept or by modulation method. The first classification gives us two protocol groups, averaging systems and avoidance systems. The averaging systems reduce the interference by averaging the interference over a wide time interval. The avoidance systems reduce the interference by avoiding it for a large part of the time.

Classifying by modulation gives us five protocols, direct-sequence (or pseudo-noise), frequency hopping, time hopping protocols based on chirp modulation and hybrid methods. Of these, the first (DS) is an averaging SS protocol while the hybrid protocols can be averaging protocols depending on whether DS is used as part of the hybrid method. All the other protocols are of the avoidance type. Table 3.1 summarizes both ways of classification.

Table 3.1
Classifying spread-spectrum MA protocols

	DS	TH	FH	Chirp	Hybrid
Averaging	x				x
Avoidance		x	x	x	x

In the following sections, the CDMA protocols are discussed where a division has been made that is based on the modulation technique.

3.2.1 Direct Sequence (DS)

In the DS-CDMA protocols the modulated information-bearing signal (the data signal) is directly modulated by a digital code signal. The data signal can be either an analog signal or a digital one. In most cases it will be a digital signal. What we often see in the case of a digital signal is that the data modulation is omitted and the data signal is directly multiplied by the code signal and the resulting signal modulates the wideband carrier. It is from this direct multiplication that the direct-sequence CDMA protocol gets its name.

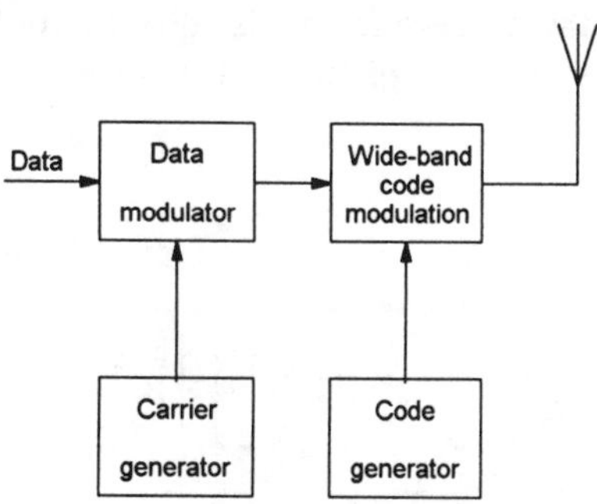

Figure 3.3 Block diagram of a DS-SSMA transmitter.

In Figure 3.3 a block diagram of a DS-CDMA transmitter is given. The binary data signal modulates an RF carrier. The modulated carrier is then modulated by the code signal. This code signal consists of a number of code bits or "chips" that can be either +1 or -1. To obtain the desired spreading of the signal, the chip rate of the code signal must be much higher than the chip rate of the information signal. For the code modulation various modulation techniques can be used but usually some form of phase shift keying (PSK) like binary phase shift keying (BPSK), differential binary phase shift keying (D-BPSK), quadrature phase shift keying (QPSK) or minimum shift keying (MSK) is employed.

If we omit the data modulation and use BPSK for the code modulation, we get the block diagram given in Figure 3.4.

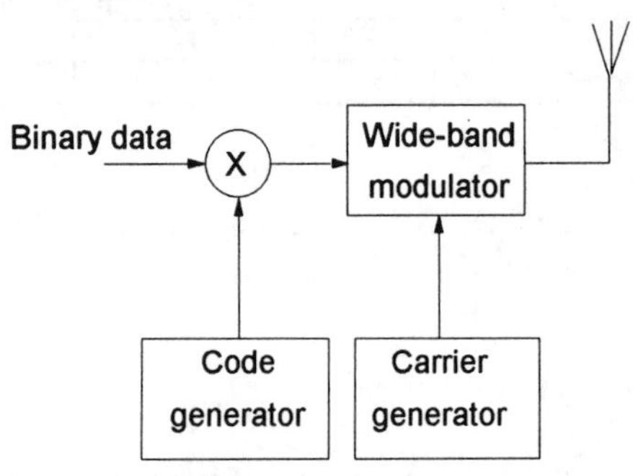

Figure 3.4 Modified block diagram of a DS-SS transmitter.

The DS-SS signal resulting from this transmitter is shown in Figure 3.5. In this figure, 10 code signals per information signal are transmitted (the code chip rate is 10 times the information chip rate) so the processing gain is equal to 10. In practice, the processing gain will be much larger (in the order of 10^2 to 10^3).

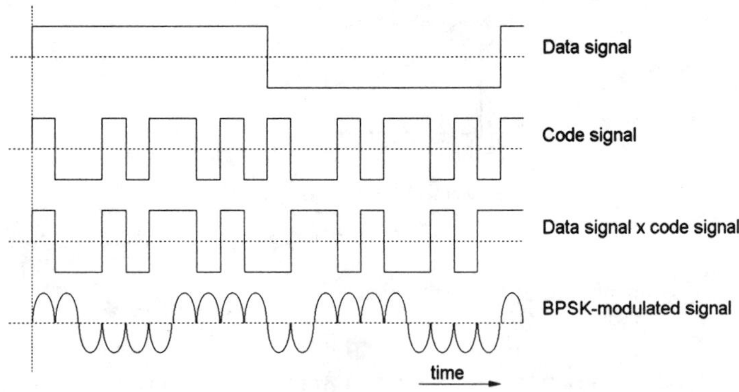

Figure 3.5 Generation of a BPSK-modulated SS signal.

After transmission of the signal, the receiver (which can be seen in Figure 3.6) uses coherent demodulation to despread the SS signal, using a locally generated code sequence. To be able to perform the despreading operation, the receiver must not only know the code sequence used to spread the signal, but the codes of the received signal and the locally generated code must also be synchronized. This synchronization must be accomplished at the beginning of the reception and maintained until the whole signal has been received. The synchronization/tracking block performs this operation. After despreading a data modulated signal results and after demodulation the original data can be recovered.

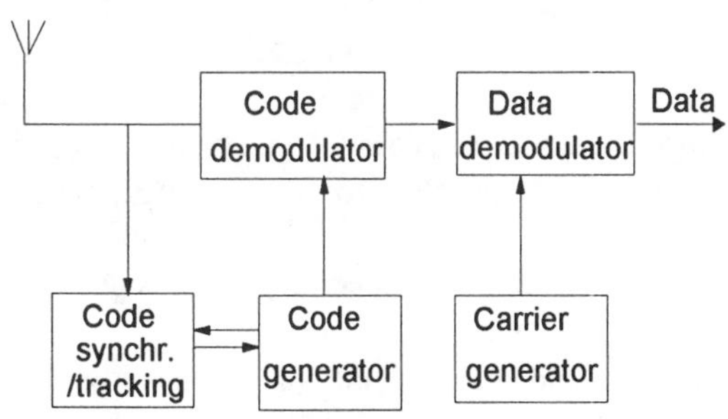

Figure 3.6. Receiver of a DS-SS signal.

In the previous section a number of advantageous properties of spread-spectrum signals were mentioned. The most important of those properties from the viewpoint of CDMA protocols is the multiple access capability, the multipath interference rejection, the narrowband interference rejection, and, with respect to secure/private communication, the low probability of interception (LPI). We explain these four properties for the case of DS-CDMA.

- Multiple access: If multiple users use the channel at the same time, there will be multiple DS signals overlapping in time and frequency. At the receiver coherent demodulation is used to remove the code modulation. This operation concentrates the power of the desired user in the information bandwidth. If the cross-correlation between the code of the desired user and the code of the interfering user is small, coherent detection will only put a small part of the power of the interfering signals into the information bandwidth.
- Multipath interference: If the code sequence has an ideal autocorrelation function, then (as can be seen in Section 3.3) the correlation function is zero outside the interval $[-T_c, T_c]$ where T_c is the chip duration. This means that if the desired signal and a version that is delayed for more than $2T_c$ are received, coherent demodulation will treat the delayed version as an interfering signal putting only a small part of the power in the information bandwidth.

- Narrowband interference: The coherent detection at the receiver involves a multiplication of the received signal by a locally generated code sequence. However, as we saw at the transmitter, multiplying a narrowband signal with a wideband code sequence spreads the spectrum of the narrowband signal so that its power in the information bandwidth decreases by a factor equal to the processing gain.
- LPI: Because the direct sequence signal uses the whole signal spectrum all the time, it will have a very low transmitted power per hertz. This makes it very difficult to detect a DS signal.

Apart from the above-mentioned properties, the DS-CDMA-protocols have a number of other specific properties that we can divide into advantageous (+) and disadvantageous (–) behavior:

\+ The generation of the coded signal is easy. It can be done by a simple multiplication.

\+ Since only one carrier frequency has to be generated, the frequency synthesizer (carrier generator) is simple.

\+ Coherent demodulation of the spread-spectrum signal is possible.

\+ No synchronization among the users is necessary.

– It is difficult to acquire and maintain the synchronization of the locally generated code signal and the received signal. Synchronization has to take place within a fraction of the chip time.

– For correct reception the locally generated code sequence and the received code sequence must be synchronized within a fraction of the chip time. This combined with the nonavailability of large contiguous frequency bands practically limits the spread bandwidth to 10-20 MHz.

– The power received from users close to the base station is much higher than that received from users further away. Since a user continuously transmits over the whole bandwidth, a user close to the base will constantly create a lot of interference for users far from the base station, making their reception impossible. This near-far effect can be solved by applying a power control algorithm so that all users are received by the base station with the same average power. However this control proves to be quite difficult.

3.2.2 Frequency Hopping (FH)

In the frequency hopping CDMA protocols, the carrier frequency of the modulated information signal is not constant but changes periodically. During time intervals T the carrier frequency remains the same but after each time interval the carrier hops to another (or possibly the same) frequency. The hopping pattern is decided by the code signal. The set of available frequencies the carrier can attain is called the hop-set.

The frequency occupation of an FH-SS system differs considerably from a DS-SS system. A DS system occupies the whole frequency band when it transmits, whereas an

FH system uses only a small part of the bandwidth when it transmits but the location of this part differs in time.

Suppose an FH system is transmitting in frequency band 2 during the first time period (see Figure 3.7). A DS system transmitting in the same time period spreads its signal power over the whole frequency band so the power transmitted in frequency band 1 will be much less than that of the FH system. However the DS system transmits in frequency band 1 during all time periods while the FH system only uses this band part of the time. On average both systems will transmit the same power in the frequency band.

The difference between the FH-SS and the DH-SS frequency usage is illustrated in Figure 3.7.

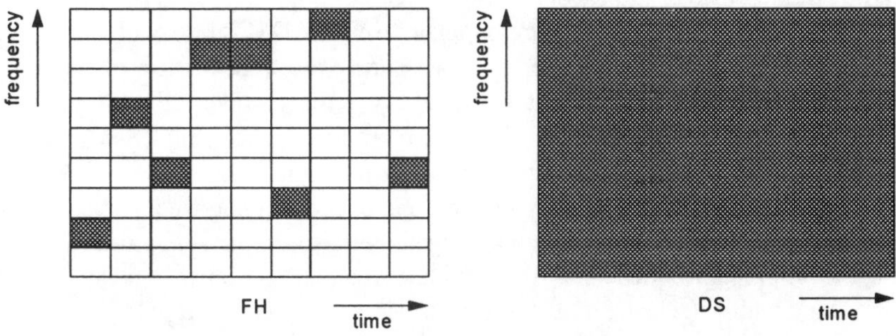

Figure 3.7 Time/frequency occupancy of FH and DS signals.

The block diagram for an FH-CDMA system is given in Figure 3.8.

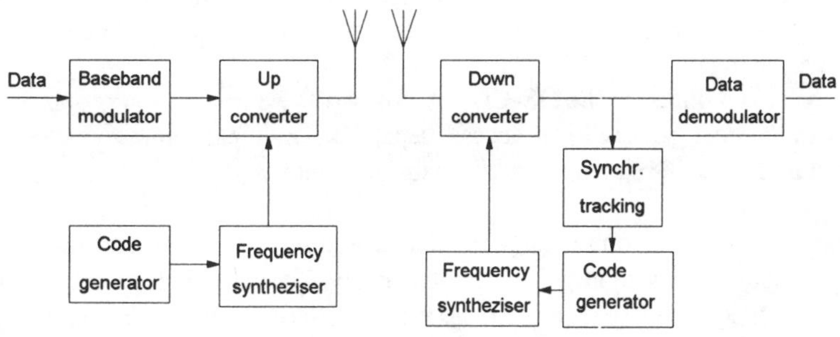

Figure 3.8 Block diagram of an FH-CDMA transmitter and receiver.

The data signal is baseband modulated on a carrier. Several modulation techniques can be used for this but it does not really matter for the application of frequency hopping. Usually FM modulation will be used for analog signals and GSK modulation for digital signals. Using a fast frequency synthesizer that is controlled by the code signal, the carrier frequency is converted up to the transmission frequency.

The inverse process takes place at the receiver. Using a locally generated code sequence the received signal is converted down to the baseband modulated carrier. The data is recovered after (baseband) demodulation. The synchronization/tracking circuit ensures that the hopping of the locally generated carrier synchronizes to the hopping pattern of the received carrier so that correct despreading of the signal is possible.

Within the frequency hopping CDMA protocols a distinction is made that is based on the hopping rate of the carrier. If the number of hops is (much) greater than the data rate, one speaks of a fast frequency hopping (F-FH) CDMA protocol. In this case the carrier frequency changes a number of times during the transmission of one bit, so that one bit is transmitted in different frequencies. If the number of hops is (much) smaller than the data rate, one speaks of slow frequency hopping (S-FH) CDMA protocols. In this case multiple bits are transmitted at the same frequency.

The occupied bandwidth of the signal on one of the hopping frequencies depends not only on the bandwidth of the information signal but also on the shape of the hopping signal and the hopping frequency. If the hopping frequency is much smaller than the information bandwidth (which is the case in slow frequency hopping), then the information bandwidth is the main factor that decides the occupied bandwidth. If however the hopping frequency is much greater than the information bandwidth, the pulse shape of the hopping signal will decide the occupied bandwidth at one hopping frequency. If this pulse shape is very abrupt (resulting in very abrupt frequency changes), the frequency band will be very broad, limiting the number of hop frequencies. If we make sure that the frequency changes are smooth, the frequency band at each hopping frequency will be about $1/T_h$ times the frequency bandwidth, where T_h is equal to the hopping frequency. We can make the frequency changes smooth by decreasing the transmitted power before a frequency hop and increasing it again when the hopping frequency has changed.

As has been done for the DS-CDMA protocols, we discuss the properties of FH-CDMA with respect to multiple access capability, multipath interference rejection, narrowband interference rejection, and probability of interception.

- Multiple access: It is quite easy to visualize how the F-FH and S-FH CDMA protocols obtain their multiple access capability. In the F-FH protocol one bit is transmitted in different frequency bands. If the desired user is the only one to transmit in most of the frequency bands, the received power of the desired signal will be much higher than the interfering power and the signal will be received correctly.

  In the S-FH protocol multiple bits are transmitted at one frequency. If the probability of other users transmitting in the same frequency band is low enough, the desired user will be received correctly most of the time. For those times that interfering users transmit in the same frequency band, error-correcting codes are used to recover the data transmitted during that period.

- Multipath interference: In the F-FH CDMA protocol the carrier frequency changes a number of times during the transmission of one bit. Thus, a particular signal frequency will be modulated and transmitted on a number of carrier frequencies. The multipath effect is different at the different carrier frequencies. As a result, signal frequencies that are amplified at one carrier frequency will be attenuated at another carrier frequency and vice versa. At the receiver the responses at the different hopping frequencies are averaged thus reducing the multipath interference. This is not as effective as the multipath interference rejection in a DS-CDMA system but it still gives quite an improvement.
- Narrowband interference: Suppose a narrowband signal is interfering on one of the hopping frequencies. If there are G_p hopping frequencies (where G_p is the processing gain), the desired user will (on the average) use the hopping frequency where the interferer is located $1/G_p$ percent of the time. The interference is therefore reduced by a factor G_p.
- LPI: The difficulty in intercepting an FH signal lies not in its low transmission power. During a transmission, it uses as much power per hertz as a continuous transmission would. But the frequency at which the signal is going to be transmitted is unknown and the duration of the transmission at a particular frequency is quite small. Therefore, although the signal is more readily intercepted than a DS signal, it is still a difficult task to perform.

Apart from the above-mentioned properties, the FH-CDMA protocols have a number of other specific properties that we can divide into advantageous (+) and disadvantageous (–) behavior:

+ Synchronization is much easier with FH-CDMA than with DS-CDMA. With FH-CDMA synchronization has to be within a fraction of the hop time. Since spectral spreading is not obtained by using a very high hopping frequency but by using a large hop-set, the hop time will be much longer than the chip time of a DS-CDMA system. Thus, an FH-CDMA system allows a larger synchronization error.
+ The different frequency bands that an FH signal can occupy do not have to be contiguous, because we can make the frequency synthesizer easily skip over certain parts of the spectrum. Combined with the easier synchronization, this allows much higher spread-spectrum bandwidths.
+ Because FH-CDMA is an avoidance SS system, the probability of multiple users transmitting in the same frequency band at the same time is small. If a user far from the base station transmits, it will be received by the base station even if users close to the base station are transmitting, since those users will probably be transmitting at other frequencies. Thus, the near-far performance is much better than that of DS.

- + Because of the larger possible bandwidth a FH system can employ, it offers a higher possible reduction of narrowband interference than a DS system.
- – A highly sophisticated frequency synthesizer is necessary.
- – An abrupt change of the signal when changing frequency bands will lead to an increase in the frequency band occupied. To avoid this, the signal has to be turned off and on when changing frequency.
- – Coherent demodulation is difficult because of the problems in maintaining phase relationships during hopping.

3.2.3 Time Hopping (TH)

In the time hopping CDMA protocols the data signal is transmitted in rapid bursts at time intervals determined by the code assigned to the user.

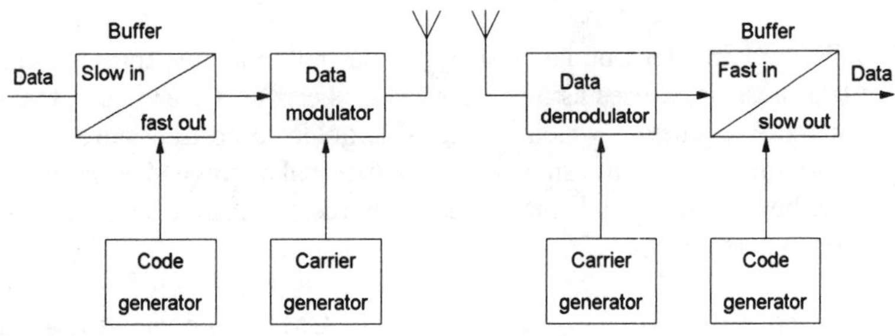

Figure 3.9 Block diagram of a TH-CDMA transmitter and receiver.

The time axis is divided into frames and each frame is divided into M time slots. During each frame the user will transmit in one of the M time slots. Which of the M time slots is transmitted depends on the code signal assigned to the user. Since a user transmits all of its data in 1, instead of M time slots, the frequency it needs for its transmission has increased by a factor M. A block diagram of a TH-CDMA system is given in Figure 3.9.

Figure 3.10 shows the time-frequency plot of the TH-CDMA systems. Comparing Figure 3.10 with Figure 3.7, we see that the TH-CDMA protocol uses the whole wideband spectrum for short periods instead of parts of the spectrum all of the time.

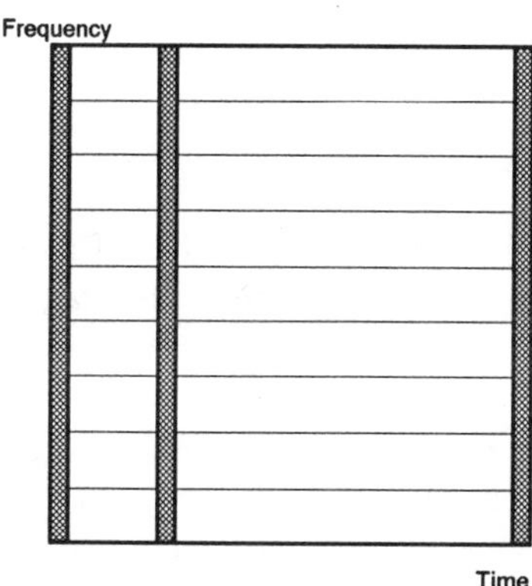

Figure 3.10 Time-frequency plot of the TH-CDMA protocol.

Following the same procedure as for the previous CDMA protocols, we discuss the properties of TH-CDMA with respect to multiple access capability, multipath interference rejection, narrowband interference rejection, and probability of interception.

- Multiple access: The multiple access capability of TH-SS signals is acquired in the same manner as that of the FH-SS signals, namely by making the probability of users' transmissions in the same frequency band at the same time small. In the case of time hopping all transmissions are in the same frequency band so the probability of more than one transmission at the same time must be small. This is again achieved by assigning different codes to different users. If multiple transmissions do occur, error-correcting codes ensure that the desired signal can still be recovered.
  If there is synchronization among the users, and the assigned codes are such that no more than one user transmits at a particular slot, then the TH-CDMA protocol reduces to a TDMA protocol where the slot in which a user transmits is not fixed but changes from frame to frame.
- Multipath interference: In the time hopping CDMA protocol, a signal is transmitted in reduced time. The signaling rate therefore increases and dispersion of the signal will now much sooner lead to overlap of adjacent bits. Therefore, no advantage is to be gained with respect to multipath interference rejection.
- Narrowband interference: A TH-CDMA signal is transmitted in reduced time. This reduction is equal to $1/G_p$ where G_p is the processing gain. At the receiver we will only receive an interfering signal during the reception of the desired signal. Thus we only receive the interfering signal $1/G_p$ percent of the time, reducing the interfering power by a factor G_p.
- LPI: With TH-CDMA the frequency at which a user transmits is constant but the times at which a user transmits are unknown, and the durations of the transmissions

are very short. Particularly when multiple users are transmitting, this makes it difficult for an intercepting receiver to distinguish the beginning and end of a transmission and to decide which transmissions belong to which user.

Apart from the above-mentioned properties, the TH-CDMA protocols have a number of other specific properties that we can divide into advantageous (+) and disadvantageous (–) behavior:

+ Implementation is simpler than that of FH-CDMA protocols.
+ It is a very useful method when the transmitter is average-power limited but not peak-power limited since the data are transmitted is short bursts at high power.
+ As with the FH-CDMA protocols, the near-far problem is much less of a problem since TH-CDMA is an avoidance system, so most of the time a terminal far from the base station transmits alone, and is not hindered by transmissions from stations close by.
– It takes a long time before the code is synchronized and the time is short in which the receiver has to perform the synchronization.
– If multiple transmissions occur, a lot of data bits are lost so a good error-correcting code and data interleaving are necessary.

3.2.4 Chirp Spread Spectrum

Although chirp spread spectrum is not yet adapted as a CDMA protocol, for the sake of completeness a short description is given here.

A chirp spread-spectrum system spreads the bandwidth by linear frequency modulation of the carrier. This is shown in Figure 3.11.

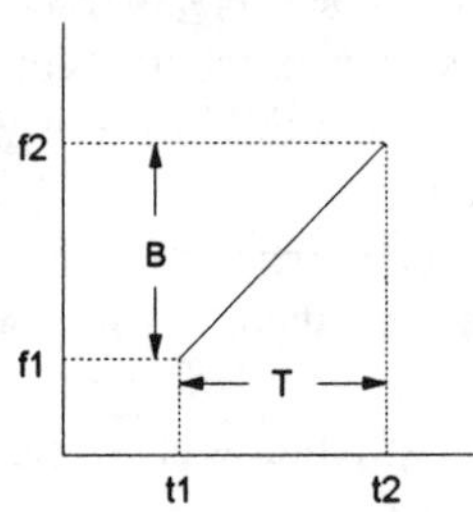

Figure 3.11 Chirp modulation.

The processing gain G_p is the product of the bandwidth B over which the frequency is varied and the duration T of a given signal waveform:

$$G_p = BT \tag{3.2}$$

3.2.5 Hybrid Systems

The hybrid CDMA systems include all CDMA systems that employ a combination of two or more of the above-mentioned spread-spectrum modulation techniques. If we limit ourselves to the direct-sequence, frequency hopping and time hopping modulations, we have four possible hybrid systems: DS/FH, DS/TH, FH/TH, and DS/FH/TH.

The idea of the hybrid system is to combine the specific advantages of each of the modulation techniques. If we take, for example, the combined DS/FH system we have the advantage of the anti-multipath property of the DS system combined with the favorable near-far operation of the FH system. Of course, the disadvantage lies in the increased complexity of the transmitter and receiver. For illustration purposes, we give a block diagram of a combined DS/FH CDMA transmitter in Figure 3.12.

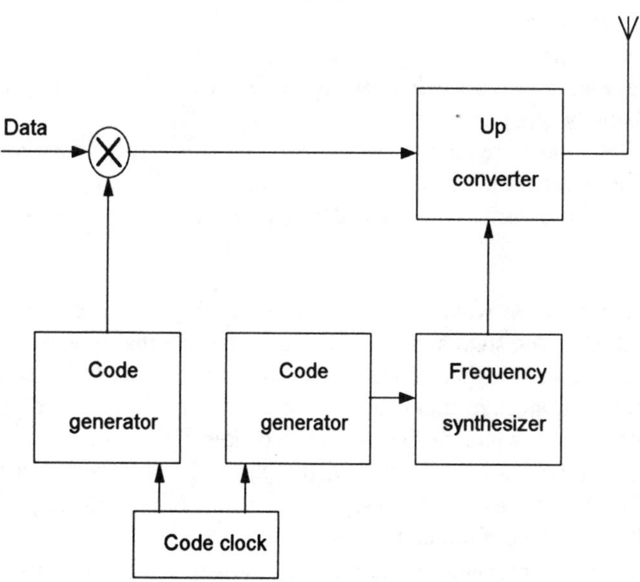

Figure 3.12 Hybrid DS-FH transmitter.

The data signal is first spread using a DS code signal. The spread signal is then modulated on a carrier whose frequency hops according to another code sequence. A code clock ensures a fixed relation between the two codes.

3.3 CODE GENERATION

For all the CDMA protocols that have been discussed in the previous sections, it is important that the codes that are assigned to the users make possible a good separation between the signal of a desired user and the signals of interfering users. Since the separation is made by correlating the received signal with the locally generated code signal of the desired user, we can translate this demand to the demand of a low cross-correlation between the codes assigned to the different users. The autocorrelation of the code is also a very important aspect because this decides how well we are able to synchronize and lock the locally generated code signal to the received signal. Also the ability of a DS-CDMA system to combat the effects of multipath interference depends on the autocorrelation function. The ideal autocorrelation function of a binary sequence would be an impulse function centered around zero.

The demands on the correlation properties are probably the most important demands for the development of suitable codes for CDMA communication. We can, however, think of a number of other demands. A reasonably complete list of the desired code properties follows:

1. The code signal must be easy to generate;
2. It must have the desired randomness properties (autocorrelation, cross-correlation, equal probability of a 0 or a 1);
3. Long periods are required to obtain a satisfactory autocorrelation. The autocorrelation function repeats itself each period;
4. To obtain secure transmission, the code signal must be difficult to reconstruct from a short segment.

There are several ways in which code sequences can be generated. One of these is the use of feedback shift registers and this method is the one that is generally employed in CDMA systems. A shift register consists of a number of cells (numbered from 1 to r) where each cell is a storage unit that, under the control of a clock pulse, moves its contents to its output while reading its new contents from its input. In the normal configurations of a feedback shift register, the input of cell m will be a function of the output of cell m−1 and the output of cell r (the last cell of the shift register). The output of the last cell forms the desired code sequence.

The function combining the outputs of cell m−1 and cell r to an input for cell m can be either a linear function or a nonlinear function. This gives us two classes of feedback shift registers, the linear feedback shift registers (linear FSRs) and the nonlinear feedback shift registers (nonlinear FSRs).

The Linear Feedback Shift Registers

The two most common realizations of linear FSRs are shown in Figures 3.13a and 3.13b. The FSR in Figure 3.13a is called a simple FSR because all the feedback signals are returned to a single input. The FSR in Figure 3.13b is called a modular FSR In this configuration the feedback is returned to two or more inputs. All the additions in the figure are module 2 additions, and c1 ... c4 are coefficients that are either 0 or 1.

A popular linear FSR is the maximum length FSR. A linear FSR with r storage elements is a maximum length FSR if it produces the longest possible sequence of 2^{r-1} before repeating itself. The popularity of these sequences is a result of the good autocorrelation and cross-correlation behavior. For the autocorrelation of an ML (maximum length) sequence of length L one can show that:

$$R(\tau) = \begin{cases} 1 & \tau = 0, L, 2L, \ldots \\ \dfrac{-1}{L} & \text{otherwise} \end{cases} \tag{3.3}$$

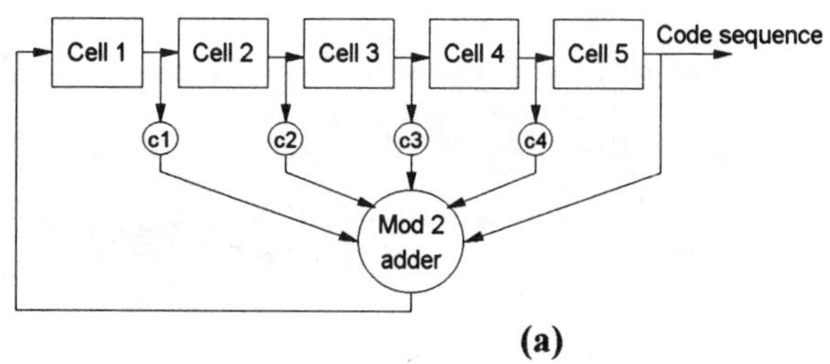

(a)

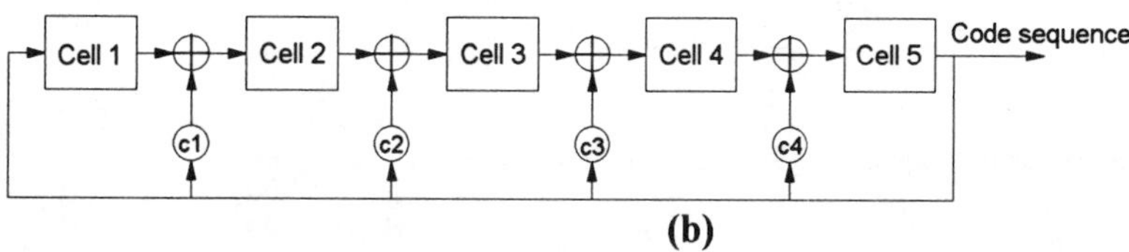

(b)

Figure 3.13 Block diagram of (a) a simple linear FSR and (b) a modular linear FSR.

As an example the autocorrelation and cross-correlation of two 31-bit ML sequences are shown in Figure 3.14. The first sequence is obtained by feeding back the outputs of cells 3 and 5 to the input of cell 1 while the second sequence is obtained by feeding back the outputs 5, 4, 3 and 2. Figure 3.15 shows the feedback shift register for the first sequence.

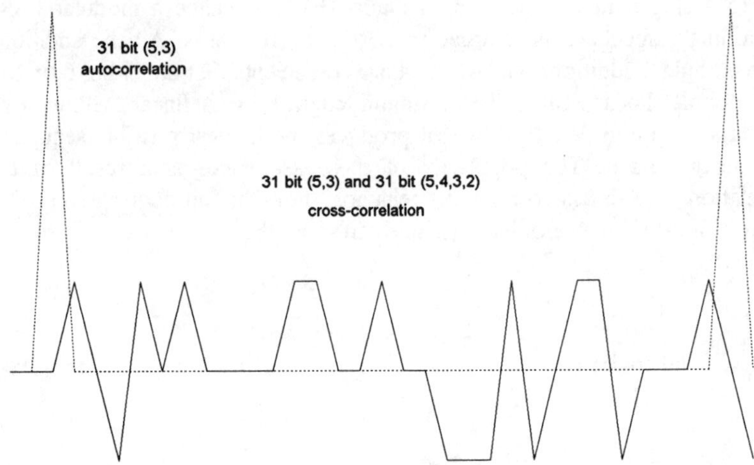

Figure 3.14 The autocorrelation and cross-correlation of maximum length code.

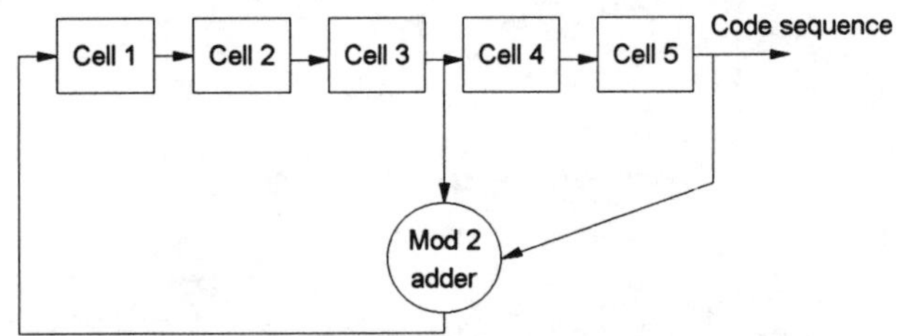

Figure 3.15 Block diagram of a (5,3) maximum FSR.

To see the advantages of ML codes with respect to non-ML codes using the same number of storage elements, we show the autocorrelation and cross-correlation of two non-ML codes that are also obtained from a 5-cell shift register. The first code is obtained by feeding back outputs 5 and 4, the second code is obtained by feeding back outputs 5 and 1. Both sequences are 21 bits. The autocorrelation of the first code and cross-correlation are shown in Figure 3.16.

A good measure of the rejection of the signals of interfering users is the ratio of the maximum cross-correlation coefficient and the autocorrelation coefficient. The smaller this ratio is, the better are the interfering users rejected. For the maximum length codes in Figure 3.15, this ratio can be calculated as 0.23 while for the non-ML codes in Figure 3.16 this ratio can be calculated to be equal to 0.43.

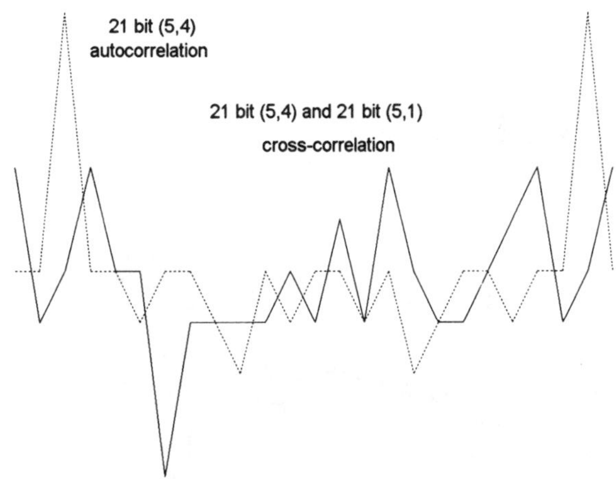

Figure 3.16 The autocorrelation and cross-correlation of non-ML codes.

The ML codes have the disadvantage that their number is limited. This was one of the reasons to search for non-ML codes that do have a satisfactory correlation behavior. One class of such codes are the Gold codes. These codes are generated by module 2 adding of the outputs of two different ML codes. Figure 3.17 shows the generation of a Gold code from two 31-bit ML sequences. By shifting the ML sequences with respect to one another, a new Gold code is produced. Thus 2^{n-1} Gold codes can be generated from 2 ML-FSR with n storage elements. To this number we can add the 2 original ML codes to obtain $2^{n-1}+1$ different codes.

For the Gold codes a formula can be derived that gives the maximum cross-correlation coefficient:

$$K = \begin{cases} 2^{(n+1)/2} + 1 & n \text{ odd} \\ 2^{(n+2)/2} + 1 & n \text{ even} \end{cases} \tag{3.4}$$

Since the maximum autocorrelation coefficient M is equal to the length of the code, we can find the following for the ratio K/M of the Gold codes from Figure 3.17:

$$\frac{K}{M} = \frac{2^{(5+1)/2} + 1}{2^5 - 1} = 0.29 \tag{3.5}$$

Comparing this to the ratio of the maximum length codes, we see that the Gold codes come quite close.

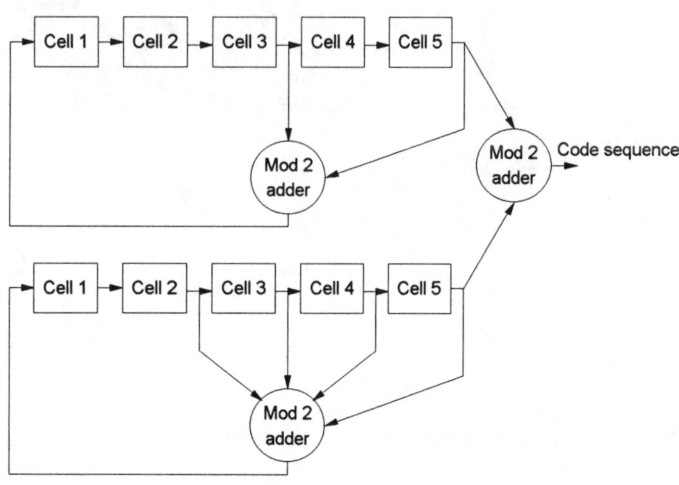

Figure 3.17 Gold code generator using (5,3) and (5,4,3,2) ML generators.

The linear FSRs have most of the desired properties we want for our codes. They are easy to generate and have reasonable randomness properties. The ML codes in particular have long periods. The property of a difficult reconstruction from a short segment is, however, not met by the linear FSRs. This is due to the linear mathematical relation in generating the code, making it easy to solve the code as a solution to a set of linear equations. One way to solve this problem is to use a linear FSR with a nonlinear output as shown in Figure 3.18. Another solution is the use of nonlinear FSRs.

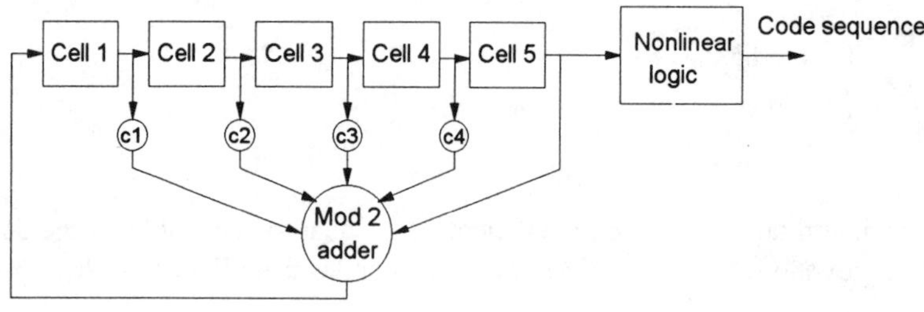

Figure 3.18 Block diagram of a linear FSR with nonlinear output.

The Nonlinear Feedback Shift Registers

The linear relation in the maximum length codes makes it quite easy to discover the code. Using a nonlinear relation in generating the code makes it much harder to break . This leads to the nonlinear feedback shift registers. These registers are more or less the same as the linear FSRs but the module 2 addition in Figure 3.13a is replaced by a nonlinear function. The block diagram of a nonlinear FSR is shown in Figure 3.19.

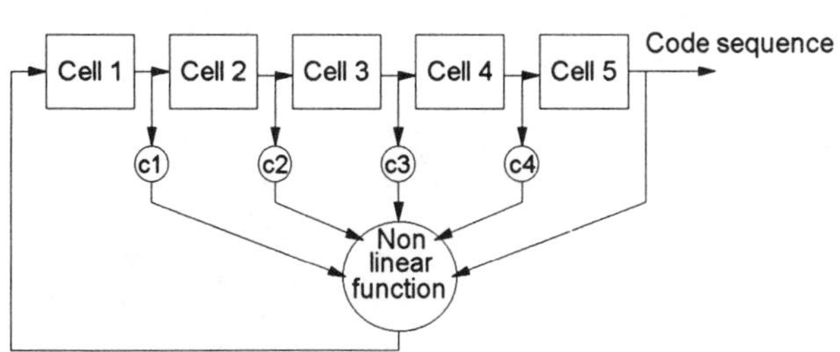

Figure 3.19 Block diagram of a nonlinear FSR.

Nonlinear codes do however have a number of disadvantages of their own. First, not much is known about the properties of the nonlinear FSRs and, second, some of the nonlinear FSRs may not generate a cycle but a transient instead, so from any initial state the register will propagate toward the all-ones state. For these reasons the nonlinear FSRs are not much used in CDMA systems.

Because the choice of a suitable code is so critical in a CDMA system, much research is still going on in trying to develop new codes. At the moment the Gold codes are among the most popular codes used because of the relative ease with which such codes can be generated and the favorable correlation properties.

REFERENCES

[1] Special issue on spread-spectrum communication, *IEEE Trans.Comm.*, Vol. COM-30, May 1982.

[2] M.K. Simon, J.K. Omura, R.A. Scholtz and B.K. Levitt, Spread-spectrum communications, Vol. I, II, III, *Comp. Sci.*, 1985.

[3] R.A. Scholtz, "The spread-spectrum concept," *IEEE Transactions on Communications*, Vol. COM-25, pp. 748-755, August 1977.

[4] D.J. Torrieri, "*Principles of Secure Communication Systems,*" Artech House, Norwood, Mass., 1985.

[5] G.R. Cooper and C.D. McGillem, "*Modern Communication and Spread-Spectrum,*" McGraw-Hill Book Company, New York, 1986.

[6] S.G. Glisic and P.A. Leppanen (ed.), "*Code Divison Multiple Access Communications,*" Kluwer Academic Publishers, Boston, 1995.

[7] A.J. Viterbi, "*CDMA Principles of Spread-Spectrum Communications,*" Addison-Wesley Publishing Company, Reading, Mass., 1995.

[8] R.C. Dixon, "*Spread-Spectrum Systems,*" John Wiley & Sons, New York, 1984.

Chapter 4
DS-CDMA Capacity with Imperfect Power Control

4.1 INTRODUCTION

In a DS-CDMA system, the requirement for power control is the most serious negative point. The power control problem arises because of the multiuser interference. All users in a DS-CDMA transmit the messages by using the same bandwidth at the same time and therefore users interfer with each other. Due to the propagation mechanism, the signal received by the base station from a user terminal close to the base station will be much stronger than the signal received from another terminal located at the cell boundary. Hence the distant users will be dominated by the close-in user. To achieve a considerable capacity, all signals, irrespective of the distance, should arrive at the base station with the same mean power. A solution to this problem is power control, which attempts to achieve a constant received mean power for each user. Therefore the performance of the transmitter power control (TPC) is one of the several dependent factors when deciding on the capacity of a DS-CDMA system.

This chapter is organized as follows: Section 4.2 describes the near-far effect model. A generalized expression has been assumed for the spatial distribution of traffic in investigating the near-far effect on the performance of a CDMA network. The traffic distribution models the practical circumstances of the traffic in the mobile communication environment for any value of path-loss exponent. This assumption yields a log-normal pdf (probability density function) of the mean power.

In Section 4.3 the performance of the reverse link (from terminal transmitter to base station receiver) is investigated. First, we study a CDMA system without any power control in the reverse link, assuming a generalized spatial user distribution which shows that the power control is inevitable for a CDMA system. Using the generalized spatial distribution, the influence of the near-far effect on the capacity of a CDMA system is investigated. It is rather impractical to assume that there is perfect power control.

Therefore, second, the imperfection effect on the reverse link capacity of a CDMA system is evaluated for both single cell and multiple cell networks in Section 4.3. Imperfection may be caused by the dynamic range of the TPC, the spatial user distributions, and the propagation statistics (such as fading and shadowing). The effect of each item on performance is not investigated explicitly. To investigate the overall effect of imperfect power control, a combined scheme is considered in terms of the standard deviation of the log-normal distribution of the received signal. The outage probability is evaluated in the presence of imperfect power control.

A mathematical model is developed to obtain the performance for the forward link (from base station transmitter to user terminal receiver) of a CDMA system in Section 4.4. The two types of power control schemes are discussed in this section, namely, carrier-to-interference (C/I) and distance-driven schemes. One of the forward power control schemes is investigated in the presence of shadowing. Finally, the forward link and reverse link capacities are compared.

4.2 NEAR-FAR EFFECT

Figure 4.1 illustrates a CDMA system with a cluster size of one.

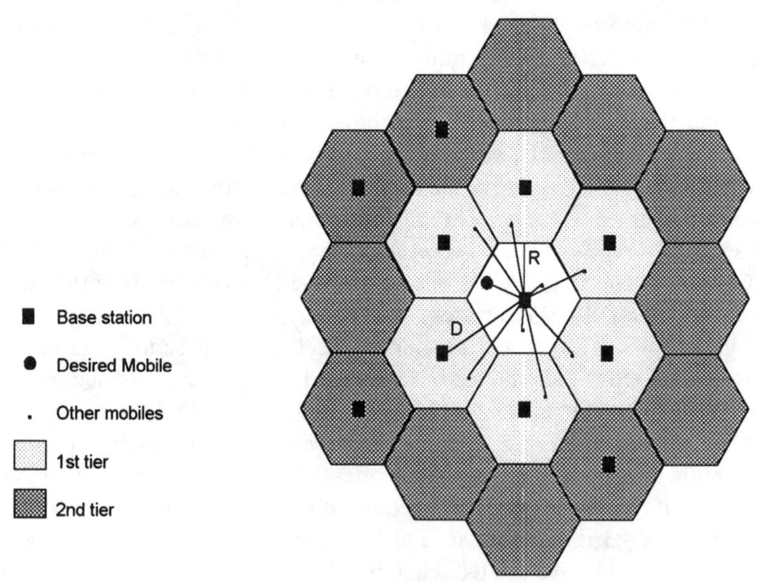

Figure 4.1 Cellular CDMA system.

The near-far effect in such a system is described by expressing the mean power at the central receiver of the home cell as

$$\xi_i \approx r^{-\gamma} \tag{4.1}$$

Here, γ is the path-loss law exponent for the mobile radio channel which lies in the range of 2 to 5 and r is defined as the (normalized) distance of a mobile transmitter from the central base station:

$$r \overset{\Delta}{=} \frac{d}{d_{max}} \tag{4.2}$$

where d is the distance from the receiving station to the mobile terminal and d_{max} is the radius of the circular area with its center at the base station, and in this circular region mobile terminals move around.

To study the near-far effect on a DS-CDMA system, it is necessary to assume a suitable spatial distribution of the generated traffic $\tilde{N}(r)$, which is defined as the number of users per unit of area at a distance r from the common receiver.

Several types of spatial distributions have been assumed in the literature for the sake of simplicity in analysis (e.g., quasi-uniform distribution [1], bell-shaped distribution [2], etc). Generally a spatial distribution is assumed which should be close to the actual model and easy to analyze. Accordingly, we have considered a spatial distribution model [3,4] which is highly suitable to study the near-far effects in a real mobile environment, for example, with Rayleigh fading and log-normal shadowing. A general spatial distribution function is assumed as

$$\tilde{N}(r) = \frac{N_t}{(2\pi)^{3/2} r^2 \sigma_i} \exp\left[\frac{(ln\xi_i)^2}{2\sigma_i^2}\right] \tag{4.3}$$

where ξ_i is the mean power, σ_i^2 is the logarithmic spatial variance, and N_t is the total traffic offered to the channel.

Figure 4.2 shows the plot $\tilde{N}(r)$ versus r for N_t=1 and spatial spread σ_i as a parameter considering γ=3 (Figure 4.2a), γ=4 (Figure 4.2b) and γ=5 (Figure 4.2c). We see from Figure 4.2 that for $\sigma_i \leq 6$ dB.

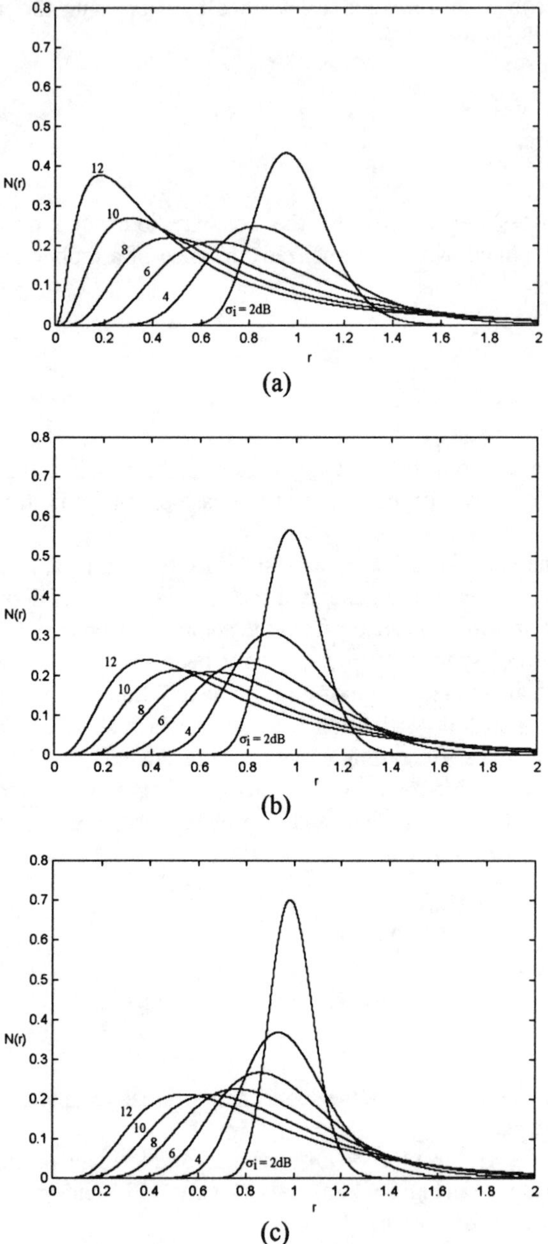

Figure 4.2 Spatial distribution of traffic: (a) path-loss exponent $\gamma = 3$; (b) $\gamma = 4$; (c) γ=5, with a total traffic $N_t = 1$ and spatial spread σ_i shown in radio cell (0< r <=1).

$\tilde{N}(r)$ simulates a traffic distribution such that most of the users' signals arrive from a considerable distance as compared to the case for $\sigma_i > 6$ dB. Now, one can easily select a suitable value for σ_i to investigate the near-far effect on mobile radio networks for a given value of γ and spatial distribution of traffic. Figure 4.2 shows that σ_i increases with the increase in γ for similar traffic distribution. By defining a spatial distribution shape factor S_i as

$$S_i \triangleq \frac{\sigma_i}{\gamma} \tag{4.4}$$

and using (4.1), (4.3) can be written as

$$\tilde{N}(r) = \frac{N_t}{(2\pi)^{3/2} r^2 S_i} \exp\left[-\frac{(ln\, r)^2}{2S_i^2}\right] \tag{4.5}$$

Now by simply varying S_i, a different user distribution can be obtained, as shown in Figure 4.3. Thus now using S_i, we need not draw distribution function $\tilde{N}(r)$ for various combinations of σ_i and γ. Therefore S_i is called a spatial distribution shape factor and it determines the shape of the spatial distribution. It can be seen from Figures 4.2 and 4.3 that there are users beyond the cell boundary (i.e., beyond r=1). As far as CDMA systems are concerned, such a spatial distribution is a realistic consideration because in a CDMA cellular system the cluster size is 1 since every cell uses the same carrier frequency or set of carrier frequencies. Hence in a cellular CDMA system, each base station receives not only interference from users located within its own cell but also from users located in the surrounding cells.

Using (4.5), the pdf of the distance between mobile users and the base station can be obtained:

$$f_r(r) = \frac{2\pi}{N_t}\tilde{N}(r)r = \frac{1}{\sqrt{2\pi} r S_i}\exp\left[-\frac{(ln^2 r)}{2S_i^2}\right] \tag{4.6}$$

Using path-loss law relation (4.1) and (4.6), the pdf of the received mean power ξ_i becomes

$$f_{\xi_i}(\xi_i) = f_r(r)\left|\frac{dr}{d\xi_i}\right| = \frac{1}{\sqrt{2\pi}\sigma_i \xi_i}\exp\left[-\frac{(ln\,\xi_i)^2}{2\sigma_i^2}\right] \tag{4.7}$$

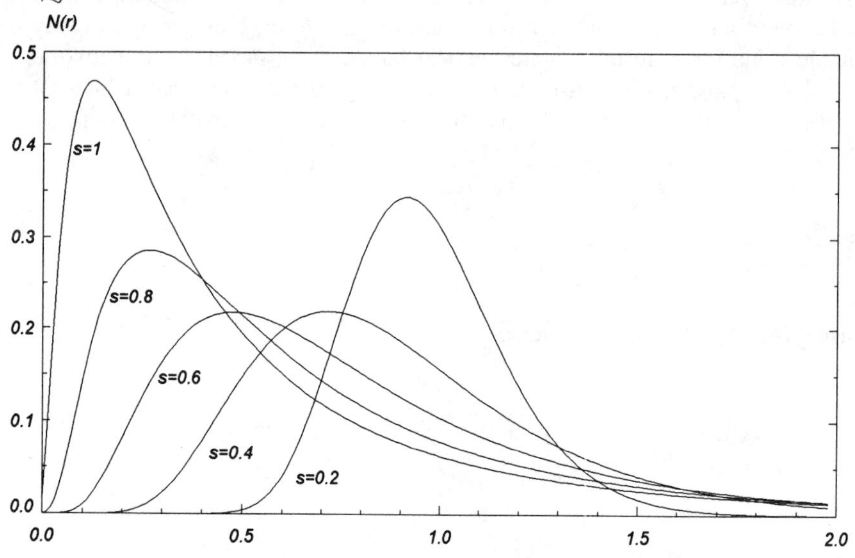

Figure 4.3 General spatial distribution function.

Equation (4.7) shows that the pdf of the received mean power is a log-normal distribution with zero logarithmic mean and logarithmic variance σ_i^2. The value of σ_i depends on the shape of $\tilde{N}(r)$ due to the factor S_i and on the path-loss γ. The reason for using the spatial user distribution given by (4.5) is fourfold.

1. It leads to log-normal pdfs for the mean power, which makes it analytically attractive because of well-known approximation techniques [5,6] to add log-normally distributed variables.
2. The effect of different values for the path-loss law exponent γ can be investigated.
3. The effect of different spatial distributions for the interferers can be investigated by simply varying parameter S_i (see Figure 4.3).
4. The near-far effect can be studied in a Rayleigh and log-normal fading environment. The distribution is realistic in the CDMA environments.

4.3 MULTIUSER INTERFERENCE IN THE REVERSE LINK

In this section the multiuser interference on the reverse link is discussed. The influence of the near-far effect on the multiuser interference is investigated. The bit energy to noise density ratio E_b/N_o is important for reliable system operation. The bit energy to noise density ratio E_b/N_o can be interpreted as the signal-to-noise ratio after correlation at the receiver. To have good communication quality, it is assumed that the C/I ratio (carrier-to-interference ratio) must be larger than the capture ratio α:

$$\frac{C}{I} \geq \alpha \quad \text{with } \alpha \triangleq G\left[\frac{E_b}{N_o}\right]_{min} \tag{4.8}$$

From [7,8] we find that the minimum requirement for E_b/N_o is 5 (+7 dB) in the case of noncoherent reception. It is reported in [8] that this value for E_b/N_o implies a bit error probability (BEP) equal to or greater than 10^{-3}, disregarding a specific modulation technique or code sequence. A BEP of 10^{-3} can be assumed to be adequate for digital voice transmission. The performance is measured in terms of outage probability, defined as the probability that the ratio of the reference user power and the multiuser power is less than the capture ratio:

$$P_{out} = Pr\left[\frac{P_d}{P_n} < \alpha\right] = \sum_{k=1}^{n} Pr\left[\frac{P_d}{P_k} < \alpha \middle| k\right] F(n,k) \tag{4.9}$$

where $\Pr(P_d \,/\, P_k < \alpha | k)$ is the conditional outage probability, P_d is the desired power, P_n is the joint interferers power caused by n interferences, P_k is the joint interference power caused by k interferers, and $F(n,k)$ is the probability that k out of n interferers are active. When using voice activity monitoring and assuming a transmitting probability a, the probability that k out of n interferers are active can be described by a binomial distribution:

$$F(n,k) = \binom{n}{k} a^k (1-a)^{n-k} \tag{4.10}$$

When no noise activity monitoring is used (a=1), then the outage probability is given by Pr $(P_d/P_k < \alpha)$. It is worth mentioning here that during conversation a speaker is not active all the time but remains silent for a certain percentage of the time. Extensive studies show that the either speaker is active only 35% to 40% of the time. Studies done in Europe suggest that the total activity due to voice and background noise is higher in a mobile environment than in a wireline and can have values between 50% and 60% [9]. It is possible to monitor the voice activity and suppress transmission for a user when no voice is present. This voice activity monitoring implies that the transmitter is not active during silent periods in human speech. Thus, the multiuser interference can be reduced. First, the single cell CDMA is discussed.

4.3.1 Single Cell

A single cell system is depicted in Figure 4.4 by a circular area. A central receiver is assumed at the base station B, and N mobile terminals are moving around it. We have considered one reference user, which implies that there are N−1 interferers.

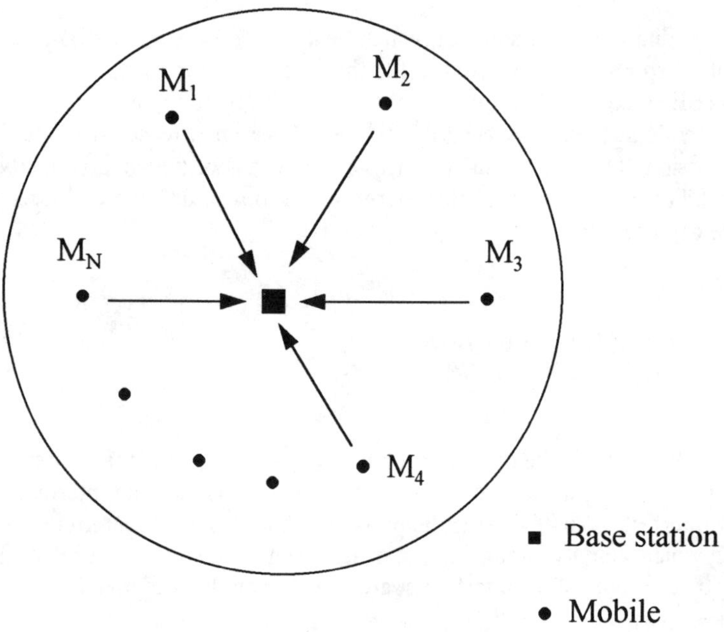

Figure 4.4 Receiver link in a single cell.

For notational convenience it is assumed that the number of interferers $n = N-1$. The base station receives the desired power $\overline{p}_d$ of the reference user and the joint interference power $\overline{p}_n(=\Sigma \overline{p}_i)$. Using (4.9) the conditional outage probability can be written as

$$P'_{out} = \int_0^{\infty} \{ \int_0^{\alpha \overline{p}_n} f\overline{p}_d(\overline{p}_d) d\overline{p}_d \} f\overline{p}_n(\overline{p}_n) d\overline{p}_n \tag{4.11}$$

Using (4.11), the conditional outage probability is evaluated because of the near-far effect. It means the pdf of $\overline{p}_d$ is a log-normal function given by (4.7) with the following substitution

$$\begin{aligned} \overline{p}_d &\rightarrow \xi_i \\ \sigma_d &\rightarrow \sigma_i \end{aligned} \tag{4.12}$$

The pdf of $\overline{p}_i$ is also a log-normal function. Therefore the pdf of $\overline{p}_n \left(=\Sigma \overline{p}_i\right)$ is also a log-normal function given by [5,6]:

$$f\,\overline{p}_n\,(\overline{p}_n) = \frac{1}{\sqrt{2\pi}\,\sigma_n \overline{p}_n} \exp\left[-\frac{\left(ln\,\overline{p}_n - m_n\right)^2}{2\sigma_n^2}\right] \tag{4.13}$$

Here m_n and σ_n^2 represent logarithmic mean and logarithmic variance, respectively. These moments can be obtained with Fenton's method for $\sigma_i < 4$ dB [5], and Schwartz and Yeh's technique for σ_i lies between 4 and 12 dB [6], assuming uncorrelated interferers. Fenton [5] showed that the sum of n stochastic independent log-normal variables is well approximated by another log-normal variable with mean:

$$m_n \underline{\Delta} \ln[\Sigma \exp(m_i + \sigma_i \, {}^2\!/\!_2 - \sigma_n \, {}^2\!/\!_2)] \tag{4.14}$$

and logarithmic variance

$$\sigma_n^2 \, \underline{\Delta} \ln[1 - \{\Sigma \exp(2m_i + \sigma_i^2)(1 - \exp\sigma_i^2)\} / \{\Sigma \exp(m_i + \sigma_i \, {}^2\!/\!_2)\}^2 \,] \tag{4.15}$$

Schwartz and Yeh [6] developed an alternative technique for the determination of m_n and σ_n^2. They derived an exact expression for the mean and variance for the sum of two log-normal variables and then used a recursive approach to obtain m_n and σ_n^2.

The conditional outage probability is determined by using (4.11)-(4.13).

$$P'_{out} = \frac{1}{\sqrt{2\pi}} \int_{l_n}^{\infty} \exp\left(-u^2\!/\!_2\right) du \; \underline{\Delta} Q(l_n) \tag{4.16}$$

where

$$l_n \underline{\Delta} (m_d - m_n - \ln \alpha) / (\sigma_d{}^2 + \sigma_n{}^2)^{1/2} \tag{4.17}$$

and $Q(l_n)$ is the Q function.

In the event of perfect power control without voice activity monitoring, the outage probability is given by

$$P_{out} = \Pr\left[(N-1) > \frac{1}{\alpha} \right] \tag{4.18}$$

because for perfect power control the average power received at the base station from the different mobiles is the same, $P_d/P_k = 1/(N-1)$.

In the case of voice activity monitoring and perfect power control, the outage probability with n interferers is given by the probability of more than n_α users being active at the same time:

$$\Pr(k > n_\alpha) = \begin{cases} 0 & k < n_a \\ \sum\limits_{k=n_\alpha}^{n} F(n,k) & k \geq n_\alpha \end{cases} \tag{4.19}$$

Here n_α is the maximum number of fully active interferers.

In practice in a power control system, the average received power at the base station may not be the same for each user signal. The performance of a power control system depends on the speed of the adaptive power control system, the dynamic range of the transmitter, the spatial distribution of the users, and the propagation statistics. All these factors influence the pdf of the received power. Because the objective of this chapter is to investigate the influence of power control imperfections on the capacity, the explicit influence of the factors mentioned above on the pdf of the average received power is not investigated. Instead, the pdf is assumed to be log-normal. The imperfection in the power control system is determined by the logarithmic variance σ^2 of the log-normal power distribution of the received signal. In the case of perfect power control the logarithmic variance is 0 dB. The outage probability due to the imperfect power control system can now be obtained in the way described above for the near-far effect.

To investigate the reverse link capacity of a single cell system without power control, two spatial distributions are considered according to (4.5) with S_i=0.58 and

S_i=0.69. Table 4.1 shows the logarithmic standard deviation of the received power for both distributions in case of γ=2 and γ=4.

Table 4.1
Logarithmic standard deviation S_i of received power for two spatial distributions (S_i) and two path-loss law exponents (γ)

	S_i = 0.58	S_i = 0.69
$\gamma = 2$	σ_i = 5 dB	σ_i = 10 dB
$\gamma = 4$	σ_i = 6 dB	σ_i = 12 dB

In Figures 4.5 and 4.6 the outage probability is plotted for the situation without power control and with perfect power control. From these figures we conclude that the performance decreases as the logarithmic standard deviation of the received power due to the near-far effect increases.

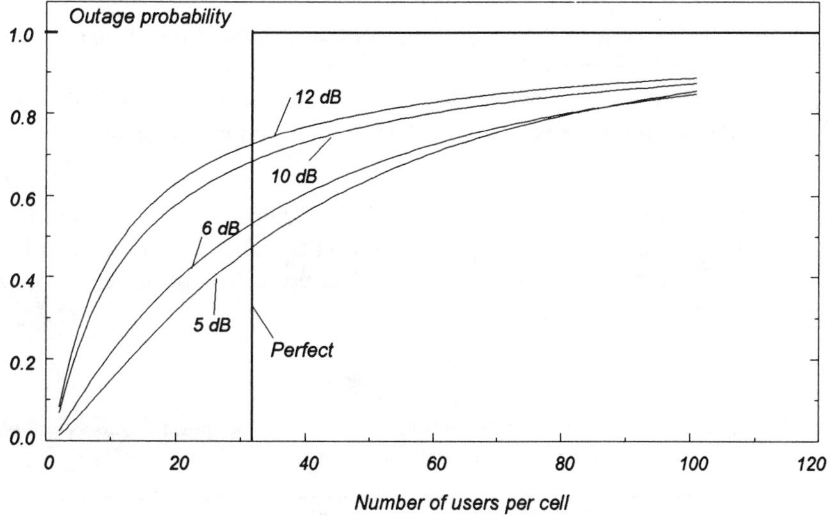

Figure 4.5 Reverse link outage probability as a function of number of users per cell for CDMA systems with perfect power control (σ_i=0 dB) and without power control (σ_i= 5, 6, 10, 12 dB). σ_i refers to the spread of the received power caused by the spatial distribution of users. Processing gain G=156, no voice activity monitoring.

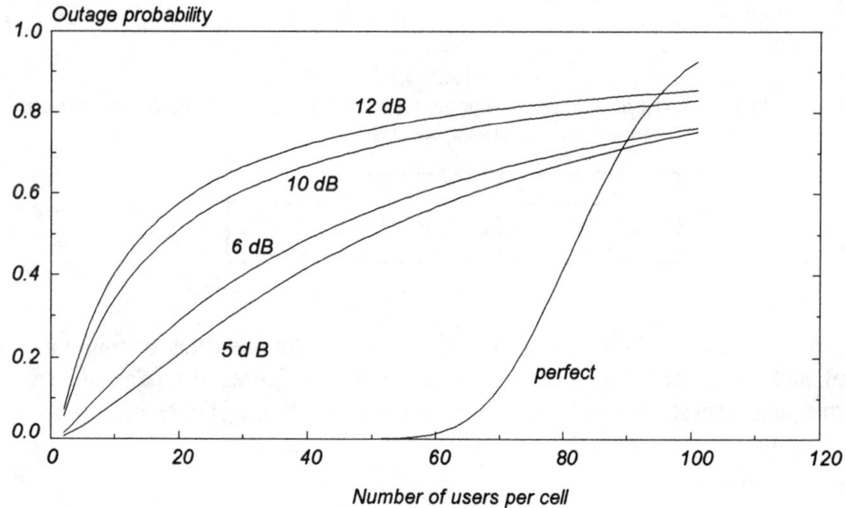

Figure 4.6 Reverse link outage probability as a function of number of users per cell for CDMA systems with perfect power control (σ_i= 0 dB) and without power control (σ_i= 5, 6, 10, 12 dB). σ_i refers to the spread of the received power caused by the spatial distribution of users. Processing gain G=156, voice activity monitoring with a=0.375.

In Table 4.2 the reverse link capacity is given for a single cell system with and without power control for given values of the processing gain G and voice activity a. It is clear from Table 4.2 that the capacity without reverse link power control is very low, which means that the power control is needed.

Table 4.2
Number of users per cell for an outage probability of 0.01. σ_i refers to the spread of the received power caused by the spatial distribution of users

	Perfect Power Control	No Power Control			
		σ_I=5 dB	σ_I=6 dB	σ_I=10 dB	σ_I=12 dB
G=100 / a=1	21	1	1	1	1
G=156 / a=1	33	1	1	1	1
G=156 / a=0.375	59	2	1	1	1

Figures 4.7 and 4.8 present the outage probability in the situation with perfect and imperfect power control. It is important to stress once more that in the model used for imperfect power control, the explicit influence of factors such as power control speed, spatial user distribution, and the like, is not considered. The results for perfect power control agree with the results obtained in [8].

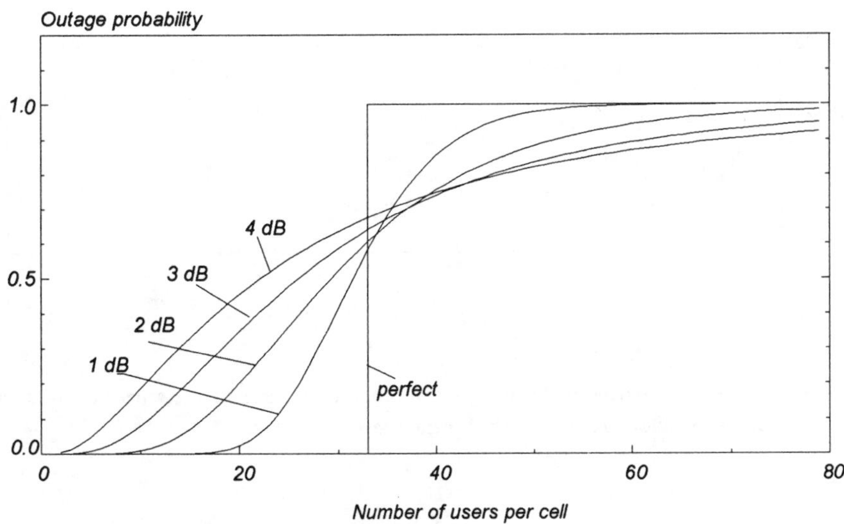

Figure 4.7 Reverse link outage probability as a function of number of users per cell for CDMA systems with perfect power control ($\sigma_i = 0$ dB) and with imperfect power control (σ = 1, 2, 3, 4 dB). σ refers to the spread of the received power caused by the imperfections of the power control system. Processing gain G=156, no voice activity monitoring (a=1).

We see from Figures 4.7 and 4.8 that the outage probability increases with the increase in the imperfection of the power control system. This is due to the random nature of the received power in the case of imperfect power control whereas in the case of perfect power control the received power was assumed to be deterministic.

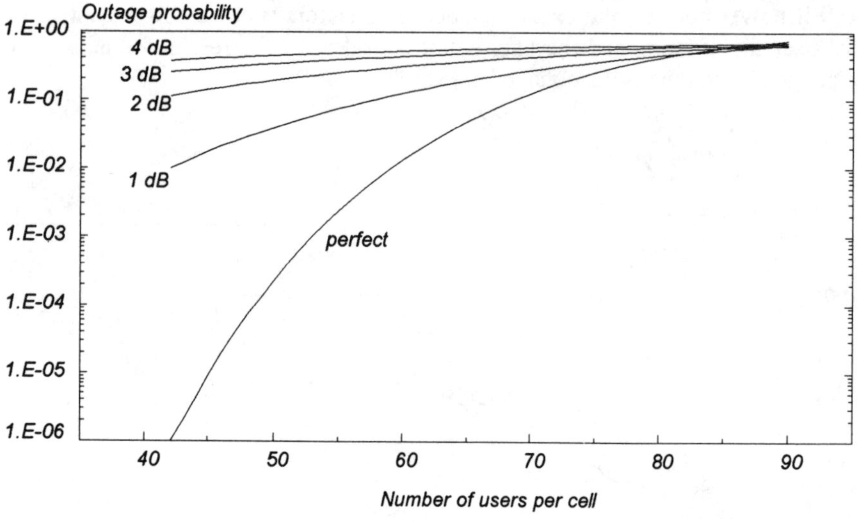

Figure 4.8 Reverse link outage probability as a function of number of users per cell with voice activity monitoring a=0.375. Other parameters as in Figure 4.7.

Table 4.3
Number of users per cell for an outage probability of 0.01. σ refers to the spread of the received power caused by the imperfections of the power control system

	Perfect Power Control	Imperfect Power Control			
		σ= 1dB	σ=2 dB	σ=3 dB	σ= 4dB
G=100 / a=1	21	12	7	5	3
G=156 / a=1	33	19	10	7	4
G=156 / a=0.375	59	43	23	12	5

Figures 4.9 and 4.10 show the effect of power control imperfections on the reverse link capacity. The value σ = 0 dB corresponds to perfect power control. The basic conclusion from Figures 4.9 and 4.10 is that the DS-CDMA system is very sensitive to power control imperfections. We further conclude that the capacity can be improved by increasing the processing gain G or decreasing the voice activity. In Table 4.3 the maximum permitted number of users per cell is given.

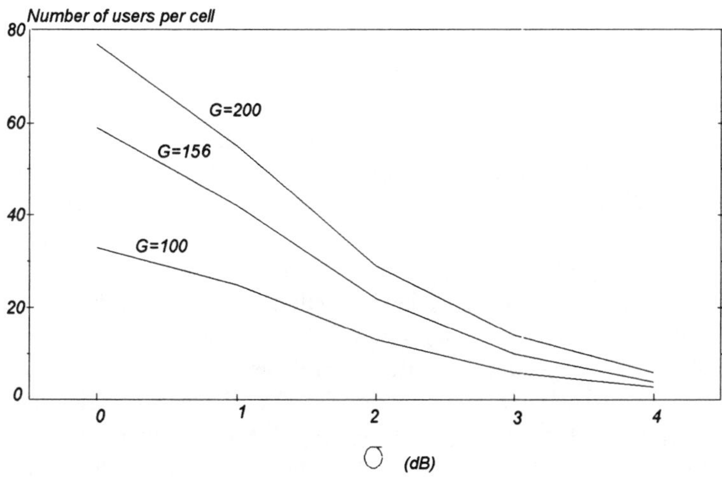

Figure 4.9 Reverse link capacity as a function of power control imperfection for three values of the processing gain G, outage probability = 0.01, and voice activity variable a=0.375.

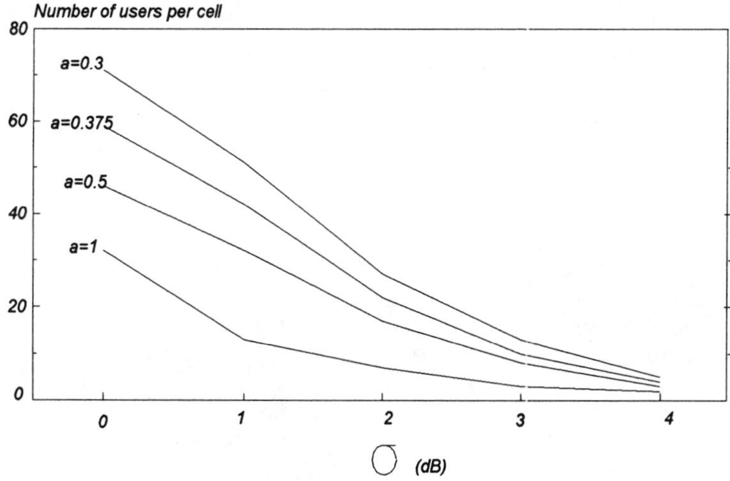

Figure 4.10 Reverse link capacity as a function of power control imperfection for four values of the voice activity variable a, outage probability = 0.01, and processing gain G=156.

4.3.2 Multiple cells

In a cellular CDMA system, each base station receives not only interference from users located in the home cell but also from users located in neighboring cells. The statistics of the power received from the reference user and the interferers in the home cell are known, but the problem remains of how to describe the interference produced by users located in surrounding cells. To solve this problem the approach of [8] is followed. In [8] the total outer cell interference is modeled as a Gaussian random variable with mean μ and logarithmic variance D^2. In [8] the linear moments were estimated by using a computer simulation and assuming shadowing with $\sigma_s = 8$ dB and sectorizing with three sectors per cell. It was found in [8] by Monte Carlo simulation that $\mu < 0.247N_s$ and $D^2 \leq 0.078N_s$ with N_s being the number of users per sector.

Without sectorizing we assume that $N = 3N_s$. Instead of using the Gaussian approximation proposed by Gilhousen et al. [8], the total outer cell interference is modeled here as a log-normal random variable with logarithmic mean m_{s1} and logarithmic variance σ^2_{s1}.

Figure 4.11 shows the Gaussian and approximated log-normal distributions for several values of N. Both distributions are very close for high values of N. Thus Figure 4.11 justifies our assumption that interference received power can be modeled as a log-normal variable. This assumption avoids the complexity in the analysis.

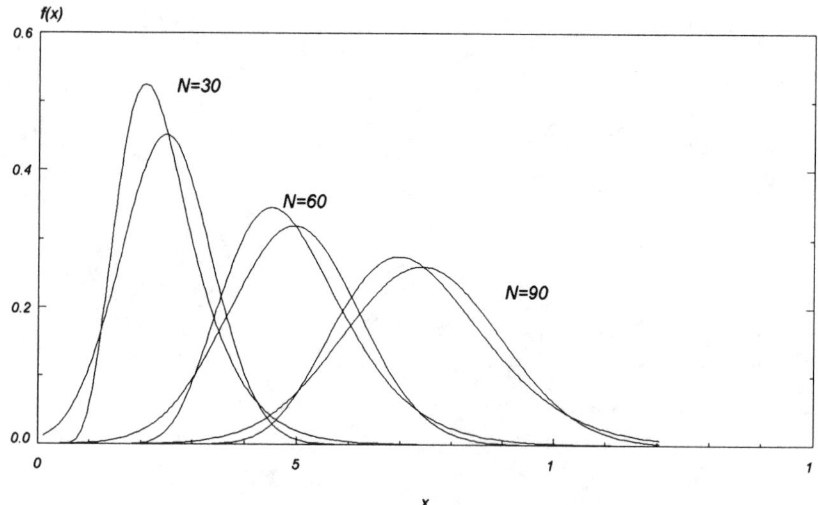

Figure 4.11 Log-normal and Gaussian approximation of outer cell interference for N=30, 60, and 90. $f(x)$ is the pdf of the random variable x.

The moments m_{s1} and σ^2_{s1} can be expressed in terms of μ and D as follows:

$$\sigma^2_{sl} = \ln\left(\frac{D^2}{\mu^2}+1\right) \quad \text{and} \quad m_{sl} = \ln\mu - \frac{1}{2}\ln\left(\frac{D^2}{\mu^2}+1\right) \tag{4.20}$$

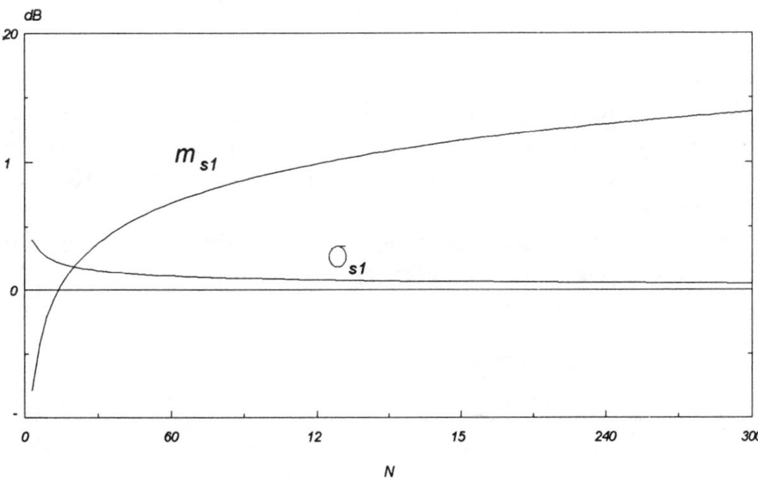

Figure 4.12 Logarithmic mean m_{s1} and logarithmic standard deviation σ_{s1} of the log-normal approximation of the outer cell interference as a function of the number of users.

Figure 4.12 depicts m_{s1} and σ_{s1} as a function of the number of users N and shows that σ_{s1} is always less than 4 dB. In the case of perfect power control the outage probability is given by

$$\Pr\left(\frac{1}{n+\overline{P}_{1,out}} < \alpha\right) = \sum_{k=0}^{n} \Pr\left(\widetilde{P}_{1,out} > \frac{1}{\alpha} - k \middle| k\right) F(n,k) \tag{4.21}$$

where $\overline{P}_{1,out}$ is the total power received due to the outer cell interference. Using the knowledge that $\overline{P}_{1,out}$ has a log-normal pdf, one obtains

$$\Pr\left(\tilde{P}_{1,out} > \frac{1}{\alpha} - k \middle| k\right) = 1 - \int_0^{\frac{1}{\alpha}-k} \frac{1}{\sqrt{2\pi}\sigma_{s1} y} \exp\left[-\frac{(\ln y - m_{s1})^2}{2\sigma_{s1}}\right] dy \tag{4.22}$$

In the case of imperfect power control, the total interference power is the sum of the interference power received from users in the home cell and from users located in surrounding cells. The outage probability can now be calculated using (4.16) after substituting the proper values of σ_d and σ_n. The proper value for σ_n is obtained by summing $N-1$ log-normally distributed interfering signals from the home cell with σ_i determined by the power control imperfections and one log-normally distributed composite interference signal from the surrounding cells with σ_{s1} determined by the number of users N.

Figures 4.13 and 4.14 show the outage probability for a multiple cell system with perfect and imperfect power control. In figures 4.15 and 4.16 the capacity is shown as a function of the power control imperfections for several values of the processing gain G and the voice activity a.

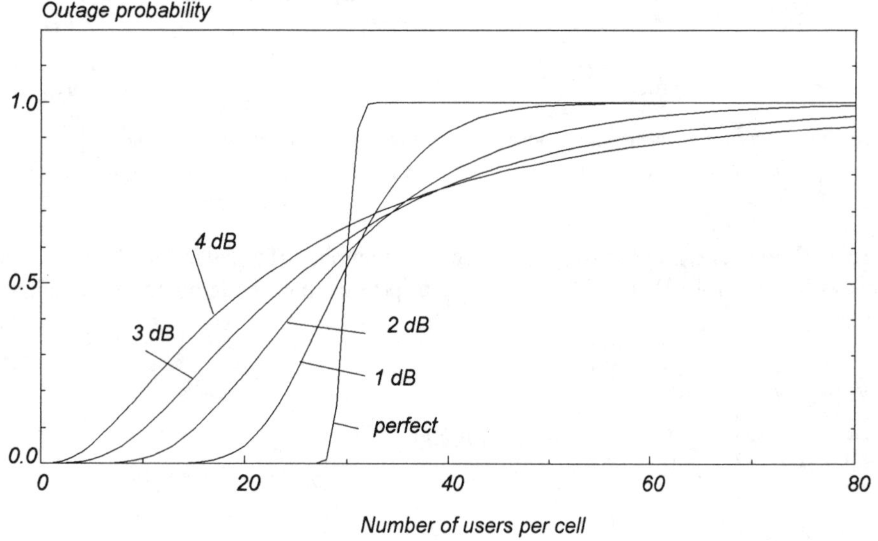

Figure 4.13 Reverse link outage probability as a function of number of users per cell for CDMA systems with perfect power control and with power control errors. Processing gain G=156, no voice activity monitoring.

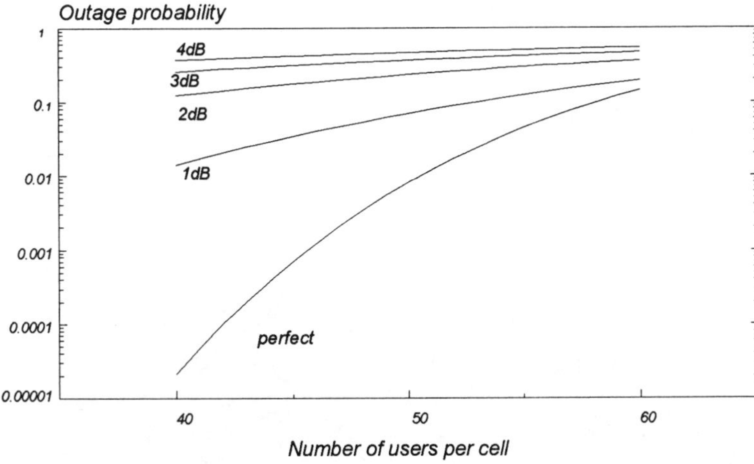

Figure 4.14 Reverse link outage probability as a function of number of users per cell for CDMA systems with perfect power control and with power control errors. Processing gain G=156,voice activity monitoring (a=0.375)

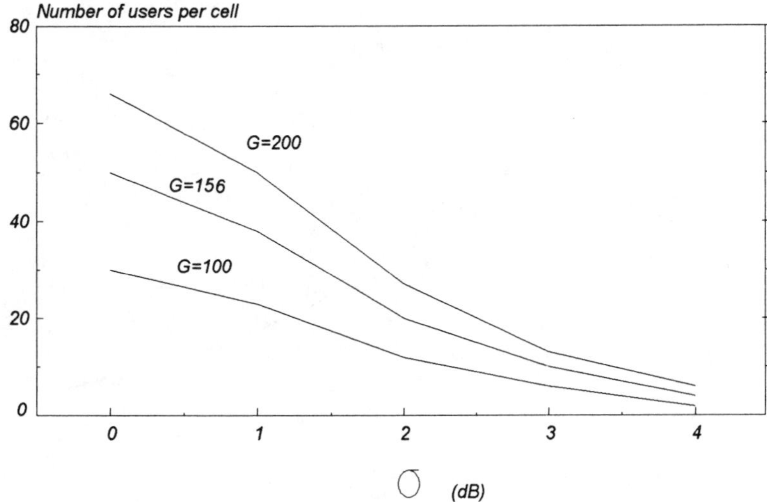

Figure 4.15 Reverse link capacity as a function of power control imperfection for three values of the processing gain G, outage probability=0.01, and voice activity variable a=0.375.

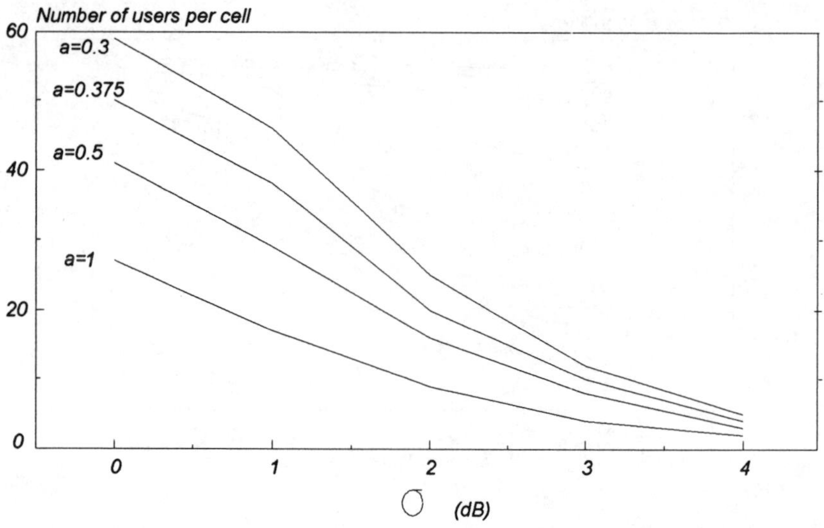

Figure 4.16 Reverse link capacity as a function of power control imperfection for four values of the voice activity variable a, outage probability=0.01, and processing gain G=156.

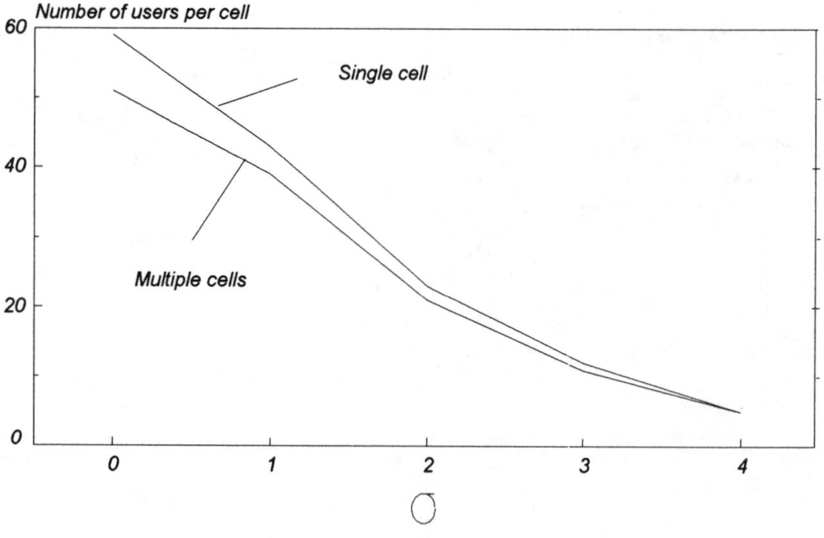

Figure 4.17 Comparison of reverse link capacity for single cell and multiple cell DS-CDMA systems.

Table 4.4 and Figure 4.17 show a comparison between the reverse link capacity for a single cell situation and a multiple cell situation. In the multiple cell situation the capacity further reduces because of the interference received from users in neighboring cells. In addition to power control and voice activity, the performance of a CDMA system can also be enhanced by introducing sectorization.

Table 4.4
Reverse link capacity (maximum number of users per cell) for DS-CDMA system with a single cell (S) and multiple cells (M)

	Perfect Power Control		Imperfect Power Control							
			σ= 1 dB		σ= 2 dB		σ= 3 dB		σ= 4 dB	
	S	M	S	M	S	M	S	M	S	M
G=100 / *a*=1	21	18	12	11	7	6	5	3	3	2
G=156 / *a*=1	33	28	19	17	10	10	7	5	4	3
G=156 / *a*=0.375	59	51	43	39	23	21	12	11	5	5

4.3.3 Sectorization

The capacity of a CDMA system can be improved by introducing sectorization. In the case of perfect sectorization, the cell is divided into D sectors. Because of perfect sectorization the number of interferers decreases with the increase in the number of sectors, implying an increase in the capacity [10].

Since practical antennas have side lobes, perfect sectorization does not exist in practice. To model the imperfect sectorization, the overlap angle v is introduced as shown in Figure 4.18. It is assumed that the receiving gain of the antennas is constant over the home sector plus a part of the adjacent sectors. It is suggested that we transform the real antenna gain patterns to this type of uniform model (Figure 4.18), analogous to the method for describing noise by the noise equivalent bandwidth.

This part is limited at the overlap angle v (measured from the sector boundaries). The overlap angle can be varied and depends on the antenna type. At angles larger than the overlap angle, the receiving antenna gain is zero. This means that when there are k interferers in the case of no sectorization; there are $(1/D+2v/360)k$ interferers in the case of D sectors. Thus the capacity with sectorization is given by

$$C_{sectorization} = \frac{360}{\frac{360}{D} + 2v} C_{nosectorization} \quad (4.23)$$

where $C_{no\ sectorization}$ where is the capacity without sectorization. Equation (4.23) shows that the sectorization increases the system capacity.

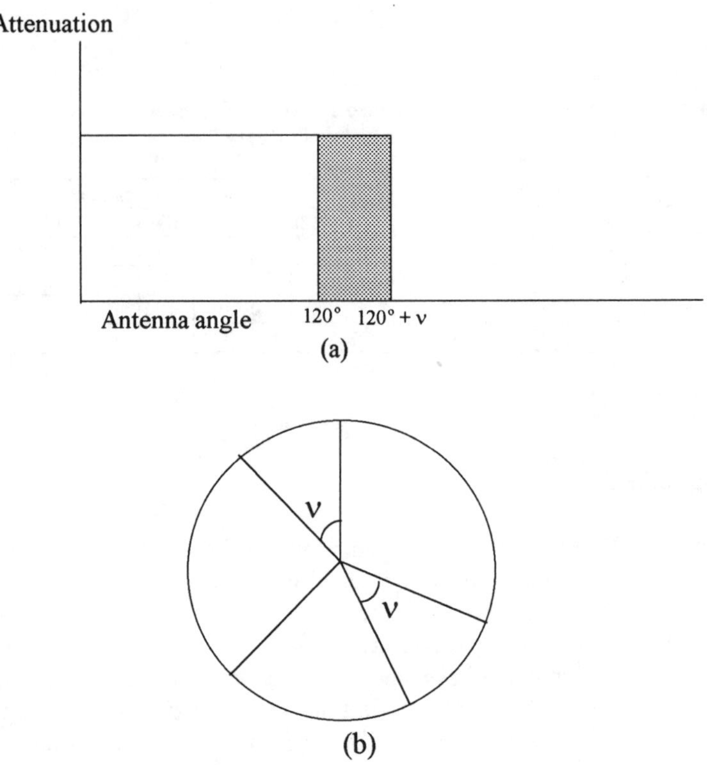

Figure 4.18 (a) Radiation pattern model for imperfect directional antenna with opening angle 120°. (b) Sector coverage with an imperfect directional antenna with opening angle 120° overlap angle ν

4.4 MULTIUSER INTERFERENCE IN THE FORWARD LINK

In a DS-CDMA system, each base station transmits signals to all active users in its service area, using one antenna. Therefore each mobile receives a composite signal from its base station consisting of the desired signal and $N-1$ interfering signals assuming N

users per cell. First, the forward link is investigated in the single cell case and then the multiple cell case is considered.

4.4.1 Single Cell

Assuming that the transmitting power used by the base station to reach a mobile is the same for each mobile, the carrier-to-interference ratio of the signal received by the reference user disregarding its position in the cell is $1/(N-1)$ with $N-1$ being the number of interferers. Obviously no forward power control is needed in a single cell system, which implies that the capacity of the system is restricted by the quality of the reverse link power control system.

4.4.2 Multiple Cells

In a multiple cell system the situation becomes more complicated than that of the single cell situation; a user located near the boundary of three cells now receives considerable interference from other base stations (Figure 4.19). Here, the carrier-to-interference ratio is computed for a reference user located at this point considering only the 12 nearest base stations as shown in Figure 4.19.

Without power control, the signal received by the mobile from the base station will be affected by both the path-loss law and log-normal shadowing. Assuming that the base station transmits a signal with power P_c to each mobile, the received desired power at the worst case position is $P_d=R^{-\gamma}\chi_d P_c$ where χ_d is a log-normal random variable with logarithmic mean m_s and logarithmic variance σ_s^2.

The expression for the power received by the reference user from base station j is given by

$$S_{r,j} = N \cdot r_j^{-\gamma} \cdot \chi_j \cdot P_c \tag{4.24}$$

where N is the number of separate signals in composite signal (determined by the number of users served by the base station), r_j is the distance between user and base station j, γ is the path-loss law exponent which is assumed to be 4, and χ_j is a log-normal random variable.

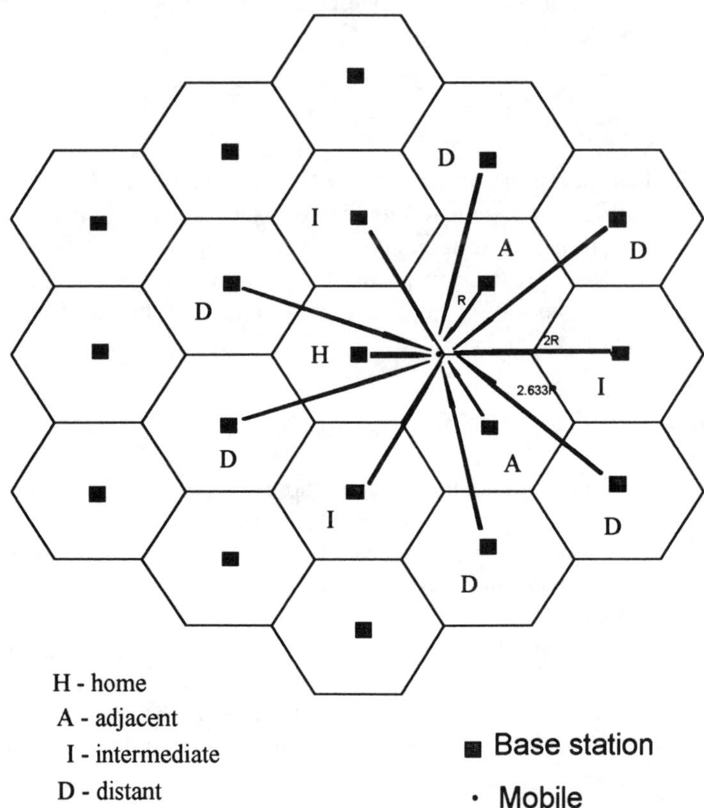

Figure 4.19 Forward link multiuser interference in cellular DS-CDMA systems.

The carrier-to-interference ratio can now be obtained using Figure 4.19:

$$C/I = \frac{\chi_1}{(N-1)\chi_1 + N\sum_{i=2}^{3}\chi_i + N(2)^{-\gamma}\sum_{i=4}^{6}\chi_i + N(2.633)^{-\gamma}\sum_{i=7}^{12}\chi_i} \tag{4.25}$$

To find the total (normalized) interference power in the denominator of (4.25), 12 random variables must be added. To do this, the distribution of $(N-1)\chi_1$, $N{\cdot}\chi_2$, $N{\cdot}\chi_3$, $N{\cdot}2^{-\gamma}\chi_4$, etc., has to be determined. Generalizing this problem implies that it is necessary to have knowledge of the distribution of the function $y{=}C{\cdot}x$ with C being a constant and

x being a log-normal variable with logarithmic mean m_s and logarithmic standard deviation σ_s. Using simple random variable transformation, it is found that y is again a log-normal random variable.

$$f(y) = \frac{1}{\sqrt{2\pi}\,\sigma_s y} \exp\left[-\frac{(\ln y - \ln C - m_s)^2}{2\sigma_s^2}\right] \tag{4.26}$$

From (4.26) it is clear that y has the same logarithmic standard deviation σ_s but a logarithmic mean m_y=ln(C)+m_s. Further the 12 random variables are assumed to be independent variables. Now the sum of the 12 random variables can be computed again by using Schwartz and Yeh's algorithm for σ_s values between 4 and 12 dB [6] and Fenton's algorithm for σ_s values of less than 4 dB [5]. Knowing now the pdf of the desired power and the total interference power received by the reference user, the outage probability as a function of the number of users N can be obtained.

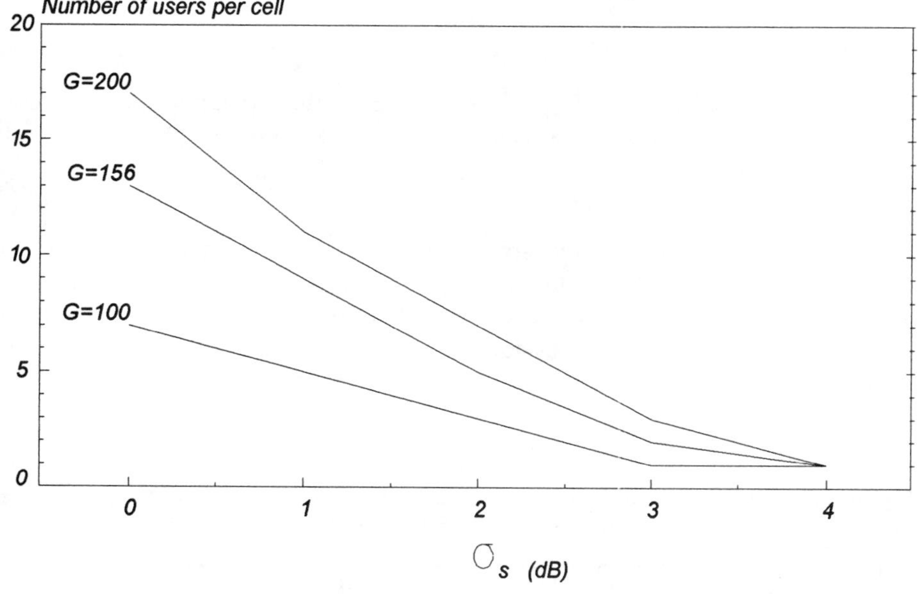

Figure 4.20 Multiple cell forward link capacity as a function of logarithmic standard deviation σ_s due to shadowing for three values of the processing gain G, outage probability=0.01, and voice activity variable a=0.375.

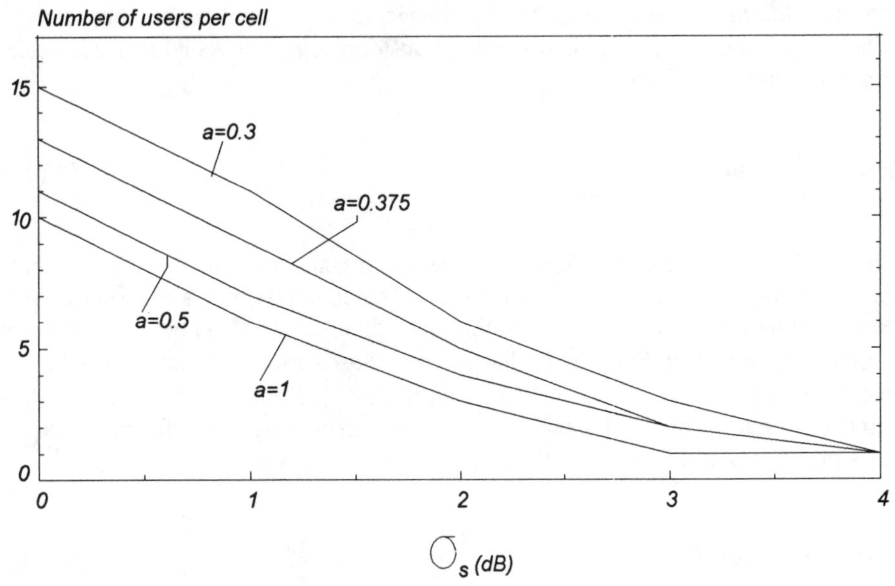

Figure 4.21 Multiple cell forward link capacity as a function of logarithmic standard deviation σ_s due to shadowing for four values of the voice activity variable a, outage probability=0.01, and processing gain G=156.

In Figures 4.20 and 4.21 the forward link capacity is shown as a function of the logarithmic standard deviation σ_s for an outage probability of 0.01 for several values of the processing gain G and the voice activity. These plots show that the forward link capacity decreases drastically for shadowed environments when not using any form of forward power control.

Table 4.5

Forward link capacity for an outage probability of 0.01 in the presence of shadowing G=156 and voice activity factor a=0.375; no power control

Forward Link Without Power Control								
Amount of shadowing (σ_s [dB])	0	1	2	3	4	6	8	10
Number of users	13	9	5	3	1	1	1	1

In Table 4.5 the forward link capacity is shown without forward power control in the presence of shadowing. From this table it can be concluded that the forward link without power control is a very heavy restriction on the capacity of a CDMA system, especially in shadowed environments. This implies that the interference problem on the forward link needs careful examination.

4.4.3 Forward Power Control

A possible solution for the multiuser interference problem on the forward link is the use of a power control scheme. The principle of forward power control is, however, less straightforward than for the reverse link. Two possible forward power control schemes are discussed here with the main objective of stressing the complexity of a forward power control scheme.

1. Distance driven [7]: With knowledge of the positions of the mobiles, it is possible to minimize the total power transmitted by each base station by transmitting at a higher power level to users located at the cell boundary and at a lower power level to users close to the base station. Distance-driven power control is most appropriate for nonshadowed environments because then the power attenuation will only depend on the distance. To offer each mobile the opportunity to measure its distance from the base station, the base station must transmit a pilot signal.

2. C/I driven [8]: This implies that the C/I ratio of each user is minimized according to the individual needs of each user. To do this, each mobile must send information about its C/I ratio to the base station. Then the base station can decide if the transmitted power to a specific user should be either increased or decreased. Using the analysis results from [7] and [8], the capacities of both power control methods are compared in Table 4.6.

Table 4.6
Maximum number of users per cell (N) with and without forward power control; G=156, a=375

	No Power Control	C/I Driven	Distance Driven
Shadowing 8 dB	1 user	38 users	—
No shadowing	13 users	—	30 users

For the C/I-driven power control scheme, the results from [8] obtained by a computer simulation performed by Gilhousen et al. are used. To perform this simulation, the following parameters were assumed: G=156, a=0.375, v=20%, σ_s = 8 dB with G being the processing gain, a being the voice activity factor, v being the portion of the transmit power devoted to the pilot signal, and σ_s being the logarithmic standard deviation of the log-normal shadowing.

From Table 4.6 it becomes clear that the capacity increases from 1 to 38 users by introducing the C/I-driven power control scheme. The forward link capacity of the distance-driven power control (Lee [7]) was calculated in the case of a nonshadowed environment. In this case the capacity increases from 13 to 30 users by introducing distance-driven power control.

Table 4.6 shows that the increase in capacity with distance-driven power control in a nonshadowed environment is less than with C/I-driven power control in a shadowed environment. Because it is expected that the performance of a distance-driven power control system in a shadowed environment becomes even worse, it is less interesting for practical purposes to compute the capacity in that case. An explanation for the higher capacity of a C/I-driven power control system is that the C/I-driven power control system minimizes the total transmitting power more carefully than a distance-driven power control scheme. In practice, some problems can occur since central control with full knowledge of the interference situation in the network is assumed and hence control information must be transmitted. This problem is also stressed by Zander in [11].

4.5 CONCLUSIONS

The influence of the near-far effect and imperfect power control on the capacity of a cellular DS-CDMA system was investigated. Capacity is defined as the maximum number of users per cell for which the outage probability is less than the threshold ratio α. To calculate the outage probability, a threshold ratio of 5 is assumed for the bit energy-to-noise density ratio in the case of noncoherent reception. For DS-CDMA systems this threshold ratio in general implies a bit error probability of better than 0.001, which is appropriate for digital speech transmission.

Two separate cases were considered:

1. Reverse link: base station receiver and mobile transmitter;
2. Forward link: base station transmitter and mobile receiver.

From the calculations, the previously existing idea that power control on the reverse link is needed in DS-CDMA systems is reconfirmed. Without any form of reverse link power control, the capacity of the system becomes very low because of the influence of the near-far effect.

Perfect reverse link power control is found to improve the capacity. The influence of imperfect power control on system performance was also investigated by describing the error of the power control system as the standard deviation of the assumed log-normal received signal. The system was found to be very sensitive to errors in the reverse link power control. If we also consider the interference received from users located in surrounding cells, the capacity decreases even more.

Multiuser interference on the forward link turns out to be a very serious problem because each user not only receives the interference from its own cell site but also from the cell sites of surrounding cells. The multiuser interference on the forward link puts a limit on the capacity especially in shadowed environments. A possible solution to circumvent this problem is forward power control. Two power control schemes were discussed briefly: C/I-driven and distance-driven power control. The C/I-driven power control scheme gives the best result, which can be explained by the fact that in this case each user receives the minimum power needed for a predefined performance (C/I ratio).

A distance-driven power control system is not very appropriate in shadowed environments because then the received power is not only a function of the distance of the user to the base station but also depends on the log-normal shadowing spread.

In practice, a forward power control scheme is more complicated than a reverse link power control scheme because central control with full knowledge of the location of individual users (distance driven) or the C/I ratio of individual users (C/I driven) is required.

When comparing the reverse link capacity to the forward link capacity of a cellular DS-CDMA system, we concluded that the forward link with perfect power control (C/I-driven or distance-driven) is more critical than the reverse link with perfect power control. In the case of perfect forward link power control and imperfect reverse link power control with logarithmic standard deviation larger than 1 dB, the reverse link is the most critical link. Imperfect forward power control was not investigated but it is expected that forward power control imperfections have a negative effect on the capacity, analogous to reverse link power control imperfections.

Finally, to compute the outage probability on the reverse link without any power control, a general spatial user distribution that leads to a log-normal probability distribution of the received power was developed. This user distribution is a generalized form of the spatial distribution proposed in [3]. It offers the possibility to investigate several user distributions by simply changing one parameter of the general user distribution. Also the effect of various path-loss law exponents (γ) and shadowing on the performance of the system can be investigated easily by using this model.

REFERENCES

[1] J.C. Arnbak and W. van Blitterswijk, "Capacity of a slotted ALOHA in Rayleigh fading channels," *IEEE J. Selected Areas Comm.*, Vol. SAC-5, pp. 261-269, February 1987.

[2] S.A. Musa and W. Wasylkiwskyj, "Cochannel interference of spread spectrum systems in a multiple user environment," *IEEE Trans.Comm.*, Vol. 26, pp. 1405-1413, October 1978.

[3] R. Prasad, "Performance analysis of mobile packet radio networks in real channels with inhibit multiple access," *IEE Proceedings-I*, Vol.138. No. 5, pp.458-464, October 1991.

[4] R. Prasad, M.G. Jansen and A. Kegel, "Capacity analysis of a cellular direct sequence code division multiple access system with imperfect power control," *IEICE Trans. Comm.*, Vol. E 76-B, pp. 894-905, August 1993.

[5] L.F. Fenton, "The sum of log-normal probability distributions in scatter transmission systems," *IRE Trans. Commun. Syst.*, Vol. CS-8, pp. 57-67, March 1960.

[6] S.C. Schwartz and Y.S. Yeh, "On the distribution function and moments of power sums with log-normal components," *Bell Syst. Tech. J.*, Vol. 61, No.7, pp. 1441-1462, September 1982.

[7] W.C.Y. Lee, "Overview of cellular CDMA," *IEEE Trans. Vehicular Technol.*, Vol. 40, No.2, pp. 291-301, May 1991.

[8] K.S. Gilhousen, I.M. Jacobs, R. Padovani, A. Viterbi, L.A. Weaver, Jr. and C.E. Wheatley III, "On the capacity of a cellular CDMA system," *IEEE Trans. Vehicular Technol.*, Vol.40, No.2, pp. 303-312, May 1991.

[9] H.J. Braun, G. Cosier, D. Freeman, A. Gilloire, D. Sereno, C.B. Southcott and A. van der Krogt, "Voice control of the Pan-European digital mobile radio system," *CSELT Tech. Rep.*, Vol. XVIII, No. 3, pp. 183-187, June 1990.

[10] M.G. Jansen and R. Prasad, "Capacity, throughput, and delay analysis of a cellular DS CDMA system with imperfect power control and imperfect sectorization," *IEEE Trans. Vehicular Technol.*, Vol. 44, pp. 67-75, February 1995.

[11] J. Zander, "Optimum power control in cellular radio systems," Internal report Royal Institute of Technology, Kista, Sweden, Report TRITA-TTT-9101.

Chapter 5
Indoor CDMA Systems

5.1 INTRODUCTION

Indoor wireless communication has significant advantages over the conventional cabling, namely, mobility of users, elimination of wiring and rewiring, drastic reduction of wiring in new buildings, flexibility in changing or creating new communication services, time and cost savings, and reduction of downtime [1-13]. Much attention is being paid to the use of DS-SS modulation for indoor wireless multiple access communication, over multipath fading channels. DS-SS modulation provides both multiple access capability and resistance to multipath fading [14-31].

In this chapter we investigate the performance of a DS-CDMA system for indoor applications. First the propagation characteristics of the indoor environment are discussed. Then, after a system description the performance model is explained. This analytical model is used to analyze the performance in terms of the bit error probability, throughput, and delay. The model is developed for DPSK modulation. The effect of two diversity schemes, selection diversity and maximal ratio combining, is considered and also the effect of forward error correcting (FEC) codes on the performance is studied.

It is shown that maximal ratio combining is superior to selection diversity. Furthermore, FEC coding can improve performance, although the maximum bit rate decreases under the restriction of fixed bandwidth and fixed processing gain. All results are obtained under the assumption of perfect power control.

5.2 PROPAGATION CHARACTERISTICS

Knowledge of propagation characteristics of a channel is the bread and butter of radio communication engineers. If a radio engineer does not possess full knowledge of the channel, he or she can never design a good radio communication system.

Therefore knowledge of indoor radio propagation characteristics is a prerequisite for the design of indoor radio communication systems. A lot of measurements have been done to obtain information concerning propagation loss, spatial distribution of power when the environment is physically static, wideband and narrowband statistics concerning the random variation of received signals at a fixed location due to any movement of the surroundings, and delay spread. Reference [1] provides a good review of these measurements. In this section, all the aspects mentioned above are briefly discussed.

5.2.1 Path Loss

In general, the spatially averaged power P_0 at a point a distance d from the transmitter is a decreasing function of d. Usually, this function is represented by a path-loss-power law of the form

$$P_0 \approx d^{-\gamma} \tag{5.1}$$

In free space the path-loss law exponent $\gamma = 2$, so the power law obeys an inverse-square law. Recently, a lot of measurements have been done to obtain information about the value of the path-loss law exponent in a practical indoor environment. As reported in [2,3], the path-loss law exponent $\gamma < 2$ if the transmitter and the receiver are positioned in the same hallway in an office building. A value of $\gamma = 1.8$ is reported in [2] for a test frequency at 910 MHz. This gain over the free-space situation is likely due to waveguiding effects in the hallway. In rooms which are located off the same hallway containing the transmitter, values of about $\gamma = 3$ have been measured and in rooms off a hallway perpendicular to the transmitter's hallway, values of 4 up to 6 have been measured. Extremely high values of γ ($\gamma = 6$) correspond to buildings with metalized partitions. In [4] path-loss results for factories are discussed for a test frequency at 1.3 GHz. Factory buildings generally have few internal partitions. Aisles are arranged in an orderly, orthogonal (intersecting) fashion and are typically flanked by metal machinery or inventory. In both, the case of a line of sight (LOS) path and the case of an obstructed (OBS) line of sight path with light clutter and with heavy clutter of the surroundings (machinery and storage racks) were distinguished. For the LOS case a value of $\gamma = 1.79$ was reported for both light and heavy clutter [4]. For the OBS case $\gamma = 2.38$ was reported for light clutter and $\gamma = 2.81$ for heavy clutter. Indoor measurement results at frequencies of 2.4, 4.75, and 11.5 GHz are reported in [5-7]. In the case of a LOS path, the path-loss law constant γ is in the range from 1.8 to 2, while in the case of an obstructed path, the path-loss law exponent γ has the values 3.3, 3.8, and 4.5 at frequencies of 2.4, 4.75, and 11.5 GHz, respectively. The value of γ also depends on the morphology of the building.

In a CDMA system, the path-loss law is responsible for the so called *near-far effect*. This implies that the average power received by the base station from a terminal located close to the base station is much higher than the average power received from a terminal located at a larger distance from the base station. In this chapter it is assumed however that the base station provides perfect average power control to combat the near-far effect. As a result, equal power is received from each terminal independent from the distance to the base station.

5.2.2 Multipath Propagation

In an indoor environment, due to the building structure and surrounding inventory, many reflections of the signal arrive at the antenna of the receiver at different times (Figure 5.1).

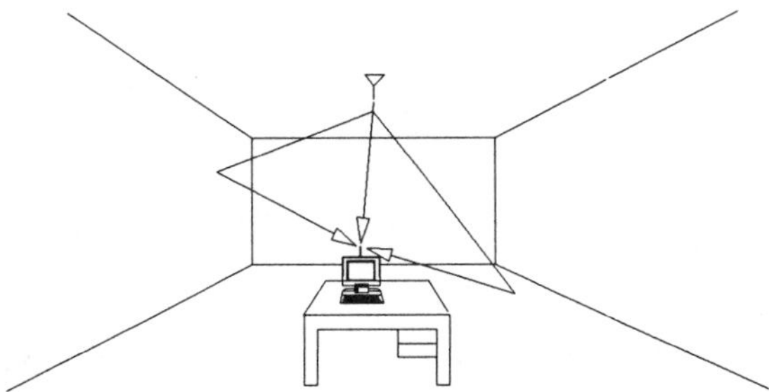

Figure 5.1 Multipath propagation.

These reflections add up and cause signal peaks or dips, dependent on the phases of the reflected signals. This implies that the indoor multipath channel can be represented by multiple paths having a real positive gain β_l, propagation delay τ_l, and phase shift γ_l, where l is the path index. The baseband complex response of the lth path is given by:

$$h_l(t) = \beta_l \exp\{j\gamma_l\}\delta(t - \tau_l) \tag{5.2}$$

where $j = \sqrt{-1}$. Figure 5.2 shows a possible impulse response with four paths.

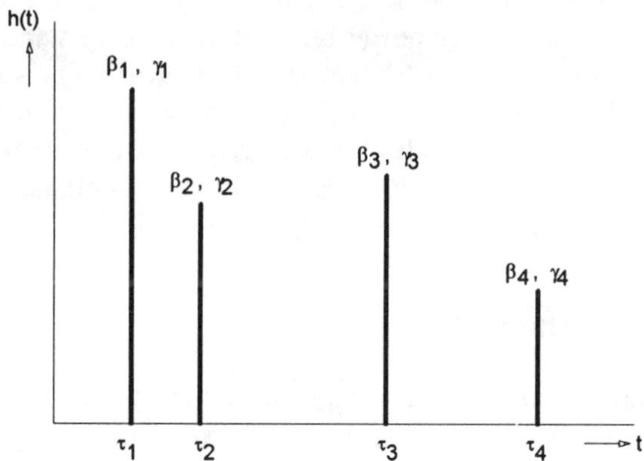

Figure 5.2 Discrete impulse response with four paths.

Now the baseband complex channel impulse response is modeled as [3]:

$$h(t,x) = \sum_{l=1}^{L} \beta_l \exp\{j\gamma_l\}\delta(t - \tau_l) \tag{5.3}$$

where $\delta(\cdot)$ is the Dirac delta function, $h(t,x)$ is the time domain response, which is also position dependent (indicated by x), and L is the number of paths ($l=1$ indicates the first arriving path).

Because of the motion of people, or other time varying environmental factors, the parameters β_l, τ_l, and γ_l are randomly changing functions of time. This type of distortion is experienced as a variation in the received signal strength and is termed fading. Besides the time selectivity of the channel, it is also important to consider the frequency selectivity of the channel. It should be emphasized that the frequency selectivity and time selectivity are viewed as two different types of distortion. The former depends on the coherence bandwidth of the channel (which is related to the multipath spread) relative to the transmitted signal bandwidth W. The latter depends on the time variations of the channel characterized by the coherence time (which is related to the Doppler spread). In the following, these aspects, i.e., coherence bandwidth, multipath delay spread, Doppler spread, and coherence time, are described in detail.

Coherence Bandwidth and Time Delay Spread

The impulse response, which can be interpreted as a power delay density function, yields useful information about the multipath characteristics of the channel. This power delay profile may be treated as a delay density function, where the delay is weighted by the signal level at that delay. An appropriate way to characterize the power delay profile is the RMS multipath delay spread denoted by T_m, which is defined as the standard deviation (i.e., the square root of the central moment or variance) of the power delay profile. Practically, the RMS delay spread T_m is determined from the measured power delay profile by calculating the standard deviation of this profile [3]:

$$T_m \underline{\Delta} \sqrt{\mathrm{E}\left[\tau^2\right] - \mathrm{E}^2[\tau]} \tag{5.4a}$$

with

$$\mathrm{E}[\tau] \underline{\Delta} \frac{\sum_l \tau_l \beta_l}{\sum_1 \beta_l^2} \tag{5.4b}$$

and

$$\mathrm{E}\left[\tau^2\right] \underline{\Delta} \frac{\sum_l \tau_l \; \beta_l}{\sum_l \beta_l^2} \tag{5.4c}$$

where β_l is the received power of the lth pulse.

Another possible way to characterize the power delay profile is by defining a maximum time delay spread, which is determined by the range of delays over which the power delay profile is essentially nonzero [8]. For measurements one can, for example, choose the maximum time delay spread to be the delay at which the power of the received pulse is 30 dB lower than the first received pulse.

Typical values for the RMS time delay spread measured in an indoor environment at a 910-MHz frequency are in the range of 50 up to 250 ns [2]. In [5]-[7], RMS delay spread values from 10 up to 20 ns are reported for a frequency band of 2.4 to 11.5 GHz. Maximum delay spreads of 300 ns have been measured in [3] and maximum delay spreads of 270 ns have been reported in [9]-[11] and other references for indoor communication in the frequency band of 850 Mhz to 4 GHz. For a detailed overview, refer to [1]. If the data bit duration is larger than the RMS delay spread, then the channel introduces a negligible amount of intersymbol interference. The reciprocal of the RMS time delay spread is a measure for the coherence bandwidth $(\Delta f)_c$ of the channel:

$$(\Delta f)_c \approx \frac{1}{T_m} \tag{5.5}$$

Figure 5.3 shows an example of a measured impulse response.

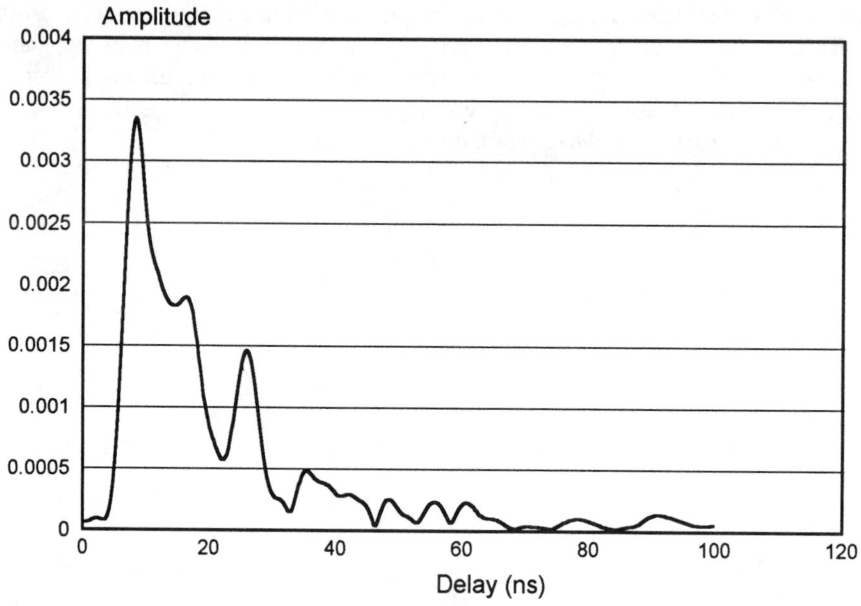

Figure 5.3 Measured power delay profile.

The *coherence bandwidth* $(\Delta f)_C$ is the bandwidth over which the signal propagation characteristics are correlated, i.e., the width of the frequency correlation function $\Phi_C(\Delta f)$. Thus two sinusoids with frequency separation larger than the coherence bandwidth are affected differently by the channel. If the coherence bandwidth $(\Delta f)_c$ is small compared to the bandwidth of the transmitted signal W, the channel is frequency selective. In this case the different frequency components in the signal are subject to different gains and phase shifts. On the other hand, if the coherence bandwidth is large compared to the signal bandwidth W, the channel is frequency nonselective and all frequency components are subject to the same attenuation and phase shift.

The frequency correlation function $\Phi_C(\Delta f)$ and the delay profile $\Phi_C(\tau)$ can be related by applying the Fourier transform:

$$\Phi_C(\Delta f) \xleftrightarrow{\text{Fourier}} \Phi_C(\tau) \tag{5.6}$$

Resuming: $W >> (\Delta f)_C \Rightarrow$ frequency selective channel

$W << (\Delta f)_C \Rightarrow$ frequency nonselective channel

The system bandwidth of a spread-spectrum system is generally larger than the coherence bandwidth, which implies that in that case the channel is frequency selective.

Coherence Time and Doppler Spread

To classify the time characteristics of the channel, the *coherence time* and *Doppler spread* are important parameters. The coherence time is the duration over which the channel characteristics do not change significantly. The time variations of the channel are evidenced as a Doppler spread in the frequency domain, which is determined as the width of the spectrum when a single sinusoid (constant envelope) is transmitted. Both the time correlation function $\Phi_c(\Delta t)$ and the Doppler power spectrum $\Phi_c(f)$ can be related to each other by applying the Fourier transform:

$$\Phi_c(\Delta t) \xleftrightarrow{Fourier} \Phi_c(f) \tag{5.7}$$

The range of values of the frequency f over which $\Phi_c(f)$ is essentially nonzero is called the Doppler spread B_d of the channel. Since $\Phi_c(f)$ is related to $\Phi_c(\Delta t)$ by the Fourier transform, the reciprocal of B_d is a measure of the coherence time $(\Delta t)_c$ of the channel. That is,

$$(\Delta t)_c \approx \frac{1}{B_d} \tag{5.8}$$

The coherence time $(\Delta t)_c$ is a measure of the width of the time correlation function. Clearly, a slow changing channel has a large coherence time or, equivalently, a small Doppler spread. The rapidity of the fading can now be determined either from the correlation function $\Phi_c(\Delta t)$ or from the Doppler power spectrum $S_c(f)$. This implies that either the channel parameters $(\Delta t)_c$ or B_d can be used to characterize the rapidity of the fading. If the bit time T_b is large compared to the coherence time, then the channel is subject to fast fading. When selecting a bit duration that is smaller than the coherence time of the channel, the channel attenuation and phase shift are essentially fixed for the duration of at least one signaling interval. In this case the channel is slowly fading or quasi-static.

Resuming: $T_b >> (\Delta t)_c \Rightarrow$ fast fading channel
$T_b << (\Delta t)_c \Rightarrow$ slow fading channel

Indoor measurements [12] showed that in any fixed location, temporal variations in the received signal envelope caused by movement of personnel and machinery are slow, having a maximum Doppler spread of 6.1 Hz. The fading statistics, which will be described in detail in the next subsection, are statistically nonstationary in the strict sense. Wide-sense-stationary characteristics, however, are exhibited for periods of time up to at least 2 seconds. Table 5.1 gives an overview of the terminology that can be used to characterize the time and frequency domain properties of fading.

Table 5.1
Terminology

Channel Models	$T_b << (\Delta t)_C$	$T_b >> (\Delta t)_C$	
$W<<(\Delta f)_C$	Nondispersive flat-flat fading	Time dispersive frequency-flat fading	Frequency nonselective channel
$W>>(\Delta f)_C$	Frequency dispersive time-flat fading	Doubly(time and frequency) dispersive	Frequency selective channel
	Time nonselective or slow fading channel	Time selective or fast fading channel	

Multiple Resolvable Paths

Since it is assumed that the signal bandwidth is much larger than the coherence bandwidth of the channel [$W>>(\Delta f)_C$], there exist a number of resolvable paths [8]. This is called the inherent diversity of spread spectrum. In fact this implies that delayed versions of the initial pulse can be grouped in clusters which can be resolved independently. As described in (5.2), each cluster has a gain denoted by β_l, a phase denoted by θ_l, and a time delay denoted by τ_l. It is reasonable to assume that the path phase θ_l of the signal at the receiver is an independent random variable uniformly distributed in [0, 2π]. The path delay τ_l is also an independent random variable and is assumed uniform over $[0,T_b]$. The path gain β_l is independent for each cluster and is described in detail in the next subsection. Because the fading rate in an indoor environment is slow, the random variables β_l, θ_l, and τ_l which describe the channel do not vary significantly over two bits.

The question remains of how to determine the maximum number of resolvable paths denoted by L in terms of communication system parameters. It is clear from Figure 5.4 that two signals must be separated by one chip time T_c in order to be resolved. Using the result of time resolution of direct-sequence spread-spectrum signals, the maximum number of paths L can be estimated from the maximum multipath delay spread T_{max} as follows:

$$L = \left\lfloor \frac{T_{max}}{T_c} \right\rfloor + 1 \tag{5.9}$$

where $\lfloor x \rfloor$ is the largest integer that is less than or equal to x.

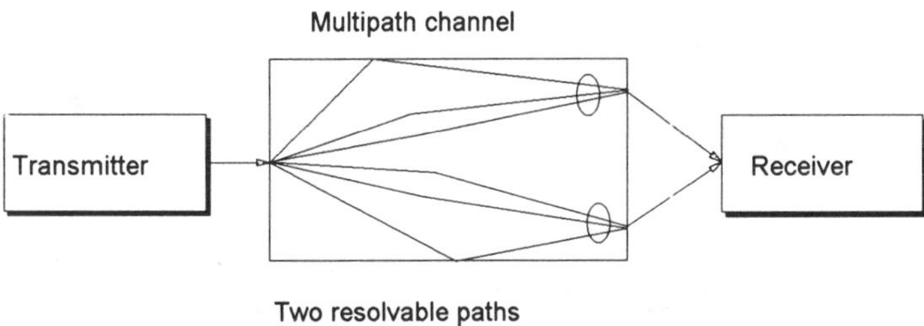

Figure 5.4 Multipath fading and multiple resolvable paths.

Actually, (5.9) is a very simple model for the number of resolvable paths. In practice, it is likely that the number of resolvable paths is a random variable. In this book however we will use the simple model.

5.2.3 Fading Characteristics

The time resolution of a DS-CDMA receiver is equal to the chip duration T_c. The reflections can now be grouped in clusters, each with a difference of less than the chip duration T_c between the arrival time of the first ray and the arrival time of the last ray. These clusters (hereafter called paths) can be resolved by the spread-spectrum receiver (inherent multipath diversity of spread-spectrum modulation). The criterion for the existence of multiple resolvable paths is that the signal bandwidth W is much larger than the coherence bandwidth of the channel [$W >> (\Delta f_c)$]. This phenomenon is described by the lowpass equivalent impulse response of the channel. For the link between the kth user and the base station this impulse response is given by

$$h_k(t) = \sum_{l=1}^{L} \beta_{lk} \exp\{j\gamma_{lk}\}\delta(t - \tau_k) \tag{5.10}$$

where l denotes the path index. Since the channel is modeled as a discrete multipath slow fading channel, there are L discrete multipath links between each user and the receive antenna of the base station, where L is determined by the time resolution T_c of the DS-CDMA system and the delay spread T_{max} of the channel according to (5.9). Each path has an independent phase denoted by θ_{lk} (uniform distribution $[0,2\pi]$), an independent time delay denoted by τ_{lk} (uniform distribution $[0,T_b]$), and an independent gain denoted by γ_{lk}. It is assumed that all path gains are described by the same pdf. This assumption is only valid in environments with relatively small delay spreads since in general we have to deal with a delay profile where the attenuation of each path is a decaying function of the delay of each path. In situations with relatively large delay spread, we have to take into account the delay profile, which means that the path gains cannot be described by the same pdf. We assume that in an indoor environment, the RMS delay spread is low enough in order to assume a constant delay profile implying that all path gains are described by the same pdf. The distribution of the path gain β_{lk} depends on the fading statistics of the channel described in this section. The situation is depicted in Figure 5.4 for the case of two resolvable clusters.

In an indoor environment, not only do a number of reflected signals arrive at the receiver antenna, but there is also often a strong (dominant) direct LOS component. This is a very important consideration with respect to the fading statistics. A distinction can be made between the spatial signal strength distribution and the signal strength distribution at the input of a fixed receiver antenna. Both distributions are discussed below.

Spatial Signal Strength Distribution

If the environment is static, there is a fixed spatial pattern of signal maxima and minima. Since a significant part of the received signal envelope is due to a constant path, the spatial signal strength distribution can be described by a Rician distribution which is confirmed by measurements [2].The Rician distribution is characterized by a parameter (R), which is the ratio of the peak power and the power received over specular paths.

Time Signal Strength Distribution

Due to the existence of a LOS component, the signal strength distribution at the input of a fixed receiver antenna under normal working conditions (moving surroundings and moving people) is also Rician distributed, which is in accordance with recent measurements done in office [2] and factory buildings [13]. From [2] it is known that the parameter R=6.8 dB corresponds to a brick building with reinforced concrete and plaster, as well as some ceramic block interior partitions. The parameter R=11 dB corresponds to a building having the same construction, but with an open-office interior floor plan and nonmetallic ceiling tiles throughout.

Rician Distribution

It is concluded that the path gain β_{lk} is an independent Rician random variable. The Rician probability density function is

$$p_\beta(r) = \frac{r}{\sigma^2}\exp\left[-\frac{r^2+s^2}{2\sigma^2}\right] I_0\left[\frac{sr}{\sigma^2}\right]$$

$$0 \le r < \infty,\ S \ge 0 \qquad (5.11)$$

$$\mathrm{R} = \frac{s^2}{2\sigma^2}$$

The shape of this Rician distribution is plotted in Figure 5.5

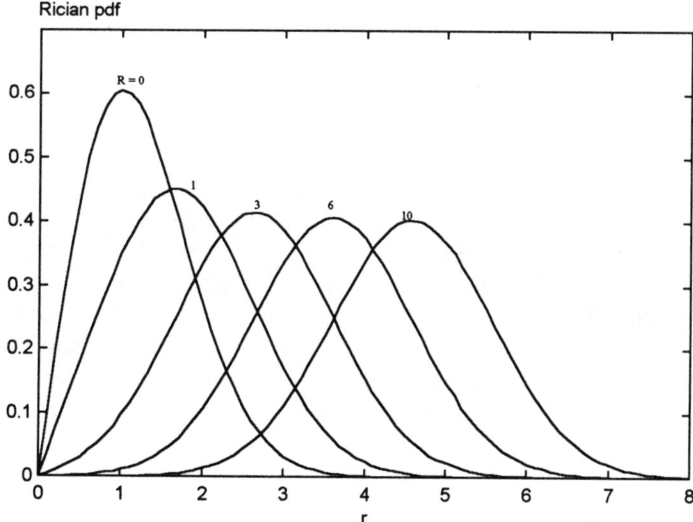

Figure 5.5 Rician probability density function.

In (5.11), $I_0(\cdot)$ is the modified Bessel function of the first kind and zero order. The parameter s is the peak value of the specular radio signal due to the superposition of the dominant LOS signal and the time invariant scattered signals reflected from walls, ceilings, and stationary inventory. The average signal power that is received over specular paths (due to moving objects inside the building) is denoted by σ^2. If there is no dominant component and the received signal consists merely of reflected versions of the original signal, then the parameter R approaches zero and the Rician distribution changes into a Rayleigh distribution:

$$p_\beta(r) = \frac{r}{\sigma^2}\exp\left[-\frac{r^2}{2\sigma^2}\right] \tag{5.12}$$
$$0 \le r < \infty$$

This shows that in fact the Rayleigh distribution is similar to a Rician distribution with direct to specular ratio R=0, i.e., s=0. Also when s, i.e., R, is large, (5.11) is approximately Gaussian. Therefore Rayleigh statistics is one extreme case of the Rician statistics, while Gaussian is another extreme case of the Rician statistics.

5.2.4 Summary

In this section we discussed the general propagation characteristics of an indoor environment. The channel is characterized as a frequency selective, slow fading channel. The most important effects encountered are path loss due to the distance between receiver and transmitter, and multipath propagation due to reflection and absorption of radio waves. The multipath effect is modeled by describing the channel as a filter with a discrete impulse response. This response consists of a number of resolvable paths, each having an independent gain, phase, and delay described by random variables. It is important to realize that a real radio channel is too complex and unpredictable for a mathematical model. For accurate results it is therefore better to resort to elaborate simulations. The model that will be used in this book however is useful to provide insight into the performance of several system configurations.

5.3 SYSTEM MODEL

In this section we describe the system model of a wireless indoor system. The system configuration is discussed and models for the transmitter, the channel, and the receiver are described. Methods to increase the performance such as diversity techniques and FEC codes are introduced. Finally, the basic principles of pseudo-noise (PN) code sequences are discussed.

5.3.1 Star Connected System Configuration

The system we consider is a star connected CDMA network with K users, as shown in Figure 5.6.

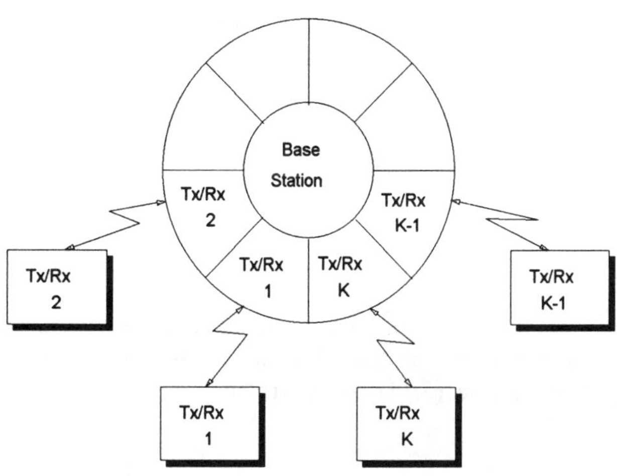

Figure 5.6 Star connected indoor wireless CDMA network.

The base station consists of a bank of spread-spectrum transmitters/receivers, one for each active user. It is assumed that the multiple access capability of the system is established by providing each user with a unique spread-spectrum code sequence. In order to have a high capacity, the cross-correlation of the spread-spectrum codes must be low. Gold codes have well-defined correlation properties and the implementation of Gold code generators is very simple. Due to these properties, Gold codes are very suitable for DS-CDMA systems. A brief description is given in Subsection 5.3.5.

5.3.2 Transmitter, Channel, and Receiver Model

Transmitter Model

The transmitter is based on direct-sequence spread-spectrum techniques with DPSK modulation. The data waveform is denoted as

$$b_k(t) = \sum_i b_k{}^i P_{T_b}(t - iT_b)$$
$$b_k{}^i \in \{0,1\} \tag{5.13}$$

Here b_k^i is the ith data bit of user k and P_{Tb} is a rectangular pulse of unit height and duration T_b. BPSK modulation of (5.13) would result in

$$s_k(t) = \sum_i A\cos(\omega_c t + \Theta_k + \Phi_k{}^i)P_{T_b}(t - iT_b)$$
$$\Phi_k^i \in \{0, \pi\} \tag{5.14}$$

Now note that DPSK is based on normal BPSK modulation with differential encoding and phase comparison detection in the demodulator. Because the phase shift of π just reverses the sign of the cosine, (5.14) can be written as:

$$s_k(t) = \sum_i Ab_k{}^i \cos(\omega_c t + \Theta_k)$$
$$b_k{}^i \in \{-1,1\} \tag{5.15}$$

With differential encoding, the coded bit is equal to the previous one if the data bit is a one; otherwise, the coded bit is made unequal to the previous one. The process starts with an arbitrary initial bit. The consequence of this method is that bit errors always occur in clusters of at least two errors. The advantage is that no permanent phase reference is required. The phase should only be stable for two subsequent bit intervals. This is especially useful in a multipath fading environment such as the indoor radio environment. Every user in the system has a unique spread-spectrum code denoted by

$$a_k(t) = \sum a_k{}^i P_{T_c}(t - iT_c)$$
$$a_k{}^i \in \{-1,1\} \tag{5.16}$$

where P_{T_c} is a rectangular pulse (chip) with duration T_C. It is assumed that each user code has length N and that $T_b/T_C = N$, so that one code sequence fits in one data bit interval. Spreading is accomplished by multiplying (5.16) by (5.15). Then the transmitted signal is denoted as:

$$s_k(t) = Aa_k(t)b_k(t)\cos(\omega_c t + \theta_k) \tag{5.17}$$

Note that in the case of DPSK, $b_k(t)$ denotes the coded bit.

A block diagram of the transmitter as described above is depicted in Figure 5.7.

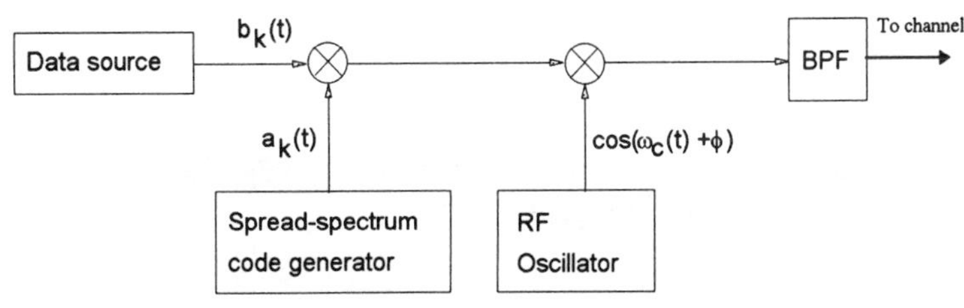

Figure 5.7 Block diagram of transmitter k.

The bandwidth expansion factor is defined as

$$B_c = \frac{T_b}{T_c} = \frac{W}{R_b} = N \tag{5.18}$$

where R_b is the data rate and W is the chip rate. Another name for the bandwidth expansion factor is the processing gain.

Channel Model

It was shown in Section 5.2 that the indoor channel can be modeled as a discrete multipath slow fading channel with the following (complex) lowpass impulse response of the bandpass channel for the link between user k and the base station:

$$h_k(t) = \sum_{l=1}^{L} \beta_{lk} \exp\{j\gamma_{lk}\}\delta(t - \tau_k) \tag{5.19}$$

where β, τ, and γ are the path gain, time delay, and phase of each path, respectively. The subscript lk refers to the lth of the kth user and j is a complex number defined as $j^2 = -1$. It is assumed that the path phase of the received signal, $(\omega_c \tau_{lk} + \gamma_{lk})$, is independently uniformly distributed over $[0,2\pi]$. The path delay is also an independently uniform variable and is assumed random over $[0, T_b]$. The path gain β_{lk} is modeled as an independent Rician random variable. The number of paths may be either fixed or randomly changing. Here fixed values of L are used according to (5.9).

Receiver Model

The receiver model consists of a matched filter, a DPSK demodulator and diversity processing components. For each unique user code a unique matched filter is required. The receiver model is depicted in Figure 5.8.

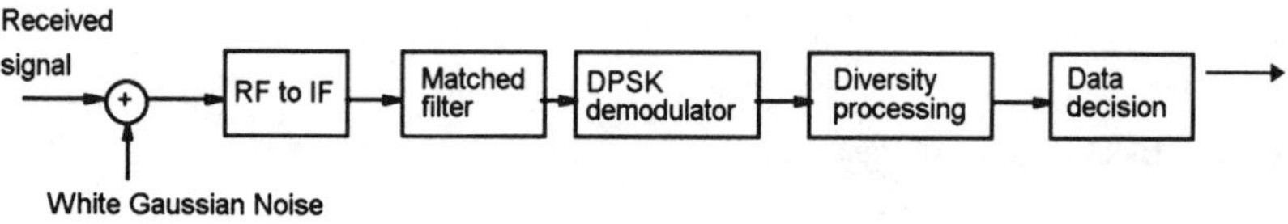

Figure 5.8 Receiver block diagram.

Applying the channel model, the received signal at the antenna for a reference user can be denoted as:

$$r(t) = \sum_{k=1}^{K} \sum_{l=1}^{L} A\beta_{lk} a_k(t - \tau_{lk}) b_k(t - \tau_{lk}) \cos(\omega_c t + \phi_{lk}) + n(t) \tag{5.20}$$

where $\phi_{lk}=(\omega_c \tau_{lk}+\gamma_{lk})$ and $n(t)$ is the normalized white Gaussian noise process with two-sided power spectral density $N_0/2$ (N_0 in the lowpass equivalent). The initial phase θ_k for each user is dropped because it might as well be included in the random path phases (γ_{lk}). Actually (5.20) is the time convolution of formulas (5.17) and (5.19) of the transmitted signal and the impulse response. Equation (5.20) can be written in the lowpass equivalent form:

$$r(t) = x(t)\cos(\omega_c t) + y(t)\sin(\omega_c t) + n(t) \tag{5.21a}$$

with

$$n(t) = n_c(t)\cos(\omega_c t) + n_s(t)\sin(\omega_c t) \tag{5.21b}$$

and

$$x(t) = \sum_{k=1}^{K}\sum_{l=1}^{L} A\beta_{lk} a_k(t-\tau_{lk}) b_k(t-\tau_{lk}) \cos(\phi_{lk}) + n_c(t) \tag{5.21c}$$

$$y(t) = \sum_{k=1}^{K}\sum_{l=1}^{L} A\beta_{lk} a_k(t-\tau_{lk}) b_k(t-\tau_{lk}) \sin(\phi_{lk}) + n_s(t) \tag{5.21d}$$

Here $x(t)$ and $n_c(t)$ are the in-phase components and $y(t)$ and $n_s(t)$ are quadrature components.

Assume that user 1 is our reference user. The lowpass response for the filter matched to the user code of the reference user is: $a_1(T_b-t)$. Time convolution of (5.21) and this matched filter response yields:

$$\begin{aligned} g_x(t) = \sum_{k=1}^{K}\sum_{l=1}^{L} A\beta_{lk} \cos(\phi_{lk}) \int_0^t a_1(s-t) a_k(s-\tau_{lk}) b_k(s-\tau_{lk}) ds \\ + \int_0^t a_1(s-t) n_c(s) ds \end{aligned} \tag{5.22a}$$

$$\begin{aligned} g_y(t) = \sum_{k=1}^{K}\sum_{l=1}^{L} A\beta_{lk} \sin(\phi_{lk}) \int_0^t a_1(s-t) a_k(s-\tau_{lk}) b_k(s-\tau_{lk}) ds \\ + \int_0^t a_1(s-t) n_s(s) \end{aligned} \tag{5.22b}$$

where $g_x(t)$ is the in-phase and $g_y(t)$ is the quadrature component of the matched filter output. Now the question of when to sample arises. Assume that the receiver can synchronize to any of the resolvable paths. So one could select the first path (no diversity), the strongest path (selection diversity), or a weighted sum of all the resolvable paths (maximal ratio combining). If the receiver synchronizes at $t=0$ to an arbitrary path, then at $t=T_b$ the matched filter output should be sampled. At the sampling point $t=T_b$, the in-phase and quadrature component of the matched filter output are

$$g_x(T_b) = \sum_{k=1}^{K}\sum_{l=1}^{L} A\beta_{lk} \cos(\phi_{lk}) \left[b_k^{-1} R_{1k}(\tau_{lk}) + b_k^{0} \hat{R}_{1k}(\tau_{lk}) \right] + \eta \tag{5.23a}$$

$$g_y(T_b) = \sum_{k=1}^{K}\sum_{l=1}^{L} A\beta_{lk} \sin(\phi_{lk})\left[b_k^{-1}R_{1k}(\tau_{lk}) + b_k^{0}\hat{R}_{1k}(\tau_{lk})\right] + \nu \tag{5.23b}$$

where b_k^{-1} and b_k^{0} are the previous and the current data bits, respectively. The correlation functions are given by

$$R_{1k}(\tau) = \int_0^{\tau} a_k(t-\tau)a_1(t)dt \tag{5.24a}$$

$$\hat{R}_{1k}(\tau) = \int_{\tau}^{T_b} a_k(t-\tau)a_1(t)dt \tag{5.24b}$$

and the noise components are

$$\eta = \int_0^{T_b} a_1(s)n_c(s)ds \tag{5.25}$$

$$\nu = \int_0^{T_b} a_1(s)n_s(s)ds \tag{5.26}$$

The noise samples η and ν are independent zero-mean Gaussian random variables with identical variance. The variance of either noise sample is the expectation of a double integral:

$$E\left[\nu^2\right] = E\left[\int_0^{T_b}\int_0^{T_b} a_1(s)a_1(q)n_c(s)n_c(q)ds\,dq\right] \tag{5.27}$$

In the case of Gaussian noise, the result of the expectation is zero if $s \neq q$ and the integral in (5.26) reduces to

$$E\left[\nu^2\right] = E\left[\int_0^{T_b} a_1^{\,2}(s)n_c^{\,2}(s)\,\mathrm{ds}\right] \tag{5.28}$$

where $a_1^2(s)=1$.

Assuming that the receiver selects the jth path of user 1, implying $\tau_{j1}=0$ and $\phi_{j1}=0$, then the complex envelope of the matched filter output $z_0=g_x(T_b)+jg_y(T_b)$ can be expressed as:

$$z_0 = A\beta_{j1}T_b b_1^0 + \sum_{k=1}^{K} A\left(b_k^{-1}X_k + b_k^0 \hat{X}_k\right) + j\sum_{k=1}^{K} A\left(b_k^{-1}Y_k + b_k^0 \hat{Y}_k\right) + (\eta_1 + j\nu_1) \tag{5.29}$$

with

$$X_1 = \sum_{\substack{l=1\\ l\neq j}}^{L} R_{11}(\tau_{l1})\beta_{l1}\cos(\phi_{l1}) \tag{5.30a}$$

$$\hat{X}_1 = \sum_{\substack{l=1\\ l\neq j}}^{L} \hat{R}_{11}(\tau_{l1})\beta_{l1}\cos(\phi_{l1}) \tag{5.30b}$$

$$Y_1 = \sum_{\substack{l=1\\ l\neq j}}^{L} R_{11}(\tau_{l1})\beta_{l1}\sin(\phi_{l1}) \tag{5.30c}$$

$$\hat{Y}_1 = \sum_{\substack{l=1\\ l\neq j}}^{L} \hat{R}_{11}(\tau_{l1})\beta_{l1}\sin(\phi_{l1}) \tag{5.30d}$$

and for $k\geq 2$

$$X_k = \sum_{l=1}^{L} R_{1k}(\tau_{lk})\beta_{lk}\cos(\phi_{lk}) \tag{5.30e}$$

$$\hat{X}_k = \sum_{l=1}^{L} \hat{R}_{1k}(\tau_{lk})\beta_{lk}\cos(\phi_{lk}) \quad (5.30f)$$

$$Y_k = \sum_{l=1}^{L} R_{1k}(\tau_{lk})\beta_{lk}\sin(\phi_{lk}) \quad (5.30g)$$

$$\hat{Y}_k = \sum_{l=1}^{L} \hat{R}_{1k}(\tau_{lk})\beta_{lk}\sin(\phi_{lk}) \quad (5.30h)$$

If the jth path is the first path, then the complete expression is used; otherwise, the paths that are received before the jth path have a negative delay with respect to the start of the jth path and the cross-correlation $R_{1k}(\tau_{lk})$ is dropped from (5.30).

If there would be only one user (K=1) and a single path (L=1), then the complex envelope of the matched filter at sampling instant n would be

$$z_n = g_x(nT_b) + jg_y(nT_b)$$
$$\Leftrightarrow \quad (5.31)$$
$$z_n = A_{lp}\exp\left[j[\phi_n - \varphi]\right]$$

where φ is an arbitrary initial phase for user 1. As described in [8, pp. 266-267], DPSK demodulation can be accomplished by taking the real part of $z_n z^*_{n-1}$ (where * denotes complex conjugation):

$$\mathrm{Re}\left[A_{lp}^{\ 2}\exp\left[j(\Phi_n - \Phi_{n-1})\right]\right] = A_{lp}^{\ 2}\cos(\Phi_n - \Phi_{n-1}) \quad (5.32)$$

It is clear that phase differences can be detected now. The same method could be used in our case by substituting (5.30) in (5.31) and thus (5.32) becomes more complex. It is assumed that z_n and z_{n-1} only differ in the data bits involved and in the Gaussian noise samples. This assumption can be made because the fading is slow compared to the data rate.

5.3.3 Diversity Techniques

In a hostile radio environment, diversity techniques can be used to improve performance. Diversity implies that a number of signals carrying the same information can be combined in order to improve the performance. Two types of diversity are considered: *selection diversity* and *maximal ratio combining*.

Selection Diversity

Selection diversity is based on selecting the strongest of a group of signals carrying the same information. With respect to a DS-CDMA system, the multiple resolvable paths can be used to accomplish selection diversity by selecting the path with the largest autocorrelation peak (output of the matched filter). This implies that the highest order of diversity that can be achieved with one antenna is equal to the number of resolvable paths. If the order of diversity that can be achieved with one antenna is too low because there are too few resolvable paths, multiple antennas can be used to increase the maximum order of diversity. The process of selection diversity is depicted in Figure 5.9.

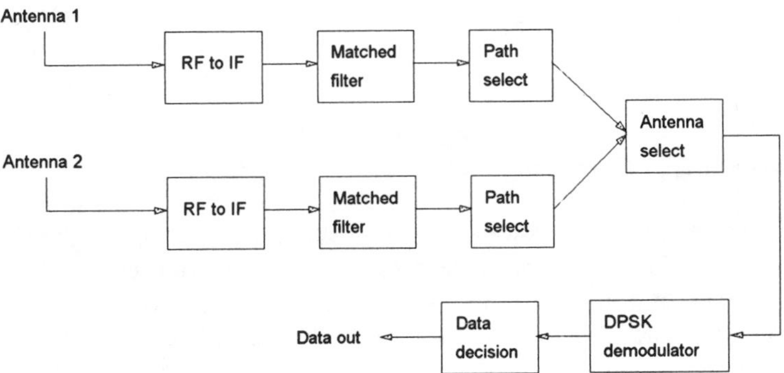

Figure 5.9 Selection diversity.

Maximal Ratio Combining

Maximal ratio combining is accomplished by summing the demodulation results of a group of signals carrying the same information and using this result as a decision variable. Maximal ratio combining with DPSK modulation differs from maximal ratio combining with BPSK modulation. In MRC/BPSK of order M the decision variable is the weighted sum of the demodulation results of M copies of the signal. The weights are taken equal to the corresponding complex-valued (conjugate) channel gain $\beta_i \exp(-j\gamma_i)$. The effect of this multiplication is to compensate for the phase shift in the channel and to weight the signal by a factor that is proportional to the signal strength. Thus a strong signal carries a larger weight than a weak signal. As already mentioned, in the hostile indoor radio channel it is very difficult to establish synchronous carrier recovery and therefore DPSK instead of BPSK can be used. Assuming that the channel parameters $\beta_i \exp(j\gamma_i)$ remain constant over two consecutive signaling intervals, then there is no need to estimate the channel parameters since the signals are automatically weighted. The price paid for this simplicity is the degradation in performance of DPSK relative to CPSK due to the presence of extra interference and noise in the DPSK demodulation loop. Figure 5.10 shows the receiver structure with maximal ratio combining.

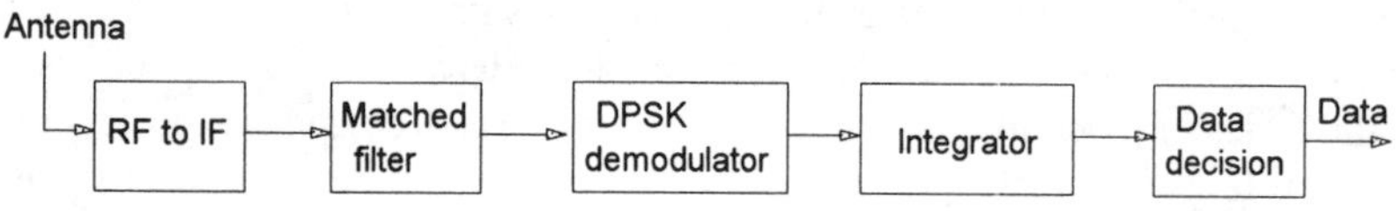

Figure 5.10 Maximal ratio combining.

5.3.4 Forward Error Correcting Codes

To enhance the performance, forward error correcting (FEC) codes can be used. In this chapter only *(n,k)* block codes are considered. An *(n,k)* forward error correcting code transforms a block of *k* data bits into a block of *n* coded bits. From coding theory it is well known that a code with a Hamming distance d_{min} can correct at least $t_c=(d_{min}-1)/2$ errors [8]. The Hamming distance depends on the type of code and on the numbers *n* and *k*. Table 5.2 shows the Hamming distance for three types of codes.

Table 5.2
Hamming distance and error correcting ability for three codes.

Code(n,k)	BCH (15,7)	Hamming (7,4)	Golay (23,12)
Hamming distance	5	3	7
Corrected errors per block	2	1	3

The probability of having *m* errors in a block of *n* bits is

$$P(m,n) = \binom{n}{m} P_e^m (1 - P_e)^{n-m} \tag{5.33}$$

with P_e being the probability of a bit error. Since the block codes can correct at least t_c errors, an upper bound for the block error probability is

$$P_{ep} = \sum_{m=t_c+1}^{n} \binom{n}{m} P_e^m (1-P_e)^{n-m} \tag{5.34}$$

and an upper bound for the bit error probability is

$$P_{ep1} = \frac{1}{n} \sum_{m=t_c+1}^{n} m \binom{n}{m} P_e^m (1-P_e)^{n-m} \tag{5.35}$$

Suppose we place a sphere of radius t_c around each of the possible transmitted code words in the code space. Codes where all these spheres are disjoint and where every received code word falls in one of these spheres are called *perfect codes*. They can correct $t_c=(d_{min}-1)/2$ errors. For these codes, (5.34) is not an upper bound but the exact block error probability. The (7,4) Hamming code and the (23,12) Golay code are examples of perfect codes [8]. Codes where all the spheres are disjoint and where every received code word is at most a distance t_c+1 from one of the possible transmitted code words are *quasi-perfect codes*. They can sometimes correct (t_c+1) errors. The block error probability for quasi-perfect codes is [8]

$$P_{eq} = \sum_{m=t_c+2}^{n} P(m,n) + \left[\binom{n}{t_c+1} - \beta_{t_c+1}\right] P_e^{t_c+1} (1-P_e)^{n-t_c-1} \tag{5.36}$$

with

$$\beta_{t_c+1} = 2^{n-k} - \sum_{i=0}^{t_c} \binom{n}{i} \tag{5.37}$$

Using the equation for the block error probability, the bit error probability for a quasi-perfect code is

$$P_{eq1} = \frac{1}{n}\sum_{m=t_c+2}^{n} mP(m,n) + \frac{t_c+1}{n}\left[\binom{n}{t_c+1} - \beta_{t_c+1}\right] P_e^{t_c+1}(1-P_e)^{n-t_c-1} \tag{5.38}$$

Like perfect codes, quasi-perfect codes are optimum on the binary symmetric channel in the sense that they result in a minimum error probability among all codes having the same block length and the same number of information bits. Therefore P_{eq1} is a lower bound for all nonperfect linear block codes. Since the (15,7) BCH code is neither a perfect code nor a quasi-perfect code, P_{ep1} and P_{eq1} are, respectively, an upper bound and a lower bound for the bit error probability with the (15,7) BCH coding.

In (5.33)-(5.38) the channel bit error probability P_e is used. If the delay spread T_{max} and the spread-spectrum code length N are given, then the number of resolvable paths L can be calculated using (5.9). Note however that in order to obtain the same data rate, the signaling rate should be increased when using FEC coding. This implies a decrease of the chip duration T_c resulting in a higher number of resolvable paths (L). Since a higher number of resolvable paths implies more multiuser interference, the channel bit error probability is higher when using FEC codes. FEC codes can therefore only be useful if they are able to decrease this (increased) channel error probability to a value that is smaller than the value of the channel error probability without FEC coding. If the channel is very noisy due to thermal noise or multiuser interference, FEC codes might worsen the performance. This occurs when the channel error probability becomes so large that the probability of having more errors in a code block than can be corrected becomes too large.

To make a fair comparison between the performance of a system with and without FEC coding it is necessary to have equal transmitted power in both cases. Since for FEC coding more bits should be transmitted, the channel SNR will be lower in that case. In fact, the SNR in the case of FEC coding is k/n times the SNR in the case of no FEC coding.

Note that due to slowly varying relative delays, memory is introduced in the channel. The theory presented above can only be applied to the channel with memory when using interleaving.

5.3.5 Pseudo-Noise Code Sequences

Code generation has been discussed in Chapter 3. It is repeated again for the completeness of this chapter.

Pseudo-noise (PN) sequences are deterministically generated sequences of two-valued symbols having a random-like structure. A special category of PN sequences is formed by maximum length shift-register sequences, or m-sequences for short. An m-sequence has length $N=2^m-1$ bits and is generated by an m-stage shift register with linear feedback as illustrated in Figure 5.11.

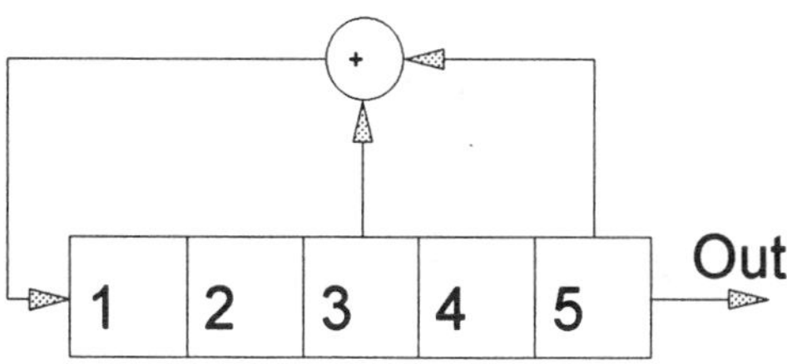

Figure 5.11 Generation of a maximum length sequence.

The sequence is periodic with period N. Each period of the sequence contains 2^{m-1} ones and $2^{m-1}-1$ zeros. Not all combinations of feedback connections are allowed. A linear feedback shift register can be described by a polynomial. Only the feedback connections described by irreducible polynomials are allowed for the generation of m-sequences. For more details on m-sequences and irreducible polynomials, refer to [31].

In DS spread-spectrum applications the binary sequence with elements $\{0,1\}$ is mapped into a corresponding binary sequence with elements $\{-1,1\}$ which is called a bipolar sequence. Gold proved that certain pairs of m-sequences of length N exhibit a three-valued cross-correlation function with values $\{-1,-t(m),\ t(m)-2\}$ where

$$
\begin{aligned}
t(m) &= 2^{(m+1)/2}+1 \quad m \text{ odd} \\
t(m) &= 2^{(m+1)/2}+1 \quad m \text{ even}
\end{aligned}
\tag{5.39}
$$

For example if $m=10$, $t(10)=2^6+1=65$ and the three possible values of the periodic cross-correlation function are $\{-1,\ -65,\ 63\}$. Two m-sequences of length N with a periodic cross-correlation function that takes on the possible values $\{-1,\ -t(m),\ t(m)-2\}$ are called preferred sequences. From a pair of preferred sequences, say A and B, we construct a set of sequences of length N by taking the modulo-2 sum of A with the N cyclically shifted versions of B or vice versa. Thus we obtain N new periodic sequences with period $N=^{2m}-1$. We may also include the original sequences A and B, and, thus, we have a total of $N+2$ sequences. The $N+2$ sequences constructed in this manner are called Gold sequences. Gold has shown that the cross-correlation function for any pair of sequences from the set of $N+2$ Gold sequences is three-valued with possible values $\{-1,\ -t(m),\ t(m)-2\}$ where $t(m)$ is given by (5.39). The peak of the autocorrelation function is of course equal to $N=2^m-1$. So for example for $m=8$ the ratio of the autocorrelation peak and the maximal cross-correlation peak is 255/33=7.33. For an m-sequence with m=8 this ratio is 255/95=2.68 [31]. In Figure 5.12 the generation of Gold sequences is depicted. The feedback connections in Figure 5.12 are those used to generate the Gold sequences of length N=127 in the performance analysis.

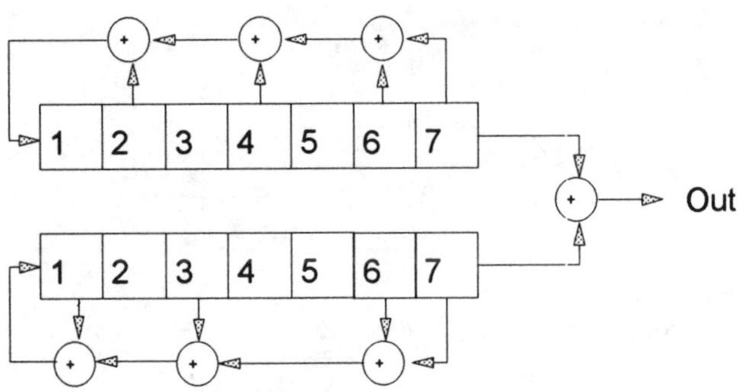

Figure 5.12 Generation of Gold sequence with length 127.

5.4 BIT ERROR PROBABILITY (BEP) ANALYSIS

In this section the performance of a star connected DS-CDMA indoor wireless network is investigated in terms of the bit error probability where we focus on the multipath effects of the indoor environment. Two diversity techniques (selection diversity and maximal ratio combining) are considered and the influence of FEC codes on the bit error probability is investigated. To make a fair comparison between systems with and without FEC coding, bandwidth efficiency is introduced.

5.4.1 BEP with Selection Diversity

To derive the bit error probability for DPSK DS SSMA with selection diversity, we start with a formula for the bit error probability of digital signals with multichannel reception, which is given in Appendix 7A.4 and derived in Appendix 4B of [8]. Formula (7A.26) in [8] is valid for $L=1$, i.e., for a single path. Since selection diversity is based on selecting the strongest of all the available paths, the formula mentioned above derives the error probability on the condition that the strongest path is selected $[P_e(\beta_{max})]$.

Let β_{j1} be the reference path. Then selection diversity is accomplished by also letting β_{j1} be the strongest path, i.e., $\beta_{j1}=\beta_{max}$. Since the fading is slow, this path will also be the strongest one for the next bit, implying that the decision variable in (5.32) is maximized. Since the data bits are equiprobable, the probability of error, P_e, is given by

$$P_e = P\left(\xi_{max} < 0 | b_1^{\,0} b_1^{\,-1} = 1\right) = P\left(\xi_{max} > 0 | b_1^{\,0} b_1^{\,-1} = -1\right) \tag{5.40}$$

where ξ_{max} is the decision variable given by (5.32) with $\beta_{j1}=\beta_{max}$.

To derive the expression for the bit error probability, it is assumed that the path delays (τ_{lk}) and the number of paths (L) are given. So the error probability given in (7.A.26) of [8] is conditional on:

- correctly chosen β_{max}
- given delay τ_{lk}
- given number of paths L.

The expression for this conditional error probability is:

$$P_e|\beta_{max},\{\tau_{lk}\},L = Q(a,b) - \frac{1}{2}\left[1+\frac{\mu}{\sqrt{\mu_0\mu_{-1}}}\right]\mathrm{I}_0(ab)\exp\left[-\frac{a^2+b^2}{2}\right] \tag{5.41}$$

$I_0(ab)$ is the modified Bessel function of zero, order which is written as:

$$I_0(ab) = \frac{1}{\pi}\int_0^{\pi}\exp[ab\cos\theta]d\theta \tag{5.42}$$

$Q(ab)$ is the Marcum Q-function, which is given by:

$$Q(a,b) = \int_b^{\infty} x\exp\left[-\frac{a^2+x^2}{2}\right]I_0(ax)dx \tag{5.43}$$

The parameters a and b are:

$$a = \frac{m}{\sqrt{2}}\left|\frac{1}{\sqrt{\mu_0}}-\frac{1}{\sqrt{\mu_{-1}}}\right| \tag{5.44a}$$

$$b = \frac{m}{\sqrt{2}}\left|\frac{1}{\sqrt{\mu_0}}+\frac{1}{\sqrt{\mu_{-1}}}\right| \tag{5.44b}$$

with

$$m = E[z_0|\beta_{max}, b_1^0] = E[z_{-1}|\beta_{max}, b_1^{-1}] \tag{5.45a}$$

$$\mu_0 = var\left[z_0|\{\tau_{lk}\}, L\right] \tag{5.45b}$$

$$\mu_{-1} = var[z_{-1}|\{\tau_{lk}\}, L] \tag{5.45c}$$

$$\mu = E[(z_0 - m)(z_{-1} - m)^*|\{\tau_{lk}\}, L] \tag{5.45d}$$

with *E[·]* denoting statistical mean and *var(·)* denoting variance. Note that the parameters m, μ, μ_0, μ_{-1} are functions of τ_{lk} and L. Using (5.29) and the assumption that $b_1^0 = b_1^{-1} = 1$, m, μ_0, μ_{-1}, and μ are obtained as

$$m = A\beta_{max}T_b b_1^{\,0} = A\beta_{max}T_b b_1^{\,-1} \tag{5.46a}$$

$$\begin{aligned}\mu_0 = {} & 2A^2\sum_{k=1}^{K}\left[\left[X_k^{\,2} + \hat{X}_k^{\,2}|\{\tau_{lk}\}, L\right]\right] \\ & + 4A^2E\left[X_1\hat{X}_1|\{\tau_{lk}\}, L\right] + 2\sigma_n^{\,2}\end{aligned} \tag{5.46b}$$

$$\mu_{-1} = 2A^2\sum_{k=1}^{K}\left[\left[X_k^{\,2} + \hat{X}_k^{\,2}|\{\tau_{lk}\}, L\right]\right] + 2\sigma_n^{\,2} \tag{5.46c}$$

$$\mu = 2A^2E\left[\sum_{k=1}^{K}\left(X_k\hat{X}_k\right) + \hat{X}_1^{\,2}|\{\tau_{lk}\}, L\right] \tag{5.46d}$$

Now the conditional bit error probability $P_e(\beta_{max}, \{\tau_{lk}\}, L)$ can be calculated. The number of resolvable paths L is a constant according to (5.9). To remove the conditioning on β_{max} and τ_{lk}, the expression for $P_e(\beta_{max}, \{\tau_{lk}\}, L)$ must be integrated over the variables β_{max} and τ_{lk}. This implies that we must know the pdfs of both β_{max} and τ_{lk}. We already mentioned that τ_{lk} is uniformly distributed over $[0,T_b]$. Deriving the probability density function of β_{max} is a little more complicated.

To derive the pdf of the strongest path (β_{max}) it is essential to note that the cumulative density function (CDF) of β_{max} is just the CDF of the pdf in (5.11) raised to the power of the order of diversity M, hence

$$p_{\beta_{max}}(r) = M\left[1 - Q\left(\frac{s}{r},\frac{r}{\sigma}\right)^{M-1}\right]\frac{r}{\sigma^2}\exp\left[-\frac{r^2+s^2}{2\sigma^2}\right]I_o\left(\frac{sr}{\sigma^2}\right) \tag{5.47a}$$

If we write (5.47a) in the binomial form, we get the final expression:

$$p_{\beta_{max}}(r) = M\sum_{i=0}^{m-1}\left[\binom{M-1}{i}(-1)^i Q^i\left(\frac{s}{\sigma},\frac{r}{\sigma}\right)\right]\frac{r}{\sigma^2}\exp\left[-\frac{r^2+s^2}{2\sigma^2}\right]I_o\left(\frac{sr}{\sigma^2}\right) \tag{5.47b}$$

where M is the order of diversity and s is the LOS component of the Rician distribution.

Since removing the conditioning on β_{max} and τ_{lk} involves extremely complicated integrals, it is useful to note that if the total number of users (K) is large, the term $2A^2E[X_1X_2+Y_1Y_2|\{\tau_{lk}\},L]$ in (5.46) can be neglected. This can be done because if K is large the contribution of this term to μ_0 becomes relatively small. This simplification makes $\mu_0=\mu_{-1}$, which means that $a=0$ in (5.44a). The conditional error probability then simplifies to

$$\begin{aligned} P_e|\beta_{max},\{\tau_{lk}\},L &= Q(0,b) - \frac{1}{2}\left[1+\frac{\mu}{\mu_0}\right]\mathrm{I}_0(0)\exp\left[-\frac{b^2}{2}\right] \\ &= \tfrac{1}{2}\left[1-\frac{\mu}{\mu_0}\right]\exp\left[-\frac{m^2}{\mu_0}\right] \end{aligned} \tag{5.48}$$

Since $I_0(0)=1$ and $Q(0,b)=\exp\left(-b^2/2\right)$, the conditioning on β_{max} can be removed as follows:

$$P_e|\{\tau_{kl}\},L=\int_0^{\infty} P_e|(\beta_{max}\{\tau_{kl}\},L)\ p_{\beta_{max}}(\beta_{max})d\beta_{max} \tag{5.49}$$

The conditioning on $\{\tau_{lk}|\neq\tau_{j1}\}$ is now removed by integrating over all $(K{*}L)-1$ identical pdfs for the path delays, all uniform over $[0,T_b]$. The parameter τ_{j1} is excluded because it is known due to the selection process of the strongest path ($\beta_{j1}=\beta_{max}$).

The method of integrating over all path delays is very time consuming. Instead it is possible to utilize the central limit theorem and conclude that μ_0 and μ are Gaussian if the product KL is sufficiently large. This is founded by the idea that μ_0 and μ are sums of a large number of random variables. This implies that the mean and the variance of μ_0 and μ have to be calculated. For convenience the condition $l\neq j$ in (5.30a) - (5.30f) is dropped, implying that user 1 has L instead of $(L-1)$ self-interference paths. The effect of this approximation is negligible if KL is large.

Combining (5.49) with the results of the Gaussian approximation yields the following expression for P_e:

$$P_e=\int_{-\infty}^{\infty}\int_{-\infty}^{\infty}\int_{0}^{\infty}\tfrac{1}{2}\left[1-\frac{\mu}{\mu_0}\right]\exp\left[-\frac{m^2}{\mu_0}\right]p_{\beta_{max}}(\beta_{max})p_{\mu}(\mu)p_{\mu_0}(\mu_0)d\beta_{max}d\mu d\mu_0 \tag{5.50}$$

where μ and μ_0 are Gaussian variables. The mean and variance of these Gaussian variables are functions of:

- The code sequences (a_k) and their correlation properties
- The number of resolvable paths (L)
- The chip time duration (T_c)
- The bit time duration (T_b)
- The Rice parameters (σ and S)
- The noise density (N_0)
- The amplitude of the received signal (A).

Variable μ can be easily removed analytically from this expression as follows:

$$\int_{-\infty}^{\infty}\left[1-\frac{\mu}{\mu_0}\right]p_{\mu}d\mu=1-\frac{E_{\tau}(\mu)}{\mu_0} \tag{5.51}$$

A double integral is left to evaluate.

To calculate the bit error probability as a function of the signal-to-noise ratio E_b/N_0 it is assumed that $\omega_c T_b = 2\pi l$, where l is an integer. This means that the transmitted power is $E_b = A^2 T_b/2$.

Figure 5.13 shows the effect of the delay spread on the performance for selection diversity with N=127 and for two different bit rates (r_b).

Computational results showed that the performance degrades with an increasing delay spread. Another important conclusion is that degradation in the performance only occurs if the increased value of the data rate or delay spread causes the number of resolvable paths to increase. This implies that the performance for a fixed code length only depends on the number of resolvable paths. The saturation of the bit error probability for high E_b/N_0 values is due to the multiuser interference. It was seen from the computations that the performance increases for higher values of the order of diversity. The Rician parameter R, which is the ratio between the direct and the reflected components in the received signal, also influences the performance.

An increase in the Rician parameter results in (1) a better wanted signal, and (2) a worse (stronger) interference signal (i.e., more interference noise). It was seen that for sufficiently high SNR (higher than 16 dB) this results in a better performance for R=11 dB as compared to R=6.8 dB. For lower SNR the increased interference noise in combination with the thermal noise plays a dominant role, which in this case causes the performance to deteriorate.

Computations were performed to compare the performance with codes of length N=127 and N=255. Tables 5.3 and 5.4 give the delay spreads that correspond to specific values of L for specific bit rates in the case of N=127 and 255, respectively. Results showed that performance is enhanced by an increase in the value of L. This happens because the codes of length 255 have better (lower) cross-correlations. Thus, the decrease in interference power per interference signal is such that even though the number of interfering signals ($KL-1$) increases, the total amount of interference power decreases. Therefore, at the cost of a doubled bandwidth, codes of 255 chips improve the performance.

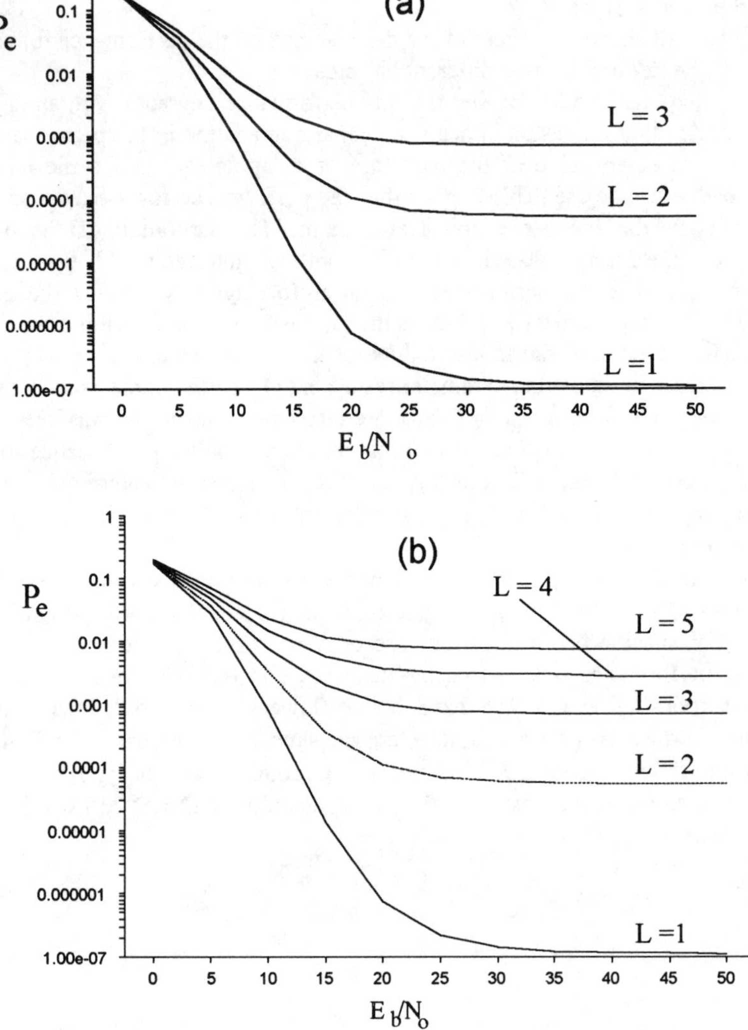

Figure 5.13 The effect of the delay spread on the bit error rate for R=6.8 dB, M=8 (selection diversity) and N=127, K=15, and (a) r_b=64 kbit/s and (b) r_b= 144 kbit/s.

Table 5.3
Delay spread (in ns) corresponding to a number of resolvable paths for three bit rates and N=127.

	32 kbit/s	64 kbit/s	144 kbit/s
L=1	T_{max}<246	T_{max}<123	T_{max}< 55
L=2	250≤T_{max}<492	123≤T_{max}<246	55≤T_{max}<110
L=3	—	246≤T_{max}<370	110≤T_{max}<165
L=4	—	—	165≤T_{max}<220
L=5	—	—	220≤T_{max}<275

Table 5.4
Delay spread (in ns) corresponding to a number of resolvable paths for three bit rates and N=255.

	32 kbit/s	64 kbit/s	144 kbit/s
L=1	T_{max}<123	T_{max}<62	T_{max}<27
L=2	123≤T_{max}<246	62≤T_{max}<124	27≤T_{max}<54
L=3	246≤T_{max}<370	124≤T_{max}<186	54≤T_{max}<81
L=4	—	186≤T_{max}<246	81≤T_{max}<108
L=5		246≤T_{max}<310	108≤T_{max}<135
L=6			135≤T_{max}<162
L=7			162≤T_{max}<189
L=8			189≤T_{max}<216
L=9			216≤T_{max}<243
L=10	—	—	243≤T_{max}<270

5.4.2 BEP with Maximal Ratio Combining

The decision variable in the case of MRC/DPSK is

$$\xi_c = \mathrm{Re}\left[\sum_{i=1}^{M}\left(AT_b\beta_{i1}b_1^0 + N_{1i}\right)\left(AT_b\beta_{i1}b_1^{-1} + N_{2i}\right)^*\right] \tag{5.52}$$

Here N_{1i} and N_{2i} are Gaussian random variables. Assuming that the multiuser interference is Gaussian, N_{1i} and N_{2i} are the sum of the Gaussian thermal noise and the multiuser interference. Since (5.52) is the sum of the demodulated signals of M paths, this method is sometimes stated as predetection combining. This form of diversity is very easily implemented with a simple integrating RC circuit. To simplify the calculations the following two assumptions are made. First, as argued in [14], it can be assumed that N_{1i} and N_{2i} are independent. It was shown in the case of selection diversity that the parameter μ, which is defined in (5.45), is nearly zero and can be ignored. This implies that $cov[N_{1j}, N_{2j}]$ is negligible compared with $var[N_{1j}]$and $var[N_{2j}]$ implying statistical independence between N_{1j} and N_{2j} . Second, it is assumed that the pairs (N_{1j}, N_{2j}) and (N_{1j}, N_{2j}) are independent for $i \neq j$. Mathematically this is not correct because the delays $\{\tau_{lk}\}_i$ are not independent of the delays $\{\tau_{lk}\}_j$. But since each set of delays is taken with reference to a different time origin (corresponding to the arrival time of the signal on the corresponding combined path) and also considering that any two resolved paths are separated by at least a chip time period, the assumption is physically reasonable. With these two assumptions, the bit error probability can be calculated using formulas (7.4.13) and (1.1.167) in [8]. If M paths are combined in such a way that the signal-to-noise-ratio per bit γ_b is given by

$$\gamma_b = \frac{E_b}{N_0} \sum_{m=1}^{M} \beta_m^2 \tag{5.53}$$

then the conditional bit error probability for DPSK is

$$P_e(\gamma_b) = \frac{1}{2^{2M-1}} \exp(-\gamma_b) \sum_{m=0}^{M-1} c_m \gamma_b^m \tag{5.54}$$

with

$$c_m = \frac{1}{m!} \sum_{i=0}^{M-1-m} \binom{2M-1}{i} \tag{5.55}$$

Then the unconditional error probability P_e is

$$P_e = \int_0^\infty P_e(\gamma_b) f_{\gamma_b}(\gamma_b) d\gamma_b \tag{5.56}$$

The problem now remains to calculate the probability density function $f_{\gamma_b}(\gamma_b)$. If we only have to deal with Gaussian noise, then γ_b is given by (5.53). In our case, however, there is also multiuser interference and the signal-to-noise-ratio per bit is given by

$$\gamma_b = \frac{E_b}{N_{ni}} \sum_{m=1}^{M} \beta_m^2 \tag{5.57}$$

where N_{ni} is the sum of the Gaussian noise and the (Gaussian) multiuser interference. Since β_l is a Rician distributed random variable, this variable can be written as

$$\beta_l = \sqrt{X_{l1}^2 + X_{l2}^2} \tag{5.58}$$

with X_{l1} and X_{l2} being independent Gaussian random variables with mean $m_x = m_{l1} = m_{l2}$ and variance $\sigma_x^2 = \sigma_{l1}^2 = \sigma_{l2}^2$. The relation with the Rician parameters S_β and σ_β is $S_\beta^2 = m_{l1}^2 + m_{l2}^2$ and $\sigma_\beta^2 = \sigma_{l1}^2 = \sigma_{l2}^2$. With this knowledge, we find that

$$\sum_{l=1}^{L} \beta_k^2 = \sum_{l=1}^{L} (X_{l1}^2 + X_{l2}^2) \tag{5.59}$$

In [8], (1.115) shows that if $y = \sum_{i=1}^{n} X_i^2$ where X_i is a Gaussian random variable with mean m_i and variance σ^2 then

$$f_y(y) = \frac{1}{2\sigma^2}\left(\frac{y}{s}\right)^{n-2/4} \exp\left(-\frac{s^2+y}{2\sigma^2}\right) I_{n/2-1}\left(\sqrt{y}\frac{s}{\sigma^2}\right) \tag{5.60}$$

with

$$s^2 = \sum_{i=1}^{n} m_i^2 \tag{5.61}$$

In our case, we have

$$z = \sum_{i=1}^{2M} X_i^2 \tag{5.62}$$

This implies that the pdf of z is

$$f_z(z) = \frac{1}{2\sigma^2}\left(\frac{z}{S_M^2}\right)^{M-1/2} \exp\left(-\frac{s_M^2+z}{2\sigma^2}\right) I_{M-1}\left(\sqrt{z}\frac{s_M}{\sigma^2}\right) \tag{5.63}$$

with $s_M^2 = \sum_{i=1}^{2M} s_i^2 = 2Ms^2$. Since $\gamma_b = \frac{E_b}{N_{ni}} z$, the pdf for γ_b is given by:

$$f_{\gamma_b}(\gamma_b) = \frac{1}{2\sigma^2 E_b/N_{ni}} \left(\frac{\gamma_b}{s_M^2 E_b/N_{ni}} \right)^{M-1/2} \exp\left(-\frac{s_M^2 + \dfrac{\gamma_b}{E_b/N_{ni}}}{2\sigma^2} \right) \tag{5.64}$$

$$I_{M-1}\left(\sqrt{\frac{\gamma_b}{E_b/N_{ni}}} \frac{s_M}{\sigma^2} \right)$$

In this equation, β_l is the path gain of the lth combined path; N_{ni} is the sum of the Gaussian noise and the multiuser interference, which is assumed to be Gaussian; $I_\alpha(x)$ is the αth order modified Bessel function of the first kind; and $S_M^2 = \dfrac{Ms^2}{\sigma^2}$ and γ_b is the sum of the signal-to-noise-ratios of the combined M paths.

Computational results showed that maximal ratio combining yields significantly better performance than selection diversity except for very low SNR. This is especially true for higher orders of diversity.

5.4.3 BEP with FEC Coding

In this section, the effects of three FEC coding schemes are considered. In the plots for a system with FEC coding the signal energy E_b stands for the transmitted energy per *information bit.* This means that E_b for a system with (15,7) BCH coding is 15/7 times the energy transmitted per code symbol.

We have calculated the coding results with M=1 for the following cases: R=6.8 dB, L=1, and N=255; R=6.8 dB, L=5 and N=255. Besides the curves for coding without diversity, Figure 5.14 also contains the curves for selection diversity with M=2, 3, and 4.

It is seen that for low values of L (such as L=1) the use of FEC coding without diversity can lead to acceptable bit error probabilities. However, for larger values of L (L=5) the use of FEC coding only is not sufficient. In that case diversity is necessary either in combination with FEC coding or not. Combining selection diversity with FEC coding is attractive when the bit error probability in the case of M=L (maximum order of diversity with one antenna) is not sufficiently low. Using FEC coding then alleviates the need to install additional antennas.

We have already seen that both longer spread-spectrum codes and FEC codes improve the performance at the expense of an increased bandwidth. It is then interesting to compare the performance for N=127 + FEC coding with the performance for N=255 without coding (Figure 5.15).

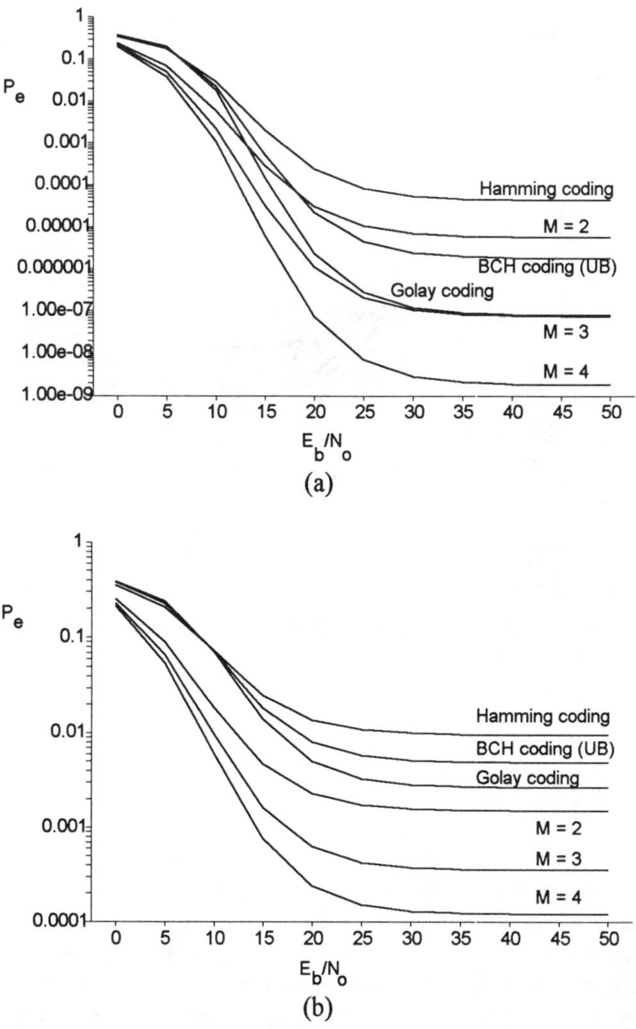

Figure 5.14 Comparison of the bit error probability using FEC coding only and selection diversity only for R=6.8 dB, N=255, and (a) L=1 and (b) L=5.

Since all three FEC codes approximately double the required signal bandwidth, the two cases mentioned above require approximately the same amount of bandwidth. It is seen that in the case of Golay coding the bandwidth efficiency of N=127 + FEC coding is higher since the performance is better than in the case of N=255 without coding for equal bandwidth. Calculations have also shown that the performance is more sensitive to the (interference) noise level if FEC codes are used. This implies that if the number of users becomes too large, the use of spread-spectrum codes of 255 chips will offer better performance than codes of 127 chips in combination with FEC codes.

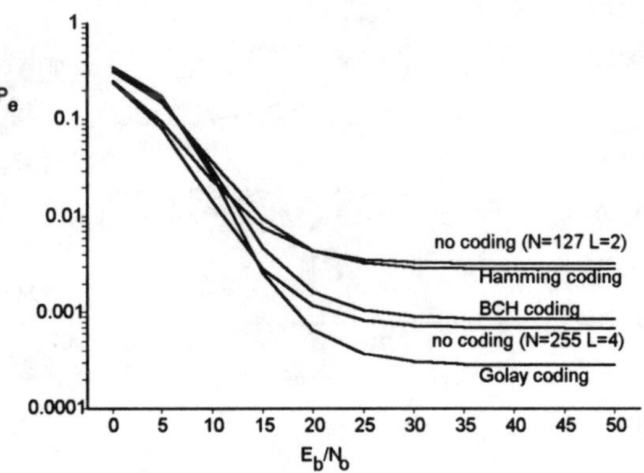

Figure 5.15 Comparison of bit error probability with r_b =64 kbit/s, T_{max}=150 ns, R=6.8 dB, and selection diversity with M=2 for N=127+FEC coding and L=2; and N=255 without FEC coding and L=4.

5.4.4 Bandwidth Efficiency

Since all three FEC codes approximately double the required signal bandwidth, the two cases mentioned above require approximately the same amount of bandwidth. It is seen that in the case of Golay coding the performance of a system with code length $N=127$ and FEC coding is better than in the case of a code length $N=255$ without coding for equal bandwidth. Since the improvement of performance is payed for by an increase in the system bandwidth, another measure of performance, *bandwidth efficiency*, is introduced. This bandwidth efficiency is defined as

$$BE \underline{\Delta} \frac{K_{max} r_b}{W} = \frac{K_{max} r_c}{N} \quad \text{(bits/Hz)} \tag{5.65}$$

where K_{max} is the maximum number of simultaneous users for which the bit error probability is less than a preset value ber_0, r_b is the bit rate, N is the spread-spectrum code length, r_c is the code rate of the FEC code used, and $W=Nr_b/r_c$ is the total bandwidth. Table 5.5 shows the bandwidth efficiency defined in (5.65) as the number of users with a given bit rate, occupying a given bandwidth with a specified bit error probability.

Table 5.5
The bandwidth efficiency as a function of the order of diversity and resolvable paths L for selection diversity with N=255, R=6.8 dB, and $ber_0 = 10^{-4}$

	L=1	L=2	L=4	L=8
M=1	—	—	—	—
M=2	0.098	0.055	0.031	0.012
M=4	0.214	0.090	0.051	0.031

In Tables 5.6a and 5.6b, a comparison is made between the performance of codes with 127 chips and 255 chips with FEC coding for selection diversity and maximal ratio combining, respectively.

Table 5.6a
Bandwidth efficiency for selection diversity with FEC coding for N=127 and N=255 for several orders of diversity

	N=127			N=255		
	Hamming (7,4)	BCH (15,7)	Golay (23,12)	Hamming (7,4)	BCH (15,7)	Golay (23,12)
M=1	0.004	0.015	0.021	0.011	0.016	0.025
M=2	0.027	0.040	0.066	0.025	0.027	0.035
M=4	0.040	0.055	0.082	0.031	0.033	0.041

Table 5.6b
Bandwidth efficiency for maximal ratio combining with FEC coding for N=127 and N=255 for several orders of diversity

	N=127			N=255		
	Hamming (7,4)	BCH (15,7)	Golay (23,12)	Hamming (7,4)	BCH (15,7)	Golay (23,12)
M=1	0.004	0.015	0.021	0.011	0.016	0.025
M=2	0.041	0.048	0.069	0.034	0.037	0.041
M=4	0.112	0.107	0.136	0.067	0.071	0.080

It is assumed that the data rate and the RMS delay spread are the same in both cases, which implies that the number of resolvable paths is doubled in the case of $N=255$ as compared to the case of $N=127$. A system with code length $N=127$, FEC coding, and diversity offers superior performance in terms of bandwidth efficiency as compared to a system with code length $N=255$, FEC coding, and diversity.

5.4.5 Monte Carlo Simulation

A Monte Carlo simulation was conducted to assess the performance, in terms of bit error probability, of a DS-CDMA radio system. Importance sampling techniques were used to accelerate the simulation [17]. The results were compared with the analytical results to verify its accuracy.

The accuracy of the calculation method is illustrated in Figure 5.16. The plot compares results obtained analytically with results obtained by computer simulation. The computer simulation was conducted without the Gaussian assumption whereas the analytical performance analysis is done using Gaussian approximation. We see that the difference between the analytical and simulation results is insignificant. Therefore, the analytical method used is a valid and fast technique to obtain the performance. Thus the Gaussian approximation can be used to obtain the performance analytically.

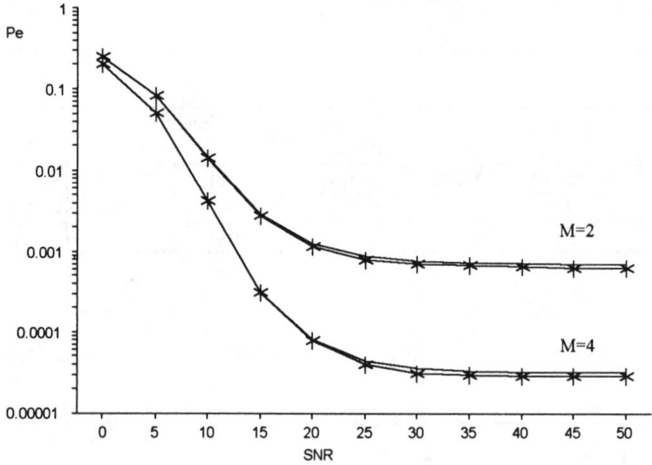

Figure 5.16 Comparison of results obtained analytically (—) and results obtained by computer simulation (-*-*-). Results are shown for $R = 6.8$ dB, $L = 4$, $N = 255$, and $M = 2.4$.

5.5 PACKET SWITCHING

Random access packet switching has already proven to be a very useful technique in an environment with high peak-to-average traffic ratios. It offers the possibility to allocate efficiently scarce radio communication channels in such a way that users with bursty traffic can share a common frequency channel without significant degradation to a single user's throughput. In ALOHA-type systems a test packet is destroyed if it is overlapped by one or more other packets. Due to this effect, in an unslotted ALOHA system without fading and capture effects, the maximum throughput defined as the number of successful packets is $^1/_2e$ (≈18%). The maximum throughput is doubled when using a slotted ALOHA system instead of an unslotted ALOHA system.

In code division multiple access systems simultaneous transmission of two or more packets is permitted (with some error rate). If the actual number of transmitting terminals exceeds the value of the maximum capacity C, then all packets are destroyed. This suggests the possibility of using CDMA systems in a random access mode of operation. In this Section, attention is focused on a star connected synchronous (slotted) DS-CDMA system where a fixed number of terminals transmit packets to a central base station.

5.5.1 Throughput and Delay Analysis

In this section a general model is described to derive expressions for these performance parameters. Two performance parameters are considered. First the throughput defined as the number of successful packets per time slot is considered as a function of the offered traffic. The second performance parameter is the packet delay as a function of the system load. The packet delay is defined as the average number of time slots used before a packet is received successfully. A very important part of the model is the traffic model, which is also described.

Throughput

The general expression for throughput is

$$S(G) \underline{\Delta} G \Pr(\text{success}) \tag{5.66}$$

where G is the average number of packets offered per time slot. The offered traffic G consists of both newly generated packets and retransmitted packets that were unsuccessfully received during a previous transmission. The packet flow of the random access CDMA system is shown in Figure 5.17.

The throughput can also be calculated as the average number of successfully received packets in the system

$$S = \sum_{k=1}^{C} kP_k P_{sk} \tag{5.67}$$

In (5.67), P_k is the probability that k packets are transmitted simultaneously and it depends on the traffic model; P_{sk} is the packet success probability if k packets are transmitted simultaneously.

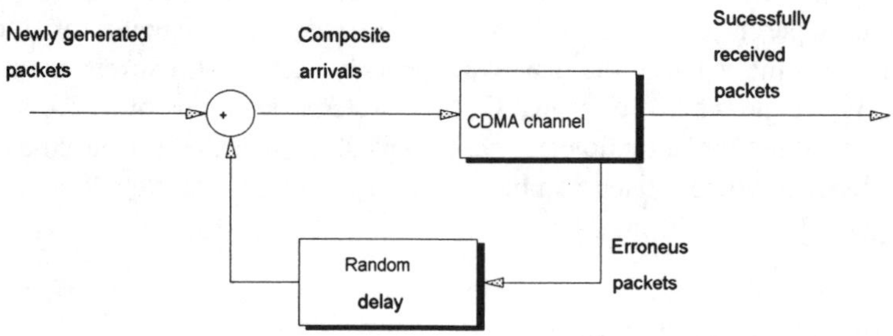

Figure 5.17 Packet flow in a random access CDMA system.

To calculate the packet success probability P_{sk} of a packet of N_p data bits, we must make an assumption about the rapidity of the fading. It was already mentioned that in an indoor environment, the channel does not change significantly within two consecutive bit intervals. This situation is defined here as *slow fading*. For the sake of completeness we also consider the situation that the fading is so fast that two consecutive bits are modeled as independent statistical variables. This situation is defined as *fast fading*. Note that in most indoor environments such as offices or warehouses, the environment can best be characterized as slow fading.

Fast Rician Fading

The probability of correctly receiving a packet of N_p data bits in the fast fading situation is

$$P_{sk} = \left[1 - P_{er}(k)\right]^{N_p} \tag{5.68}$$

where $P_{er}(k)$ is the bit error probability of the channel and k is the number of simultaneously transmitting terminals. The expression for $P_{er}(k)$ is different for selection diversity and maximal ratio combining.

For selection diversity, we found in Section 5.4 that the bit error probability is

$$P_{er,s}(\mathrm{k}) = \int_{-\infty}^{\infty} \int_{-\infty}^{\infty} \int_{0}^{\infty} P_{sd}(\beta_{max}, \mu, \mu_0) p_{\beta_{max}}(\beta_{max}) p_\mu(\mu) p_{\mu_0}(\mu_0) d\beta_{max} d\mu d\mu_0 \tag{5.69}$$

where

$$P_{sd}(\beta_{max}, \mu, \mu_0) = \frac{1}{2}\left[1 - \frac{\mu}{\mu_0}\right] \exp\left[-\frac{A^2 \beta_{max}^2 T_b^2}{\mu_0}\right] \tag{5.70a}$$

$$m = E[z_0 | \beta_{max}, b_1^0] = E[z_{-1} | \beta_{max}, b_1^{-1}] \tag{5.70b}$$

$$\mu_0 = var[z_0 | \{\tau_{lk}\}, L] \tag{5.70c}$$

$$\mu = E[(z_0 - m)(z_{-1} - m)^* | \{\tau_{lk}\}, L] \tag{5.70d}$$

The parameters z_0 and z_{-1} are the envelopes of the signal at the current sampling instant and the previous sampling instant, respectively, while the parameters μ_0 and μ are appproximated by Gaussian random variables with pdfs p_μ and $p_{\mu 0}$, respectively. The parameter μ can be easily removed as follows:

$$\int_{-\infty}^{\infty} \left[1 - \frac{\mu}{\mu_0}\right] p_\mu d\mu = 1 - \frac{E_\tau(\mu)}{\mu_0} \tag{5.71}$$

leaving a double integral to evaluate. The parameter μ_0 consists of the influence of the desired signal, multiuser interference and white Gaussian noise.

The pdf of selecting the strongest path was given by:

$$p_{\beta_{max}}(r) = M\left[1 - Q(\frac{s}{\sigma},\frac{r}{\sigma})\right]^{M-1} \frac{r}{\sigma}\exp\left[-\frac{s^2+r^2}{2\sigma^2}\right] I_0(\frac{sr}{\sigma^2}) \quad (5.72)$$

where M is the order of diversity and $Q(\frac{s}{\sigma},\frac{r}{\sigma})$ is the Marcum Q-function.

For maximal ratio combining, the bit error probability was found to be

$$P_{er,m} = \int_0^\infty P_e(\gamma_b) f_{\gamma_b}(\gamma_b) d\gamma_b \quad (5.73)$$

with the conditional bit error probability

$$P_e(\gamma_b) = \frac{1}{2^{2M-1}} \exp(\gamma_b) \sum_{m=0}^{M-1} c_m \gamma_b^m \quad (5.74)$$

where

$$c_m = \frac{1}{m!} \sum_{i=0}^{M-1-m} \binom{2M-1}{i} \quad (5.75)$$

and

$$\gamma_b = \frac{E_b}{N_{ni}} \sum_{m=1}^{M} \beta_m^2 \quad (5.76)$$

The probability density function of γ_b is

$$f_{\gamma_b}(\gamma_b) = \frac{1}{2\sigma^2 \, E_b / N_{ni}} \left(\frac{\gamma_b}{s_M^2 \, E_b / N_{ni}} \right)^{M-\frac{1}{2}} \exp\left(- \frac{s_M^2 + \frac{\gamma_b}{E_b / N_{ni}}}{2\sigma^2} \right) I_{M-1}\left(\sqrt{\frac{\gamma_b}{E_b / N_{ni}}} \frac{s_M}{\sigma^2} \right) \tag{5.77}$$

In the above equations, M is the order of diversity, β_l is the path gain of the lth combined path, N_{ni} is the sum of the Gaussian noise and assumed multiuser interference, $s_m^2 = \frac{Ms^2}{\sigma^2}$ with S being one of the Rice parameters and γ_b being the sum of the SNRs of the M combined paths. In case of maximal ratio combining, the order of diversity is chosen such that $M \leq L$ (Number of resolvable paths. The detailed derivations of these expressions can be found in Section 5.4.)

Slow Rician Fading

The expressions for the packet error probability in the case of slow Rician fading are different from the fast Rician case. For selection diversity, the bit error probability is

$$P_{er,s} = \int_{-\infty}^{\infty} \int_{-\infty}^{\infty} \int_{0}^{\infty} \{1 - P_{sd}(\beta_{max}, \mu, \mu_0)\}^{N_d} p_{\beta_{max}}(\beta_{max}) p_{\mu}(\mu) p_{\mu_0}(\mu_0) d\beta_{max} d\mu d\mu_0 \tag{5.78}$$

For maximal ratio combining, the packet success probability is

$$P_e = \int_0^{\infty} [1 - P_e(\gamma_b)]^{N_d} f_{\gamma_b}(\gamma_b) d\gamma_b \tag{5.79}$$

Here the same expressions hold as (5.79)-(5.82).

Average Delay

The average delay is defined as the average number of slot times it takes for a packet to be successfully received. Thus it is the average time duration between the packet being offered to the transmitter and the packet being successfully received. The average delay in an indoor network, assuming negligible round trip propagation delay and immediate acknowledgment, can be obtained on the same basis as was done in [30]:

$$D = 1.5 + \left[\frac{G}{CS} - 1\right]\left(\lfloor \delta_\tau + 1 \rfloor + 1\right) \tag{5.80}$$

Here it is assumed that the minimum delay for slotted systems is 1.5 slot durations. This time consists of one slot time being the average time between the time the packet is offered to the transmitter and the beginning of the next slot. The average number of retransmissions for a packet to be successfully received is $(G/CS-1)$, the delay due to each retransmission is $\lfloor \delta_\tau + 1 \rfloor + 1$, with δ_τ being the mean of the retransmission delay, which is uniformly distributed over the range from which the retransmission delay is selected. C is the user's threshold capacity.

Arrival Models

Once a model for the new packet generation process and the retransmission process is formulated, the required composite arrival distribution P_k can be found. The basic model used in this chapter assumes three operation modes for each of C terminals:

- Originating mode;
- Transmission mode;
- Backlog mode.

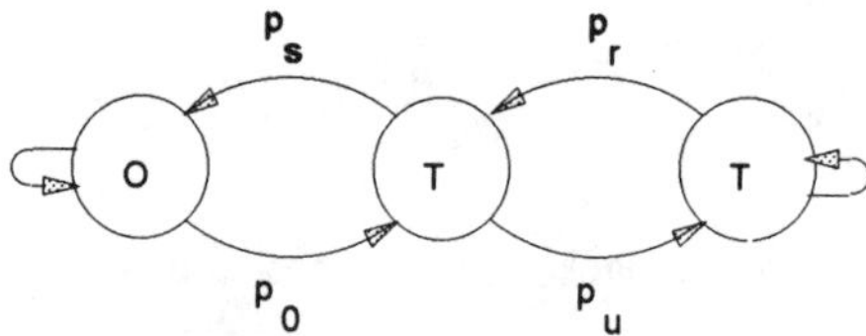

Figure 5.18 Terminal states.

In the originating mode, new packets are transmitted in any slot with probability p_0. Terminals enter the backlog mode when a transmitted packet is not received successfully. The terminals can only be in one state at a time. This implies that terminals that are in backlog mode cannot transmit new packets until the backlog packet is received correctly. In the backlog mode, packets are retransmitted with probability p_r. For completeness, the transition probabilities shown in Figure 5.18 are:

p_0: Probability of a new packet being transmitted in any slot;
p_r: Probability of a backlogged packet transmitted in any slot;
p_s: Probability of a packet successfully received in any slot;
p_u: Probability of a packet unsuccessfully received in any slot.

In general $p_r > p_0$. A simplification can be made by assuming that the generation statistics of new packets and backlogged packets are the same, implying that $p_r = p_0 = p$. In this case, the composite arrival distribution becomes binomial with parameters C and p:

$$
\begin{aligned}
P(k) &= \binom{C}{k} p^k (1-p)^{N-k} \qquad k \le N \\
&= 0 \qquad k > N
\end{aligned}
\tag{5.81}
$$

The probability p is given by

$$p = \frac{G}{C} \tag{5.82}$$

with G being the average number of new plus backlogged packets and N_s is the number of terminals in the system. After substitution, P_k is written as

$$
\begin{aligned}
P(k) &= \binom{C}{k} \left(\frac{G}{C}\right)^k \left(1-\frac{G}{C}\right)^{N_s-k} \qquad k \le C \\
&= 0 \qquad k > C
\end{aligned}
\tag{5.83}
$$

If $p_r = p_0 = p \to 0$ and $N_s \to \infty$ then the arrival model approaches a Poisson distribution with a composite finite arrival rate λ. Then the composite arrival distribution is

$$P(k) = \frac{(\lambda T)^k}{k!} \exp(-\lambda T) \tag{5.84}$$

Since the average number of transmitted new and backlogged packets is given by $G = \lambda T$, this expression can be written as

$$P(k) = \frac{G^k}{k!}\exp(-G) \tag{5.85}$$

Since the number of terminals in a slotted DS-CDMA network is finite, the binomial arrival distribution is more appropriate than the Poisson arrival distribution. In the general case when $p_r > p_0$ and C is finite, it is not possible to obtain a simple closed-form expression for the composite arrival distribution, as in the Poisson and the binomial cases. This general case is not considered in this chapter. A good description of the model is given in Chapter 9.

In Figure 5.19 normalized (with reference to C) throughput results for fast and slow Rician fading channel are depicted for an indoor, multi-picocellular surrounding. It is seen that slow fading channels yield better performance than fast fading channels.

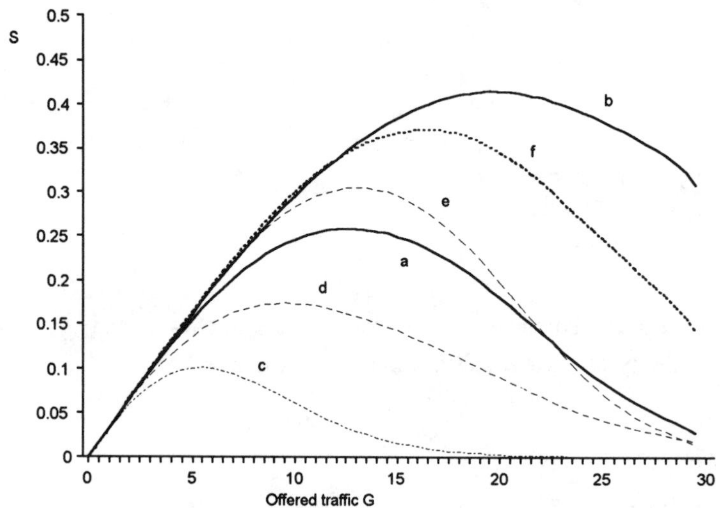

Figure 5.19 Throughput curves for fast and slow Rician fading (Multi-pico-cell environment). Maximal ratio combining, N=127, R=6.8 dB, C=30, N_d =128. M=2, L=2, (a) fast, (b) slow. M=2, L=4, (c) fast, (d) slow. M=4, L=4, (e) fast, (f) slow.

5.6 CONCLUSIONS

The performance of a star-connected DS-SS system for indoor wireless applications was evaluated considering DPSK modulation. The indoor radio channel was assumed to be of the Rician fading type. Two types of diversity, selection diversity and maximal ratio combining, were considered. The performance was assessed in terms of bit error probability and throughput. In general, we can state that diversity improves the performance significantly and that maximal ratio combining yields superior performance compared to selection diversity.

We saw that performance is quite sensitive to the value of the delay spread and the bit rate at which the data are transmitted. In general the performance is affected if an increased value of the delay spread or bit rate causes the number of resolvable paths (L) to increase.

Methods to improve the performance for a given bit rate and delay spread were investigated using diversity (selection diversity and maximal ratio combining), FEC coding, and longer spread-spectrum codes. These methods can also be combined. The drawback of diversity is that if it is necessary to have an order of diversity (M) larger than the number of resolvable paths, then the use of multiple antennas is required. In all cases additional hardware and logic are required for implementation of the system. From the results the following can be concluded.

1. Instead of installing multiple (two or more) antennas per receiver, FEC coding can be used either independent of or in combination with diversity to further improve the performance at the cost of increasing system bandwidth. The two options should be compared in terms of implementation complexity and costs.

2. FEC coding cannot completely replace diversity. The performance with FEC coding deteriorates faster than the performance for diversity if the number of paths or simultaneously active users increases.

3. Use of spread-spectrum codes of 255 chips improves the bit error probability at the cost of a higher system bandwidth as compared to spread-spectrum codes of 127 chips. However, the improvement of the bit error probability is not such that the bandwidth efficiency is improved.

4. If the interference power is below a certain limit, a system with FEC coding and spread-spectrum codes of 127 chips can compete with a system with spread-spectrum codes of 255 chips (note that these two systems require the same amount of bandwidth). This holds for both bit error probability and bandwidth efficiency. If the number of paths is too large or the number of active users crosses a certain limit, the system with spread-spectrum codes of 255 chips but without FEC codes yields better performance.

The results presented in this chapter are valid under the assumption of an independent identically distributed power delay profile, perfect power control within a cell, and perfect synchronization and tracking at the receiver.

REFERENCES

[1] H. Hashemi, "The indoor propagation channel," *Proceedings of the IEEE*, Vol. 81, pp. 943-968, July 1993.

[2] R.J.C. Bultitude, "Measurement, characterization and modeling of indoor 800/900 MHz radio channels for digital communications," *IEEE Comm. Mag.*, Vol. 25, pp. 5-12, June 1987.

[3] A.M. Saleh and R.A. Valenzuela, "A statistical model for indoor multipath propagation," *IEEE J. Selected Areas in Comm.*, Vol. SAC-5, pp. 128-137, February 1987.

[4] T.S. Rappaport and C.D. McGillem, "UHF fading in factories," *IEEE J. Selected Areas Comm.*, Vol.7, pp. 40-48, January 1989.

[5] G.J.M. Janssen and R. Prasad, "Propagation measurements in an indoor radio environment at 2.4 GHz, 4.75 GHz an 11.5 GHz," *Vehicular Technolgoy Society 42nd VTS Conference Frontiers of Technology*, Denver, Colorado, pp. 617-620, 10-13 May 1992.

[6] G.J.M. Janssen, P.A. Stigter and R. Prasad, "A model for BER evaluation of indoor frequency selective channels using multipath measurement results at 2.4, 3.75 and 11.5 GHz," *Proceedings 1994 International Zurich Seminar on Digital Communications*, Zurich, Switzerland, pp. 344-355, March 1994.

[7] G.J.M. Janssen, P.A. Stigter and R. Prasad, "Wideband indoor channel measurements and BER analysis of frequency selective multipath channels at 2.4, 4.75 and 11.5 GHz," Accepted for publication in *IEEE Trans. Commun.*

[8] J.G. Proakis, "*Digital Communications,*" McGraw-Hill, New York, 2nd edition, 1989.

[9] D.M.J. Devarsivatham, "Time delay spread measurements of wideband radio signals within buildings," *Electron. Lett.*, Vol.20, pp. 950-951, 1984.

[10] D.M.J. Devarsivatham, "Time delay spread and signal level measurements of 850 MHz radio waves in building environments," *IEEE Trans. Antennas and Propagation*, Vol. AP-34, pp. 1300-1305, November 1986.

[11] D.M.J. Devarsivatham, C. Banarjee, M.J. Krain and D.A. Rappaport, "Multi-frequency radiowave propagation measurements in the portable radio-environment," *Proc. ICC'90*, Vol. 4, pp. 335.1.1-335.1.7, April 1990.

[12] S.J. Howard and K. Pahlavan, "Performance of a DFE modem evaluated from measured indoor radio multipath profiles," *Proc. ICC'90*, Atlanta, Vol. 4, pp. 335.2.1-335.2.5, April 1990.

[13] T.S. Rappaport, "Characterization of UHF multipath radio channels in factory buildings," *IEEE Trans. Antennas and Propagation*, Vol. 37, pp. 1058-1069, August 1989.

[14] M. Kahverad and B. Ramamurthi, "Direct sequence spread-spectrum with DPSK modulation and diversity for indoor wireless communication," *IEEE Trans. Comm.*, Vol. COM-35, pp. 224-236, February 1987.

[15] H.S. Misser, C.A.F.J. Wijffels and R. Prasad, "Throughput analysis of CDMA with DPSK modulation and diversity in indoor Rician fading radio channels," *Electron. Lett.*, Vol. 27, pp. 601-602, March 1991.

[16] H.S. Misser and R. Prasad, "Spectrum efficiency of amobile cellular radio system using direct sequence spread-spectrum," *Proc.-Vol. III, Computer and Communication, 1991 IEEE Region 10 International Conference (Tencon'91)*, New Delhi, India, pp. 91-94,August 28-30, 1991.

[17] H.S. Misser, A. Kegel and R. Prasad, "Monte Carlo simulation of direct sequence spread-spectrum for indoor radio communication in a Rician fading channel," *IEE Proceedings-I*, Vol. 139, pp. 620-624, December 1992.

[18] S.A. Musa and W. Wasylkiwskyj, "Cochannel interference of spread-spectrum systems in a multiple user environment," *IEEE Trans. Comm.*, Vol. COM-26, pp. 1405-1411, October 1978.

[19] J.M. Musser and J.N. Diagle, "Throughput analysis of an asynchronous code division multiple access (CDMA) system," *Proceedings of ICC*, Philidelphia, PA, pp. 2F.2.1-2F.2.7, July 1982.

[20] R. Prasad, H.S. Misser and A. Kegel, "Performance analysis of direct sequence spread-spectrum multiple access communications in an indoor Rician-fading channel with DPSK modulation," *Electron. Lett.*, Vol. 26, pp. 1366-1367, August 1990.

[21] R. Prasad, "Throughput analysis of slotted code division multiple access for indoor radio channels," *Proc. IEEE International Symposium Spread-Spectrum Techniques and Applications*, Kings's College, London, pp. 12-17, 24-26 September 1990.

[22] R. Prasad, H.S. Misser and A. Kegel, "Indoor radio communication in a Rician channel using direct-sequence spread-spectrum multiple access with selection diversity," *Proc. IEEE International Symposium Spread-Spectrum Techniques and Applications*, King's College, London, pp. 6-11, 24-26 September 1990.

[23] R. Prasad, "Capacity of a slotted code division multiple access system for wireless in-house network communications," *Proc.-Volume III, Computer and Communication, IEEE Tencon'91*, New Delhi, India, pp. 86-90, 28-30 August 1991.

[24] R. Prasad and C.A.F.J. Wijffels, "On performance comparison of slotted CDMA in macro-, micro-, and pico-cellular environment," *European Cooperation in the Field of Scientific and Technical Research, COST 231TD(91)-72*, Leidschendam, 25-27 August 1991.

[25] R. Prasad and C.A.F.J. Wijffels, "Performance analysis of a slotted CDMA system for indoor wireless communication using a Markov chain model," *Proc. IEEE Global Telecommunications Conference*, Phoenix, Arizona, Vol.3, pp. 1953-1957, 2-5 December 1991.

[26] H.S. Misser and R. Prasad, "Bit error probability evaluation of a microcellular spread-spectrum multiple access system in a shadowed Rician fading channel," *Proc. 42nd VTS Conference*, Denver, Colorado, pp. 439-442, 10-13 May 1992.

[27] R. Prasad, C.A.F.J. Wijffels and K.L.A. Sastry, "Performance analysis of slotted CDMA with DPSK modulation diversity and BCH-coding in indoor radio channels," *Archiv für Elektronik und Uebertragungstechnik*, Volume 46, No. 6, pp. 375-382, November 1992.

[28] K.L.A. Sastry and R. Prasad, "Throughput analysis of CDMA with DPSK and BPSK modulation and diversity in indoor radio channels," *Proceedings Third IEE Conference on Telecommunications*, Edingburgh, U.K., pp. 90-94, 17-20 March 1991.

[29] R. Prasad, H.S. Misser and A. Kegel, "Performance evaluation of direct-sequence spread-spectrum multipath-excess for indoor wireless communication in a Rician fading channel," *IEEE Trans. Commun.*, Vol. 43, pp. 581-592, February/March/April 1995.

[30] C.A.F.J. Wijffels, H.S. Misser and R. Prasad, "A micro-cellular CDMA system over slow and fast Rician fading radio channels with forward error correctly coding and diversity," *IEEE Trans.Vehicular Technol.*, Vol. 42, pp. 570-580, November 1993.

[31] R.C. Dixon, "*Spread-Spectrum Systems,*" John Wiley & Sons, New York, 1976.

Chapter 6
Outdoor CDMA Systems

6.1 INTRODUCTION

Outdoor DS-CDMA systems show some basic differences with indoor DS-CDMA systems. First, in outdoor systems the service area is much larger than for indoor systems. This implies that a high number of base stations are needed to provide communication with an acceptable quality in the total service area. The objective often is to service as many users per unit area as possible with a prespecified performance using a limited system frequency band. To cover the total service area, a cellular concept is proposed. Another difference is the fact that the indoor cells are often (physically) separated by walls or ceilings with metal shielding effects, while outdoor cells are physically separated by the radio propagation characteristics.

In narrowband communications, the cellular concept implies that frequencies allocated to the service are reused in a regular pattern of areas covered by one base station, called cells. To ensure that the mutual interference between users remains below a harmful level, adjacent cells use different frequencies. The total number of adjacent cells using the total system bandwidth is called the cluster size. The closest distance between the centers of two cells using the same frequency (in different clusters) is determined by the choice of the cluster size C_u and the layout of the cell cluster. This distance is often referred to as *reuse distance*. The reuse distance R_u normalized to the size of a hexagon is related to the cluster size:

$$R_u = \sqrt{3C_u} \tag{6.1}$$

When using wideband spread-spectrum communication, adjacent cells are allowed to use the same frequency band, provided that all users have unique code sequences. This implies that the cluster size C_u of a cellular DS-CDMA system is equal to 1 and the normalized reuse distance R_u is $\sqrt{3}$. A user moving into the service area of another base station does not need to change its receiving or transmitting frequency. This process is called a *soft handover*. This complicates the network management task of the base station. Figure 6.1 shows a graphical representation of the cellular concept.

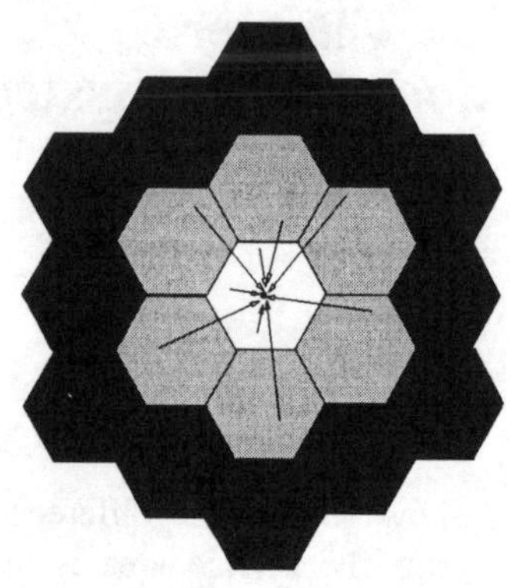

Figure 6.1 Cellular concept.

In general, a distinction is made between two types of outdoor cells. First we have the conventional *macrocells*, which are of large size (2 to 20 km) with the antenna radiating large power (0.6 to 10 W) from the tops of tall buildings. These cells are used in low user density areas, for instance, in rural areas. In high-density areas such as city centers, highways and airport areas, smaller cells called *microcells* are needed in order to have a higher user capacity per unit area. The size of microcells (0.4 to 2 km) is much less than the size of macrocells. Their antennas are at street lamp elevations operating at relatively low power (less than 20 mW).

In a multiple cell system each base station receives not only interference from users located within its own cell but also from users located in the surrounding cells, as is shown in Figure 6.1. To simplify the performance evaluation, an approximation is made by assuming that every cell causes equal average interference power. If the average interference power of the home cell is P_i, the total multiuser interference power P_t can be approximated by [1]

$$P_t \approx P_i(1+6R_u^{-\gamma}) \tag{6.2}$$

where γ is the path-loss law exponent and R_u is the (normalized) reuse distance. It is assumed here that the interference from only adjacent cells from the first tier are considered in (6.2). The interference due to the second and higher order tiers is neglected. In practice, a number of other effects occur in a multiple cell system. First, each terminal is assumed to be power controlled by its home base station. Calculating now the total interference injected by all users of this cell into another cell invokes an assumption on the spatial distribution of the terminals in the cell area and becomes a fairly complicated task. A second complication is caused by the fact that propagation effects such as path-loss law and fading have to be considered, implying that the total interference power received from other cells can perhaps best be modeled as a statistical variable. Using (6.2), however, simplifies the performance analysis and can lead to instructive results. The model described in this chapter can eventually be adapted to a refined model for the multiuser interference.

6.2 PROPAGATION CHARACTERISTICS

In an outdoor environment we have to deal with propagation mechanisms similar to those in the indoor environment.

Path Loss

In contrast to the indoor environment, in a macrocellular outdoor system there is often no line of sight (LOS) component, implying that the received signal is composed of reflections. Due to this effect, the path-loss law exponent γ is between 3 and 5 (typically 4). In a microcellular environment, however, close to the base station the path-loss law exponent $\gamma = 2$ and at the cell boundary, the path-loss law exponent $\gamma = 4$. A detailed description of this microcellular path-loss law model is given in [2]. As in the previous chapter, we assume here that the receiver provides perfect power control although this is very difficult to establish in practice in this type of environment.

Multipath Propagation

In the previous chapter it was shown that the delay spread is a very important measure in order to determine the number of resolvable paths. Typical values for the delay spread are 0.2 μs for microcells and 0.8 μs for macrocells [3].

Fading Characteristics

Due to the multipath propagation, fast signal fluctuations occurs. In a microcellular environment the signal envelope is described by a Rician distributed statistical variable. The Rician probability density function is

$$p_\beta(r) = \frac{r}{\sigma^2}\exp\left[-\frac{r^2+s^2}{2\sigma^2}\right] I_0\left[\frac{sr}{\sigma^2}\right]$$

$$0 \le r < \infty, \; s \ge 0 \tag{6.3}$$

$$\mathrm{R} = \frac{s^2}{2\sigma^2}$$

where r is the envelope of the received signal, $I_0(\cdot)$ is the modified Bessel function of the first kind and zero order, and s is the peak value of the specular radio signal due to the superposition of the dominant LOS signal and the time invariant reflected scattered signals. The average signal power that is received over specular paths (due to moving objects inside the building) is denoted by σ^2. The Rician distribution is characterized by the parameter R. Typical values for R in the microcellular environment are 7 and 12 dB [4-6]. In a macrocellular environment there is no dominant component and the received signal consists merely of reflected versions of the original signal. Then the parameter R approaches zero and the Rician distribution changes into a Rayleigh distribution:

$$p_\beta(\mathrm{r}) = \frac{r}{\sigma^2}\exp\left[-\frac{r^2}{2\sigma^2}\right] \tag{6.4}$$

$$0 \le r < \infty$$

Shadowing

In addition to fast signal fluctuations caused by multipath reflections, in an outdoor radio environment often slow signal fluctuations exist due to obstruction of the signal by hills, buildings, etc. This phenomenon is known as *shadowing*. The slow fluctuations can be modeled by a statistical variable with a log-normal pdf:

$$p_s(P_0) = \frac{1}{\sqrt{2\pi}\sigma_s P_0} \exp\left[-\frac{(\ln(P_0) - m_s)}{2\sigma_s^2}\right] \tag{6.5}$$

where σ_S and m_S are the logarithmic standard deviation and area mean power, respectively. Typical values for σ_S are between 4 and 8 dB.

In fact, (6.3) and (6.4) are conditional on the average power P_0, which is determined by (6.5). For a Rayleigh distribution, the average power $P_0=\sigma^2$, for a Rician distribution, $P_0=\sigma^2+s^2/2$. Note that we use the notation σ_s^2 for the logarithmic variance due to shadowing and σ^2 for the received multipath power; $s^2/2$ is the power of the dominant component.

6.3 SYSTEM MODEL

A star connected topology is assumed for all communications in each cell of the cellular network. This implies that all communications are established via a central base station. Since CDMA systems are interference limited, it is necessary to keep a mobile station which is close to the base station from causing an overwhelming amount of interference. For this reason the base station is assumed to provide average power control.

6.3.1 Transmitter, Channel, and Receiver Model

In this chapter we consider both BPSK and DPSK modulation. It is known that BPSK gives better error performance than DPSK but since coherent detection is very difficult to establish in a mobile radio environment, DPSK is often preferred. This subsection focuses mainly on BPSK modulation, since DPSK modulation was discussed in detail in Chapter 5.

Transmitter

The transmitted signal of the kth user is

$$s_k(t) = A a_k(t) b_k(t) \cos(\omega_c t + \theta_k) \tag{6.6}$$

where

$$\begin{aligned} a_k(t) &= \sum a_k{}^i P_{T_c}(t - iT_c) \\ a_k{}^i &\in \{-1, 1\} \end{aligned} \tag{6.7}$$

and

$$b_k(t) = \sum_i b_k^{\,i} P_{T_b}(t - iT_b)$$
$$b_k^{\,i} \in \{0,1\} \tag{6.8}$$

The index k refers to the kth user. The parameter b_k is the data waveform, b_k^j is the jth bit of the data waveform, a_k^i is the ith chip of the direct sequence code, P_T is a rectangular pulse of unit height and duration T, and T_b and T_c are the bit and chip duration, respectively. It is assumed that $T_b / T_c = N$, where N is the length of the direct sequence code.

Channel Model

The complex lowpass equivalent of the radio channel's impulse response is given by:

$$h_k(t) = \sum_{l=1}^{L} \beta_{lk} \exp\{j\gamma_{lk}\}\delta(t - \tau_k) \tag{6.9}$$

where β, τ, and γ are the path gain, time delay, and phase of each path, respectively. The subscript lk refers to the lth of the kth user and j is an imaginary number defined as $j^2 = -1$. It is assumed that the path phase of the received signal, $(\omega_c \tau_{lk} + \gamma_{lk})$, is an independent random variable uniformly distributed over $[0,2\pi]$. The path delay is also an independent random variable and is assumed random over $[0,T_b]$. The path gain β_{lk} is modeled as an independent Rician random variable. The number of paths may be either fixed or randomly changing. Here fixed values of L are used according to

$$\mathrm{L} = \left\lfloor \frac{T_{max}}{T_c} \right\rfloor + 1 \tag{6.10}$$

where T_{max} is the delay spread.

The path gain β consists of a fast and a slow component. In the case of a microcellular environment the pdf can be obtained from the Rician and log-normal pdf as follows:

$$p_\beta(r) = \int_0^\infty \frac{r}{\sigma^2} \exp\left[-\frac{r^2 + s^2}{2\sigma^2}\right] I_0\left[\frac{rs}{\sigma^2}\right] \frac{1}{\sqrt{2\pi}\sigma_s P_0} \exp\left[-\frac{\left(\ln(P_0) - m_s\right)^2}{2\sigma_s^2}\right] dP_0 \tag{6.11}$$

where $P_0=\sigma^2+s^2/2$. To simplify the integral it is assumed that only the direct LOS signal is affected by slow log-normal shadowing, which implies that $P_0=s^2/2$. This assumption is reasonable since the direct LOS component is much stronger than the total interference power received over specular paths (depending on the Rice factor). Then (6.11) changes into

$$p_\beta(r)=\int_0^\infty \frac{r}{\sigma^2}\exp\left[-\frac{r^2+2P_0}{2\sigma^2}\right]I_0\left[\frac{r\sqrt{2P_0}}{\sigma^2}\right]\frac{1}{\sqrt{2\pi}\sigma_s P_0}\exp\left[-\frac{\left(\ln(P_0)-m_s\right)^2}{2\sigma_s^2}\right]dP_0 \tag{6.12}$$

In a macrocellular environment, the average power $P_0=\sigma^2$ and the Rice factor $R=s^2/2\sigma=0$, so the composite pdf can be computed from (6.11) and is given by

$$p_\beta(r)=\int_0^\infty \frac{r}{P_0}\exp\left[-\frac{r^2}{2P_0}\right]\frac{1}{\sqrt{2\pi}\sigma_s P_0}\exp\left[-\frac{\left(\ln(P_0)-m_s\right)^2}{2\sigma_s^2}\right]dP_0 \tag{6.13}$$

Receiver

The receiver consists of a matched filter for a particular spread-spectrum code and a BPSK or DPSK demodulator. Since the DPSK case has been analyzed extensively in the previous chapter, we now focus on the BPSK case.

The output of the matched filter can be written

$$z=A\beta_{j1}T_b b_1{}^0+\sum_{k=1}^{K}A\left(b_k{}^{-1}X_k+b_k{}^0\hat{X}_k\right)+\eta \tag{6.14}$$

with

$$X_1=\sum_{\substack{l=1\\ l\neq j}}^{L}R_{11}(\tau_{l1})\beta_{l1}\cos(\phi_{l1}) \tag{6.15a}$$

$$\hat{X}_1 = \sum_{\substack{l=1 \\ l\neq j}}^{L} \hat{R}_{11}(\tau_{l1})\beta_{l1}\cos(\phi_{l1}) \tag{6.15b}$$

and for $k\geq 2$

$$X_k = \sum_{l=1}^{L} R_{1k}(\tau_{lk})\beta_{lk}\cos(\phi_{lk}) \tag{6.15c}$$

$$\hat{X}_k = \sum_{l=1}^{L} \hat{R}_{1k}(\tau_{lk})\beta_{lk}\cos(\phi_{lk}) \tag{6.15d}$$

Here b_k^{-1} and b_k^0 are the previous and the current data bits, respectively. They can be either 1 or −1 with equal probability. The parameters τ_{lk} and ϕ_{lk} are, respectively, the difference in arrival time and phase between the jth path of the reference user and the lth path of the kth user. It is assumed that ϕ_{jk} is uniformly distributed over $[0,2\pi]$; η is the thermal noise component with power spectral density $N_0/2$ and A is the signal amplitude. The partial correlation functions are given by

$$R_{1k}(\tau) = \int_0^{\tau} a_k(t-\tau)a_1(t)dt \tag{6.16a}$$

$$\hat{R}_{1k}(\tau) = \int_{\tau}^{T_b} a_k(t-\tau)a_1(t)dt \tag{6.16b}$$

6.3.2 Sources of Transmission Errors

The total received undesired power consists of thermal white noise and multiuser interference. Both aspects are discussed in this subsection.

Thermal White Noise

This type of noise is caused by the random motion of electrons in electronic components and conducting media in general. Only white Gaussian noise is considered here. It is assumed that the thermal noise component η has power spectral density $N_0/2$.

Multiuser Interference

Multiuser interference consists of self-interference, interference from the terminals in the same cell, and interference from terminals in cochannel cells. The self-interference is caused by the multiple reflections in the channel impulse response. In the performance analysis only the interference from the mobiles in the home cell and the mobiles in the six nearest cochannel cells is considered. If the signals transmitted by users in the home cell reach the base station with average amplitude A, then the signals from the cochannel cells arrive at the base station with average amplitude

$$A_{cc} = A\sqrt{R_u^{-\gamma}} \tag{6.17}$$

Here γ is the path-loss law constant and R_u is the normalized reuse distance.
Using (6.14) and (6.15) the interference power can be written

$$\begin{aligned}\sigma_{int}^2 &= \sum_{l=2}^{L} A^2 E\left\{\left[b_1^{-1}R_{11}(\tau_{l1}) + b_1^0 \hat{R}_{11}(\tau_{l1})\right]^2\right\} E\left\{\left[\beta_{l1}\cos(\phi_{l1})\right]^2\right\} \\ &+ K(A^2 + 6A_{cc}^2)\sum_{l=1}^{L} E\left\{b_k^{-1}R_{1k}(\tau_{lk}) + b_k^0 \hat{R}_{1k}(\tau_{lk})\right\} E\left\{\left[\beta_{lk}\cos(\phi_{lk})\right]^2\right\}\end{aligned} \tag{6.18}$$

For Gold codes the following equation holds [7]:

$$E\left\{\left[b_k^{-1}R_{1k}(\tau_{1k}) + b_k^0 \hat{R}_{1k}(\tau_{1k})\right]^2\right\} = \frac{2T_b^2}{3N} \tag{6.19}$$

with N being the length of the Gold code. Using (6.19), equation (6.18) can be reduced to

$$\sigma_{int}^2 = \frac{2T_b^2}{3N}\left(A^2 \sum_{l=2}^{L} E\left\{\left[\beta_{l1}\cos(\phi_{l1})\right]^2\right\} + K(A^2 + 6A_{cc}^2)\sum_{l=1}^{L} E\left\{\left[\beta_l \cos(\phi_l)\right]^2\right\}\right) \tag{6.20}$$

Microcell

In the microcellular case for $l=1$ the path gain is Rician distributed. When a Rician distributed variable is multiplied by the sine or cosine of a uniformly distributed phase, then a Gaussian variable with mean $s/\sqrt{2}$ and variance σ^2 results. This implies that

$$E\left\{\left[\beta_{lk}\cos(\phi_{lk})\right]^2 \mid s\right\} = \sigma^2 + \frac{s^2}{2} \tag{6.21}$$

By averaging over $s^2/2$ using (6.5) it can be shown that

$$E\left\{\left[\beta_{lk}\cos(\phi_{lk})\right]^2\right\} = \sigma^2 + \exp\left(m_s + \frac{\sigma_s^2}{2}\right) \tag{6.22}$$

Equation (6.20) can now be written

$$\sigma_{int}^2 = \frac{2T_b^2}{3N}\left[\sigma^2 + \exp\left(m_s + \frac{\sigma_s^2}{2}\right)\right]\left\{(L-1)A^2 + LK(A^2 + 6A_{cc}^2)\right\} \tag{6.23}$$

Macrocell

In the macrocellular case for $l=1$ the path gain is Rayleigh distributed. When a Rayleigh distributed variable is multiplied by the sine or cosine of a uniformly distributed phase then a Gaussian variable with mean 0 and variance $\sigma^2=P_0/2$ is the result. This implies that

$$E\left\{\left[\beta_{lk}\cos(\phi_{lk})\right]^2 \mid P_0\right\} = \frac{P_0}{2} \tag{6.24}$$

By averaging over P_0 using (6.5) it can be shown that

$$E\left\{\left[\beta_{lk}\cos(\phi_{lk})\right]^2\right\} = \frac{1}{2}\exp\left(m_s + \frac{\sigma_s^2}{2}\right) \tag{6.25}$$

Equation (6.20) can now be written

$$\sigma_{\text{int}}^2 = \frac{T_b^2}{3N}\exp\left(m_s + \frac{\sigma_s^2}{2}\right)\left\{(L-1)A^2 + LK(A^2 + 6A_{cc}^2)\right\} \tag{6.26}$$

If we approximate the multiuser interference by a Gaussian random process, the interference power of (6.22) and (6.26) can be added to the thermal noise power.

6.4 BEP ANALYSIS

In this section we derive expressions for the bit error probability. For the DPSK case we only show the analytical results. For a detailed derivation of the expressions for the bit error probability in this case, refer to Chapter 5. The expressions for the BPSK case are given in detail.

6.4.1 BPSK Modulation

General

The starting point for the derivation of the bit error probability is the bit error probability of a BPSK signal in an additive white Gaussian noise (AWGN) channel. As mentioned in the previous section the multiuser interference is assumed to be Gaussian distributed. This means that the interference power can be added to the thermal noise power. In (6.14) it was shown that the output of the matched filter for a single cell system can be written

$$z = A\beta_{j1}T_b b_1^0 + \sum_{k=1}^{K} A\left(b_k^{-1}X_k + b_k^0\hat{X}_k\right) + \eta \tag{6.27}$$

where the first term is the desired signal part and the sum of the last two terms represents both the multiuser interference and the thermal noise and is described by a Gaussian variable with standard deviation σ_{tot} given by

$$\sigma_{tot}^2 = \sigma_{thermal}^2 + \sigma_{int}^2 \tag{6.28}$$

The multiuser interference consists of a multiple cell system of both interference from the home cell and interference from the adjacent cells. For a microcell, the variance of the multiuser interference σ_{int}^2 is given by (6.22) and for a macrocell it is given by (6.26). The variance of the thermal noise is given by

$$\sigma_n^2 = N_0 T_b \tag{6.29}$$

Assuming that the data bits −1 and 1 are equiprobable, the bit error probability P_e can be expressed as

$$P_e = P(z < 0 | b_1^0 = 1) \tag{6.30}$$

Neglecting for time being the fading effects, the bit error probability for a BPSK signal in an AWGN channel is

$$P_e = \frac{1}{2}\operatorname{erfc}\left(\frac{ArT_b}{\sqrt{2}\sigma_{tot}}\right) \tag{6.31}$$

where erfc $(\cdot)$ is the complementary error function, given by $\left(2/\sqrt{\pi}\right)\int_x^\infty \exp\left(-t^2\right)dt$.

No Diversity

Fading can be incorporated by averaging (6.31) over the pdf given in (6.12) and (6.13). In the microcellular case (Rician fading), the expression for the bit error probability is

$$\begin{aligned} P_e = \frac{1}{2}\int_0^\infty \int_0^\infty & \operatorname{erfc}\left(\frac{ArT_b}{\sqrt{2}\sigma_{tot}}\right)\frac{r}{\sigma^2}\exp\left[-\frac{r^2 + 2P_0}{2\sigma^2}\right] I_0\left[\frac{r\sqrt{2P_0}}{\sigma^2}\right] \\ & \cdot \frac{1}{\sqrt{2\pi}\sigma_s P_0}\exp\left[-\frac{\left(\ln(P_0) - m_s\right)^2}{2\sigma_s^2}\right] dP_0 dr \end{aligned} \tag{6.32}$$

In the macrocellular case (Rayleigh fading), the expression for the bit error probability is

$$P_e = \frac{1}{2}\int_0^\infty \int_0^\infty \operatorname{erfc}\left(\frac{\mathrm{A} r T_b}{\sqrt{2}\sigma_{tot}}\right) \frac{r}{P_0} \exp\left[-\frac{r^2}{2P_0}\right] \cdot \frac{1}{\sqrt{2\pi}\sigma_s P_0} \exp\left[-\frac{\left(\ln(P_0) - m_2\right)^2}{2\sigma_s^2}\right] dP_0 dr \tag{6.33}$$

This expression can be simplified to

$$P_e = \int_0^\infty \frac{1}{2}\left[1 - \sqrt{\frac{\frac{\mathrm{A}^2 T_b}{\sigma_{tot}^2} P_0}{1 + \frac{\mathrm{A}^2 T_b}{\sigma_{tot}^2} P_0}}\right] \frac{1}{\sqrt{2\pi}\sigma_s P_0} \exp\left[-\frac{\left(\ln(P_0) - m_s\right)^2}{2\sigma_s^2}\right] dP_0 \tag{6.34}$$

Selection Diversity

To obtain the bit error probability with selection diversity, the pdf of the strongest signal (β_{max}) must be derived.

The cdf (cumulative distribution function) of the M signals subject to both Rician fading and log-normal fading is then:

$$F_\beta(r) = \int_0^\infty \left[1 - Q\left(\frac{\sqrt{2P_0}}{\sigma}, \frac{r}{\sigma}\right)\right] \frac{1}{\sqrt{2\pi}\sigma_s P_0} \exp\left[-\frac{\left(\ln(P_0) - m\right)^2}{2\sigma_s^2}\right] dP_0 \tag{6.35}$$

where

$$Q(a,b) = \int_0^\infty x \exp\left[-\frac{a^2 + x^2}{2}\right] I_0(x) dx \tag{6.36}$$

The cdf of β_{max} is the cdf of (6.35) raised to the power of M (order of diversity). The pdf of β_{max} can now be obtained by differentiating the cdf with respect to r:

$$p_{\beta_{max}}(r) = M\left\{\int_0^\infty \left[1-Q\left(\frac{\sqrt{2P_0}}{\sigma},\frac{r}{\sigma}\right)\right]\frac{1}{\sqrt{2\pi}\sigma_s P_0}\exp\left[-\frac{\left(\ln(P_0)-m_s\right)^2}{2\sigma_s^2}\right]dP_0\right\}^{M-1}$$
$$\cdot\int_0^\infty \frac{r}{\sigma^2}\exp\left[-\frac{r^2+2P_0}{2\sigma^2}\right]I_0\left[\frac{r\sqrt{2P_0}}{\sigma^2}\right]\frac{1}{\sqrt{2\pi}\sigma_s P_0}\exp\left[-\frac{\left(\ln(P_0)-m_s\right)^2}{2\sigma_s^2}\right]dP_0 \tag{6.37}$$

The bit error probability can now be obtained by averaging (6.31) over (6.37). The cdf of the M signals subject to both Rayleigh fading and log-normal fading can be easily obtained from (6.35). In the Rician case $P_0=s^2/2$. In the Rayleigh case $s=0$ and $P_0=\sigma^2$. Since $Q(0,r/\sigma)$ is simply $\exp(-b^2/2)$, (6.35) changes into:

$$F_\beta(r) = \int_0^\infty \left[1-\exp\left(-\frac{r^2}{2P_0}\right)\right]\frac{1}{\sqrt{2\pi}\sigma_s P_0}\exp\left[-\frac{\left(\ln(P_0)-m_s\right)^2}{2\sigma_s^2}\right]dP_0 \tag{6.38}$$

The cdf of β_{max} is the cdf of (6.38) raised to the power of M (order of diversity). The pdf of β_{max} can now be obtained by differentiating the cdf with respect to r:

$$p_{\beta_{max}}(r) = M\left\{\int_0^\infty \left[1-\exp\left(-\frac{r^2}{2P_0}\right)\right]\frac{1}{\sqrt{2\pi}\sigma_s P_0}\exp\left[-\frac{\left(\ln(P_0)-m_s\right)^2}{2\sigma_s^2}\right]dP_0\right\}^{M-1}$$
$$\cdot\int_0^\infty \frac{r}{P_0}\exp\left[-\frac{r^2}{2P_0}\right]\frac{1}{\sqrt{2\pi}\sigma_s P_0}\exp\left[-\frac{\left(\ln(P_0)-m_s\right)^2}{2\sigma_s^2}\right]dP_0 \tag{6.39}$$

6.4.2 Results

Bit error probability was calculated for microcells for three values of the cluster size: C_u=1, 3, and 7 with a Rician parameter of 7 dB and a standard deviation for the logarithmic distribution of 2 dB.

For C_u=1 the entire system bandwidth is used in one cell. This means that the longest possible direct-sequence codes can be used (i.e., the largest possible interference rejection). However, the distance between the cochannel cells is at the same time minimized, i.e., the largest possible interference power is present. As the cluster size increases, the maximal length of the direct-sequence codes decreases (less interference rejection) but the distance between the cochannel cells is at the same time increased (less interference power).The advantage of having the longest possible direct-sequence code (C_u=1) cannot compensate for the drawback of having the largest possible interference power. It is also seen that the performance can only be improved up to a certain limit by increasing the cluster size. Note that the number of users per cell is kept constant for all cluster sizes.

Computational results were obtained for the macrocellular systems to evaluate the spectrum efficiency, defined as the maximum number of users per cell that can be supported by the system with a prespecified bit error probability. Denoting the maximum number of users as K_{max}, the spectrum efficiency of the system is defined as

$$SE \underline{\Delta} \frac{K_{max} r_b}{W S_A} \tag{6.40}$$

where W is the total bandwidth occupied by the system, S_A is the area of the cell, and r_b is the data rate. Table 6.1 shows the values for the spectrum efficiency for a bit error probability threshold of 5×10^{-3}, T_{max}=3 μs, S_A=1 km^2, E_b/N_0=36 dB and σ_S= 8 dB.

Table 6.1

Spectrum efficiency for a bit error probability threshold of 5×10^{-3}, T_m=3 μs, S_A=1 km^2, E_b/N_0=36 dB, and σ_s= 8 dB

Cluster Size	Spectrum Efficiency
1	1.47×10^{-2}
3	9.3×10^{-3}
4	5.8×10^{-3}
7	2.7×10^{-3}

Clearly the spectrum efficiency decreases with an increasing cluster size. A cluster size of one yields the maximum spectrum efficiency for a macrocellular system.

6.5 THROUGHPUT AND DELAY ANALYSIS

As mentioned in Chapter 5, it is possible to transmit packets with data in a CDMA system. Let us assume that the number of users transmitting data packets simultaneously within a specific time slot is k. In a CDMA system, as soon as the number of active users rises above the user's threshold capacity C, fatal collisions occur and all the packets are destroyed. The packets are lost due to excessive bit errors when the number of simultaneous users exceeds the threshold capacity C. The lost packets are rescheduled and retransmitted after a sufficient time delay. Each user transmits a specific code to identify the user at the base station of a single cell system.

The (normalized) throughput is defined as the average number of suceccesfully received packets per time slot, normalized to the user's capacity C and is given by

$$S \underline{\Delta} \frac{1}{C} \sum_{k=1}^{C} k P_k P_{sk} \tag{6.41}$$

where P_k is the probability of a packet being transmitted with k–1 other packets and $P_{sk}(k)$ is the packet success probability. Assuming a binomial arrival distribution of the offered traffic G, P_k is written as

$$P_k = \binom{C}{k} \left[\frac{G}{C}\right]^k \left[1 - \frac{G}{C}\right]^{C-k} \tag{6.42}$$

The offered traffic is defined as the average number of transmissions (newly generated packets plus retransmitted packets) per time slot by k users. A binomial arrival distribution of the offered traffic seems to be an appropriate assumption for a CDMA system because there are a finite number of users sharing a direct frequency band and the performance of the system for a certain chip length depends on the number of users (PN codes). With the binomial arrival model it is further assumed that the probability that a new packet is generated is the same as the probability of retransmission of a packet.

The expression for the packet success probability $P_{sk}(k)$ depends on the rapidity of the fading. Two situations are considered in this chapter: slow fading and fast fading.

6.5.1 Throughput

Slow Fading

In a slow fading channel it is assumed that the channel characteristics are constant for a very long period compared to a signaling interval. Hence it is assumed that all bits of a packet are received with the same average power. In the case of selection diversity and DPSK modulation, the packet success probability is given by

$$P_e = \int_{-\infty}^{\infty} \int_{-\infty}^{\infty} \int_{0}^{\infty} \left[1 - \tfrac{1}{2}\left[1 - \frac{\mu}{\mu_0}\right] \exp\left[-\frac{m^2}{\mu_0}\right]\right]^{N_p} \cdot p_{\beta_{\max}}(\beta_{\max}) p_\mu(\mu) p_{\mu_0}(\mu_0) d\beta_{\max} d\mu d\mu_0 \tag{6.43}$$

and in the case of maximal ratio combining, the packet success probability is given by

$$P_e = \int_0^\infty [1 - P_e(\gamma_b)]^{N_p} f_{\gamma_b}(\gamma_b) d\gamma_b \tag{6.44}$$

where N_p is the number of bits per packet.

Fast Fading

In the case of fast fading, the channel variations are fast relative to the signaling interval, implying that each signaling symbol undergoes fading independently. The packet success probability is given by

$$P_{sk}(k) = [1 - P_e(k)]^{N_p} \tag{6.45}$$

where $P_e(k)$ is the bit error probability, which depends on the modulation (DPSK or BPSK) and on the type of diversity (selection diversity or maximal ratio combining).

6.5.2 Delay

The average delay of the system is defined as the number of slot times it takes for a packet to be successfully received. The average delay (in slots), assuming immediate acknowledgment, is given by

$$D = (1.5 + d) + \left[\frac{G}{CS} - 1\right]\left(\lfloor \delta_\tau + 1 \rfloor + 1 + 2d\right) \tag{6.46}$$

Here it is assumed that the minimum delay for slotted systems is 1.5 slot durations. This time consists of one slot time being the average time between the time the packet is offered to the transmitter and the beginning of the next slot. The average number of retransmissions for a packet to be successfully received is *(G/S–1)*, the delay due to each retransmission is $\lfloor \delta_\tau + 1 \rfloor + 1$, with δ_τ being the mean of the retransmission delay, which is uniformly distributed over the range from which the retransmission delay is selected.

6.5.3 Results

This section presents the results for cellular systems using DPSK modulation without considering the shadowing effect.

In Figure 6.2 a comparison is given between single and multiple cell systems. As expected, multiple cell systems have an inferior throughput performance compared to single cell systems, due to the interference introduced by the surrounding cells. It is also seen that the best performance is for the single cell system with 16 resolvable paths and the worst is for the multiple cell system with 32 resolvable paths.

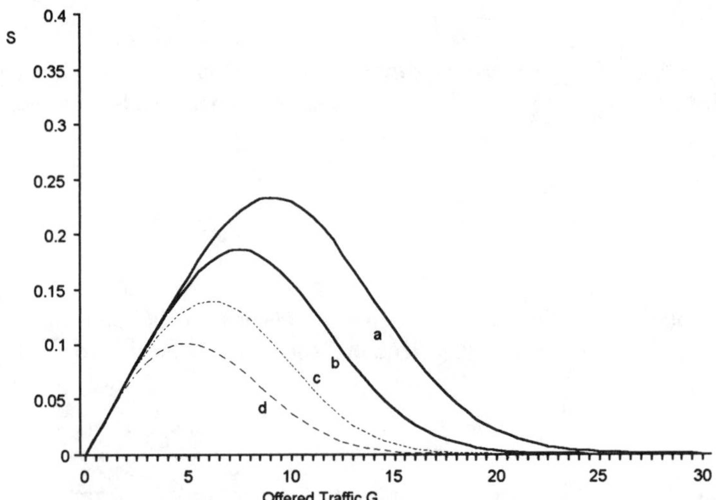

Figure 6.2 A comparison between a single cell system and a multiple cell system for fast Rician fading, microcell, selection diversity, N=255, R=12 dB, C=30, N_p=42, M=2. (a) L=16, single cell, (b) L=16, multiple cells, (c) L=32, single cell, (d) L=32, multiple cell.

Figure 6.3 shows a comparison of the performance of systems using both selection diversity and maximal ratio combining. It is clear that MRC is superior to SD.

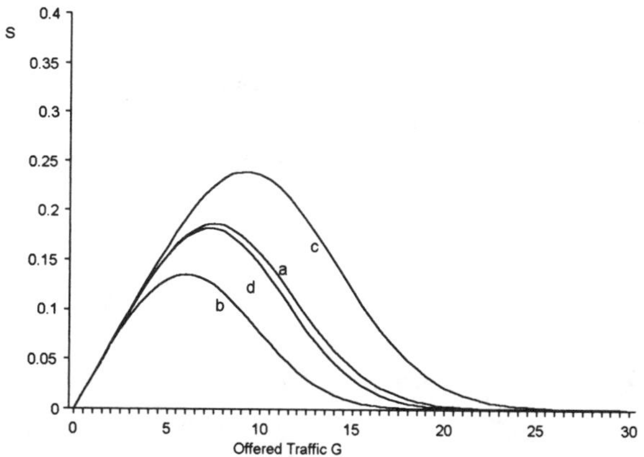

Figure 6.3 Throughput curves for two diversity methods (maximal ratio combining and selection diversity) and fast Rician fading. Multiple microcells, N=255, R=12 dB, C=30, N_p=42, M=2. (a) SD, L=16, (b) SD, L=24,(c) MRC, L=16, (d) MRC, L=24.

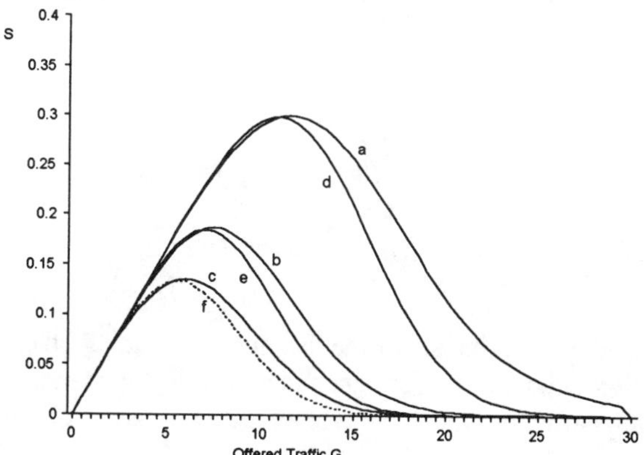

Figure 6.4 Influence of FEC coding on the throughput for different numbers of resolvable paths and fast Rician fading. Multiple microcells, selection diversity, N=255, R=12 dB, C=30, N_p=42, M=2. No coding, (a) L=8, (b) L=16, (c) L=24 and BCH (15,7) FEC coding, (d) L=16, (e) L=32, (f) L=48.

In Figure 6.4 the influence of FEC coding on throughput is shown. A comparison of curves b and d in Figure 6.4 shows that there is significant improvement in the performance of the system due to FEC coding.

The throughput-delay characteristics for various system configurations are shown in Figure 6.5. The throughput increases and the delay decreases with the decrease in L for the system with and without coding. Coding enhances the performance (i.e., high throughput and low delay; curves b and c).

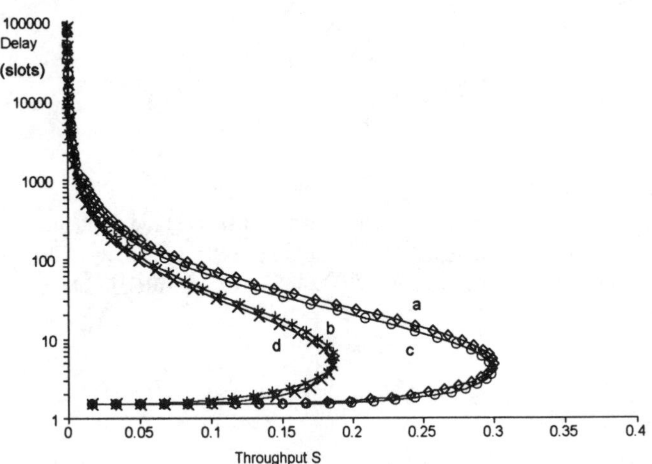

Figure 6.5 Throughput-delay curves for fast Rician fading. The delay is given in slot times. Multiple microcells, selection diversity, N=255, R=12 dB, C=30, N_p=42, M=2. No coding, (a) L=8, (b) L=16 and BCH (15,7) FEC coding, (c) L=16, (d) L=32.

Figure 6.6 compares the performance of the fast and slow fading channels. It is seen that slow fading channels yield better performance than fast fading channels. The effect of the packet length is also shown in Figure 6.6. As expected, a larger packet length results in decreased throughput.

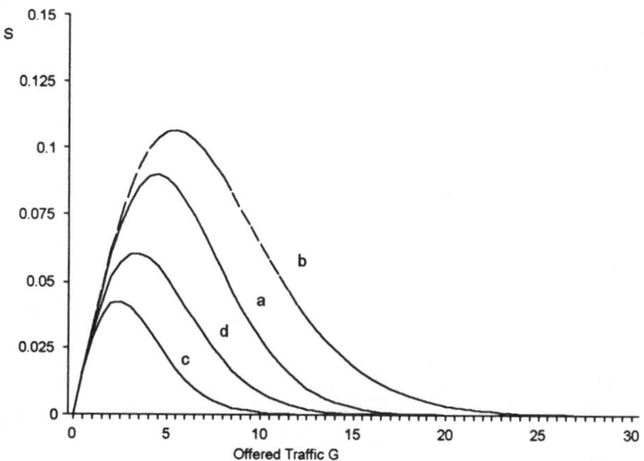

Figure 6.6 A comparison between fast and slow Rician fading channels and the effect on the throughput. Multiple microcells, maximal ratio combining, N=127, R=12 dB, C=30, M=2, L=6. No coding, (a) fast, N_p=128, (b) slow, N_p=128, (c) fast, N_p=1024, (d) slow, N_p=1024.

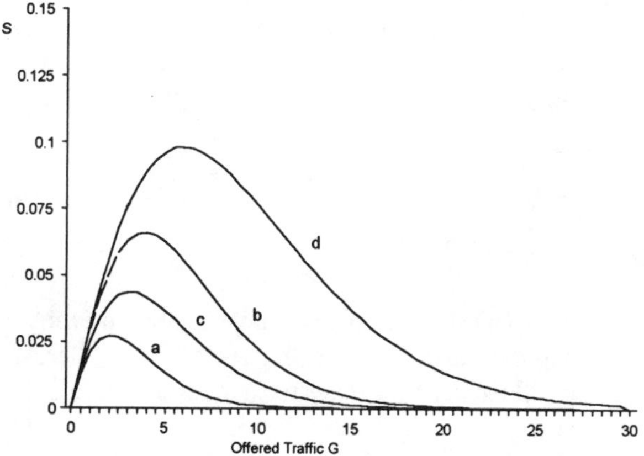

Figure 6.7 Throughput curves for fast and slow Rayleigh fading (multi-macrocell environment). Maximal ratio combining, N=127, R=0, C=30, N_p =32, M=4, (a) fast, L=32, (b) fast, L=16, (c) slow, L=32, (d) slow, L=16.

In Figure 6.7, the throughput results are shown for fast and slow Rayleigh fading channels in a macrocellular environment. As in microcells, macrocellular slow fading channels also yield a higher throughput than fast fading channels do.

In Figure 6.8 a comparison is made between macro-, micro- and picocellular systems. As expected, picocells yield the best performance due to the smallest delay spread.

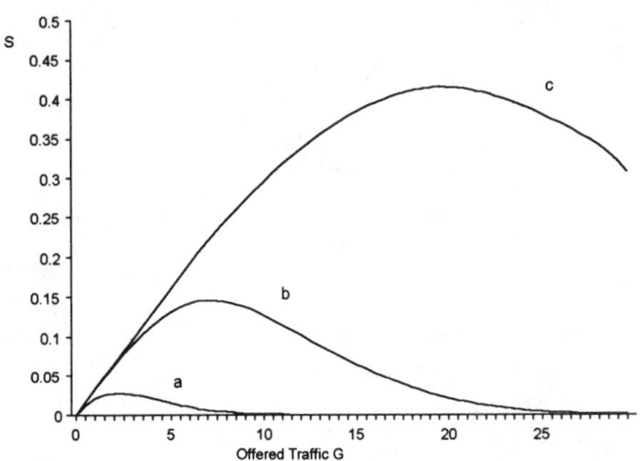

Figure 6.8 Throughput curves for slow fading in multiple macro-, multiple micro-, and multiple picocells. Maximal ratio combining, N=127, C=30, r_b=32 kbit/s, N_p=128, (a) macrocell, R=0, M=4, L=32 (7.63 μs<T_{max}<7.87 μs), (b) microcell, R=7 dB, M=4, L=8 (1.72 μs<T_{max}<1.97 μs), (c) picocell, R=6.8 dB, M=2, L=2 (246 ns<T_{max}<492 ns).

6.6 CONCLUSIONS

The performance of outdoor CDMA systems was investigated in terms of bit error rate, spectrum efficiency, throughput, and delay characteristics. Both micro- and macrocells were considered, and the influence of diversity techniques and forward error correcting codes on the performance were investigated for BPSK and DPSK modulation.

It has been shown that for both microcells and macrocells, better performance is obtained in slow fading channels than in fast fading channels. In a slow fading channel, the channel characteristics do not change significantly during two consecutive bits, whereas in a fast fading channel the rapidity of the fluctuations is such that two consecutive bits can be assumed to be independently fading. In both cases, the performance increases for decreasing packet lengths.

The main differences between microcells and macrocells are seen in terms of propagation characteristics, the delay spread, and the probability distribution of the envelope fading. In microcells, the delay spread is generally less than in macrocells. The envelope fading in microcells can be described by a Rician distribution whereas for a macrocell, modeling by a Rayleigh distribution is more appropriate. It is shown that the performance of a DS-CDMA system decreases as the delay spread increases. Furthermore, better performance is obtained in a Rician channel than in a Rayleigh channel. The performance further increases with increasing Rician factor (stronger LOS signal). From this we conclude that the performance of a DS-CDMA system in a single microcell is generally better than in a single macrocell.

In cellular systems, the optimal cluster size (in terms of maximal spectrum efficiency) is different for micro-, and macrocells. In macrocells the highest spectrum efficiency is obtained when using the entire system bandwidth in each cell ($C_u=1$). This implies that it is more effective to use longer code sequences than to separate cochannel cells. In microcells however, longer code sequences cannot compensate for the larger amount of interference when using a cluster size $C_u=1$. As the cluster size increases, the performance improves but eventually starts to deteriorate beyond a certain value of the cluster size.

Finally, results presented in this chapter show that performance is increased by using diversity techniques and forward error correcting codes. The performance increases with the order of diversity and selection diversity yields better performance than maximal ratio combining. BPSK modulation gives better results than DPSK modulation.

The results presented in this chapter are valid under the assumptions of a uniform power delay profile, perfect power control within a cell, and perfect synchronization and tracking at the receiver.

REFERENCES

[1] C.A.F.J. Wijffels, H.S. Misser and R. Prasad, "A micro-cellular CDMA-system over slow and fast Rician fading channels with forward error correcting coding and diversity," *IEEE Trans. Vehicular Technol.*, Vol. 42, pp. 570-580, November 1993.

[2] H. Harley, "Short distance attenuation measurements at 900 Mhz and 1.8 Ghz using low antenna heights for microcells," *IEEE J. Selected Areas Comm.*, Vol. 7, pp. 5-10, January 1989.

[3] T.S. Rappaport, S.Y. Seidel and R. Singh, "Path-loss and multipath delay statistics in four European cities for 900 MHz cellular and micro-cellular communications," *Electron. Lett.*, Vol. 26, pp. 1713-1714, 27 September 1990.

[4] S.T.S. Chia, R. Steel, E. Green and A. Baron, "Propagation and bit error rates measurements for a microcellular system," *J. Inst. Elect. Radio Eng.*, Vol. 57, pp. 255-266, Nov/Dec. 1987.

[5] R.J.C. Bultitude and G.K. Bedal, "Propagation characteristics on microcellular urban mobile radio channels at 910 MHz," *IEEE J. Selected Areas Comm.*, Vol.7, pp. 31-39, January 1989.

[6] R. Prasad and A. Kegel, "Effects of Rician faded and log-normal shadowed signal on spectrum efficiency," *IEEE Trans. Vehicular Technol.*, Vol. 42, pp. 274-281, August 1993

[7] M.B. Pursley, "Performance evaluation for phase coded spread-spectrum multiple access communication, part I system analysis," *IEEE Trans. Comm.*, Vol. COM-25, pp. 795-799, August 1977.

Chapter 7
Mobile Satellite CDMA Systems

7.1 INTRODUCTION

In this chapter the application of CDMA for a mobile satellite system is discussed. First a brief reason for considering a satellite link for mobile services using CDMA techniques is given. Then a description of the system model is given. Since the propagation characteristics of a satellite radio channel are in a number of respects different from the indoor and outdoor land mobile radio channels, another model is needed in order to investigate the performance of a mobile satellite system. The model developed here is valid for BPSK modulation, but can be extended for other types of modulation. It offers the possibility of investigating the performance of a mobile satellite CDMA system in terms of the bit error probability, throughput, and delay. Two diversity techniques are considered, and the effect of FEC coding on performance is also investigated. Also the influence of several propagation effects on the performance parameters is considered. We show that the mobile satellite system is strongly affected by log-normal shadowing due to blockage of the signal path between mobile terminal and satellite.

Land-mobile satellite systems seem to be a solution for realizing global personal communication networks [1].They can provide a wide range of services (e.g., position finding, radio paging, interconnection to the public switched telephone network, the possibility of private networks, voice communication, and data transmission [2]) to numerous types of terminals (e.g., land vehicles, aircraft, marine vessels, remote data collection and control sites, and portable terminals [1]). The applications of spread-spectrum modulation in the field of land-mobile satellite communications offer code division multiple access, resistance to multipath fading and a low peak-to-average power ratio. In addition, the properties of low probability of interception (LPI), antijamming resistance, and message privacy and security are attractive in some applications. Furthermore, as shown in [3,4], spread-spectrum CDMA systems can provide greater capacity than FDMA for mobile satellite communications.

The two most important propagation aspects of the satellite radio channel are shadowing and multipath fading. An appropriate model was proposed in [5-7], which assumes that the line of sight (LOS) component under foliage attenuation (shadowing) is log-normally distributed and that the multipath effect is Rayleigh distributed. Additionally, these two random processes are assumed to be correlated. In the next section this statistical propagation model is described in detail.

7.2 SYSTEM MODEL

The system under consideration consists of a satellite transceiver serving as base station and a number of mobile terminals on the earth surface communicating with the satellite transceiver (Figure 7.1). In this section, we describe the channel model and the receiver model.

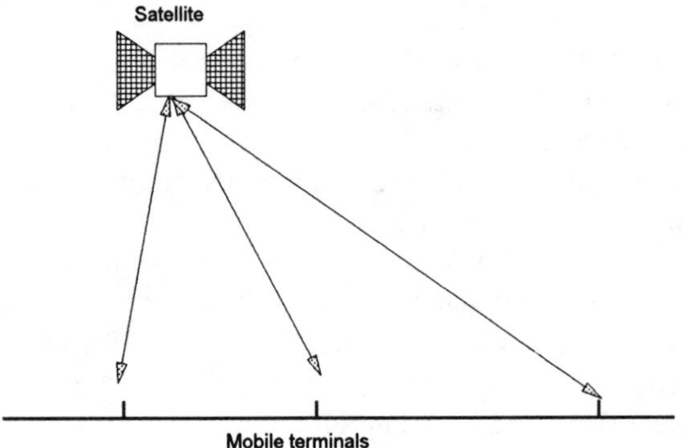

Figure 7.1 Satellite system model.

7.2.1 Channel Model

A statistical propagation model for a narrowband channel in rural and suburban environments was developed in [5-7]. The received signal is assumed to be the sum of a multipath signal with a Rayleigh distributed envelope and a shadowed LOS signal with a log-normal envelope distribution (Figure 7.2).

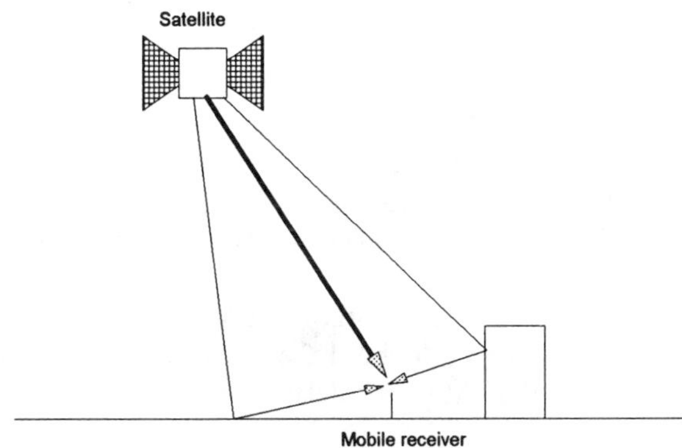

Figure 7.2 Radio channel with strong direct LOS and sum of multipath signals.

The resulting probability distribution of the received signal envelope r is given by [5-7]:

$$p_\beta(r) = \int_o^\infty \frac{r}{\sigma^2} \exp\left[-\frac{r^2+z^2}{2\sigma^2}\right] I_0\left[\frac{rz}{\sigma^2}\right] \frac{1}{\sqrt{2\pi}\sigma_s z} \exp\left[-\frac{(\ln(z)-m_s)^2}{2\sigma_s^2}\right] dz \qquad (7.1)$$

where $I_0(\cdot)$ is the modified Bessel function of the first kind and the zeroth order, σ^2 is the average scattered power due to multipath, m_s is the mean value due to shadowing, and σ_s^2 is the variance due to shadowing.

The probability density function of the received signal phase ϕ was found to be approximately Gaussian [6]:

$$p_\phi(\phi) = \frac{1}{\sqrt{2\pi\sigma_\phi^2}} \exp\left[-\frac{(\phi - m_\phi)}{2\sigma_\phi^2}\right] \qquad (7.2)$$

where m_ϕ and σ_ϕ^2 are the mean and the variance of the received signal phase, respectively.

The above model is valid for a narrowband system. If spread-spectrum modulation is used with a chip duration of less than the delay spread of the channel, the multipath power is partially reduced by the correlation operation in the receiver. The envelope and the phase distribution functions remain the same, but the values of σ^2 and σ_ϕ are reduced.

The impulse response of the wideband channel can be written as:

$$h(t) = \sum_{l=1}^{L} \beta_l \delta(t - \tau_l) \exp(j\phi_l) \tag{7.3}$$

where β_l, τ_l, and ϕ_l are the gain, time delay, and phase of the lth path, respectively. The first path is the LOS and therefore its propagation statistics are described by (7.1) and (7.2). The other paths have a Rayleigh path gain distribution and uniformly distributed phase since the direct LOS is suppressed by the correlation operation. The parameters of the various path distributions can be found if the power delay profile is known. From measurements [8] it is known that the power delay profile can be modeled as

$$P(\tau) = \frac{1}{T_m} \sigma^2 \exp(-\frac{1}{T_m} \tau) \tag{7.4}$$

where T_m is the delay spread. For a rural environment, a typical value of T_m is 0.65 µs [8].

Due to the correlation operation, the multipath power σ^2 is reduced. For path l it can be approximated as

$$\sigma_l^2 = \sigma^2 \left[1 - \exp\left(-\frac{T_c}{T_m} \right) \right] \exp\left[-(l-1)\frac{T_c}{T_m} \right] \tag{7.5}$$

where T_c is the chip duration.

Also, the phase variance of the first path will be decreased since it is determined by the amount of multipath power and the statistics of the LOS propagation. Using (4.5.19) of [9], the phase distribution function $p(\phi)$ of a log-normally shadowed Rician signal can be derived and given as [10]:

$$p(\phi) = \frac{1}{\sqrt{8\pi^3 \sigma_s^2}} \int_0^\infty \exp\left[\frac{-z^2}{2\sigma^2} - \frac{(\ln(z) - m_s)^2}{2\sigma_s^2} \right] \cdot \frac{\left(1 + G\sqrt{\pi} \exp(G^2)[1 + \text{erf}(G)]\right)}{z} dz \tag{7.6}$$

where

$$G \triangleq \frac{z \cos(\phi)}{\sqrt{2}\sigma} \tag{7.7}$$

and erf(·) is the error function: $\mathrm{erf}(G) \triangleq 2/\sqrt{\pi}\int_{o}^{G}\exp\left(-t^2\right)dt$. Now the phase variance in the case of spread-spectrum modulation can be determined by:

$$\sigma_\phi^2 = \int_{-\pi}^{\pi}\phi^2 p(\phi)d\phi \tag{7.8}$$

Note that the mean phase m_ϕ is zero because $p(\phi) = p(-\phi)$. Equations (7.6) and (7.7) are used only to determine the effect of spread-spectrum modulation on the phase variance. The combined effect of phase variation and envelope fading [6] is evaluated by using (7.2) as an approximation for (7.6).

7.2.2 Receiver Model

The total received signal for binary phase shift keying (BPSK) modulation is:

$$r(t) = \sum_{k=1}^{K}\sum_{l=1}^{L} A\beta_{lk}a_k(t-\tau_{lk})b_k(t-\tau_{lk})\cos\left[(\omega_c+\omega_{lk})t+\phi_{lk}\right]+n(t) \tag{7.9}$$

where l and k denote the path and user number, respectively, and A is the transmitted signal amplitude, which is assumed to be constant and identical for all users. For user k, $\{a_k\}$ is the spread-spectrum code, $\{b_k\}$ is the data sequence, $\omega_c+\omega_{lk}$ is the carrier plus Doppler angular frequency, ϕ_{lk} is the carrier phase, and $n(t)$ is white Gaussian noise with two-sided spectral density $N_0/2$. The instantaneous path amplitude is denoted as β. The signal is converted to baseband and correlated with a particular user code. Assuming that the receiver locks on the first path of user 1, a signal sample of the correlation output can be written as:

$$z = A\beta_{11}\cos(\phi_{11})T_b b_1^{\,0} + \sum_{k=1}^{K} A\left(b_k^{\,-1}X_k + b_k^{\,0}\hat{X}_k\right) + \eta \tag{7.10}$$

with

$$X_1 = \sum_{l=2}^{L} R_{11}(\tau_{l1})\beta_{l1}\cos(\phi_{l1}) \tag{7.11a}$$

$$\hat{X}_1 = \sum_{l=2}^{L} \hat{R}_{11}(\tau_{l1})\beta_{l1}\cos(\phi_{l1}) \tag{7.11b}$$

and for $k \geq 2$

$$X_k = \sum_{l=1}^{L} R_{1k}(\tau_{lk})\beta_{lk}\cos(\phi_{lk}) \tag{7.11c}$$

$$\hat{X}_k = \sum_{l=1}^{L} \hat{R}_{1k}(\tau_{lk})\beta_{lk}\cos(\phi_{lk}) \tag{7.11d}$$

Here b_k^{-1} and b_k^{0} are the previous and current data bits, respectively. They can be either 1 or–1 with equal probability; τ_n and ϕ_n are, respectively, the difference in arrival time and phase between the jth path of the reference user and the lth path of the kth user. It is assumed that ϕ_n is uniformly distributed over [0,2π]; η is a zero mean Gaussian variable with variance N_0T_b; and A is the signal amplitude. The partial correlation functions are given by

$$R_{lk}(\tau) = \int_{0}^{\tau} a_k(t-\tau)a_1(t)dt \tag{7.12a}$$

$$\hat{R}_{lk}(\tau) = \int_{\tau}^{T_b} a_k(t-\tau)a_1(t)dt \tag{7.12b}$$

Equation (7.10) can be written as

$$z = A\beta_{11}\cos(\phi_{11})T_b b_1^{\ 0} + \sum_{k=1} I_k + \eta \tag{7.13}$$

where I_k consists of the cross-correlations with interfering users and paths. If the number of interfering signals is large, i.e., $KL >> 1$, then the sum of these signals will be approximately Gaussian distributed. Using [11-13], a closed-form expression for the variance of the interference σ_{int}^2 can be derived. After correlation, the interference from other users and paths is:

$$\sum_{k=1} I_k = \sum_{k=1} A\left(b_k^{-1}X_k + b_k^{\ 0}X_k\right) \tag{7.14}$$

In this equation, the carrier phase ϕ_{lk} is Gaussian distributed for $l=k=1$, but uniformly distributed for the other users and paths, because transmitters are assumed to have arbitrary phases. The β_{lk} term has a shadowed Rician distribution for $l=1$ and a Rayleigh distribution otherwise. Assuming that both K and L are much larger than 1, the variance of (7.14) can be approximated by:

$$\sigma_{int}^2 \cong \sum_{k=1}^{K} \sum_{l=1}^{L} A^2 E\left\{\left[b_k^{-1} R_{lk}(\tau_{lk}) + b_k^0 \hat{R}_{lk}(\tau_{lk})\right]^2\right\} E\left\{\left[\beta_{lk} \cos\phi_{lk}\right]^2\right\} \tag{7.15}$$

The second moment of $\beta\cos\phi$ can be calculated as follows: If $l=1$, then β has a shadowed Rician distribution, which can be viewed as a Rician distribution with a variable Rice parameter $z^2/2\sigma^2$. The product of this Rician variable with the cosine of a uniformly distributed variable gives a Gaussian variable with a mean of $z/\sqrt{2}$ and a variance of σ^2. Thus $E[(\beta_{1k}\cos\phi_{lk})^2]=\sigma^2+z^2/2$, on the condition that z is constant. This condition can be removed by integrating over the independent log-normal distribution of z:

$$E\left[(\beta_{1k} \cos\phi_{1k})^2\right] = \int_0^\infty \left(\sigma_l^2 + \frac{z^2}{2}\right) p(z)dz \tag{7.16}$$

where σ_l^2 is the scattered multipath power for path l. Since z has a log-normal distribution, we know that

$$\int_0^\infty \sigma_l^2 p(z) = \sigma_l^2 \int_0^\infty p(z) = \sigma_l^2 \tag{7.17}$$

Furthermore,

$$\int_0^\infty \frac{z^2}{2} p(z)dz = \frac{1}{2} E\{z^2\} \tag{7.18}$$

If we define the logarithmic moments of z as

$$m_s = E\{\ln(z)\} \tag{7.19}$$

and

$$\sigma_s^2 = E\left\{(\ln(z))^2\right\} - E\{\ln(z)\} \tag{7.20}$$

then the moments of the variable z are [14]

$$\mu = E\{z\} = \exp\left(m_s + \frac{\sigma_s^2}{2}\right) \tag{7.21}$$

and

$$D^2 = E\{z\} - E^2\{z\} = \exp\left(2m_s + \sigma_s^2\right)\left[\exp\left(\sigma_s^2\right) - 1\right] \tag{7.22}$$

From this it can be concluded that

$$\frac{1}{2}E\left\{z^2\right\} = \frac{1}{2}\exp\left[2m_s + 2\sigma_s^2\right] \tag{7.23}$$

Using (7.17), (7.18), and (7.23), (7.16) simplifies to:

$$E\left[\left(\beta_{1k}\cos\phi_{1k}\right)^2\right] = \sigma_l^2 + \frac{1}{2}\exp\left[2\sigma_s^2 + 2m_s\right] \tag{7.24}$$

If $l>1$ then β has a Rayleigh distribution. In that case, $z=0$ and

$$E\left[\left(\beta_{lk}\cos\phi_{lk}\right)^2\right] = \sigma_l^2 \tag{7.25}$$

The variance of the cross-correlations can be calculated for Gold codes as described in [15]:

$$E\left\{\left[b_k^{-1}R_{lk}(\tau_{lk}) + b_k^0\hat{R}_{lk}(\tau_{lk})\right]^2\right\} = \frac{2T_b^2}{3N} \tag{7.26}$$

where N is the Gold code length, which is assumed here to be equal to T_b/T_C. Now, by substituting (7.24)-(7.26) in (7.15), a closed-form expression for the interference power is obtained:

$$\sigma_{int}^2 = \frac{2KA^2T_b^2}{3N}\left(\sigma^2 + \frac{1}{2}\exp\left[2m_s + 2\sigma_s^2\right]\right) \tag{7.27}$$

where K is the total number of users, N is the length of the spreading code, and

$$\sigma^2 = \sum_{l=1}^{L}\sigma_l^2 \tag{7.28}$$

In (7.27) it is assumed that all users have the same power and the same amplitude distribution function. This is valid for the downlink, but it is an approximation for the uplink.

7.3 PERFORMANCE ANALYSIS

The bit error probability (BEP) is considered to be a basic performance measure for digital systems. This bit error probability has been derived in the following subsections for systems with and without diversity. The derivations are made based on BPSK modulation but can be easily extended to other types of modulation.

7.3.1 BEP Analysis Without Diversity

Assuming that the data bits −1 and 1 are equiprobable, the bit error probability P_e can be expressed as:

$$P_e = P\left(z < 0 \middle| b_0^1 = 1\right) \tag{7.29}$$

The decision variable z can be written as

$$z = Z_X + Z_Y \tag{7.30}$$

where Z_X is the desired component with pdf denoted by $P_X(X)$ and Z_Y is the total noise plus interference described by a Gaussian pdf denoted by Z_Y with mean 0 and variance σ_t^2:

$$\sigma_t^2 \triangleq N_0 T_b + \sigma_{int}^2 \tag{7.31}$$

where N_0 is the power density of the Gaussian noise, T_b is the duration of a data bit, and σ_{int}^2 is the interference power.

From (7.13) it is clear that z_x consists of β with a shadowed Rician distribution, multiplied by the cosine of Gaussian distributed phase. Using [9], the total distribution function of $z_x=\beta\cos(\phi)$ can be written as:

$$p_X(x) = 2\int_{|x|}^{\infty} p_\beta(r) p_\phi\left[\arccos\left(\frac{x}{r}\right)\right]\frac{dr}{\sqrt{r^2-x^2}} \tag{7.32}$$

Substituting (7.1) and (7.2) in (7.32), the expression for the pdf of Z_X is

$$p_X(x) = \int_{|x|}^{\infty}\int_{0}^{\infty}\frac{r}{\sigma^2\sqrt{\pi^2\sigma_s^2\sigma_\phi^2}}\exp\left[-\frac{(\ln(z)-m_s)^2}{2\sigma_s^2}-\frac{(r^2+z^2)}{2\sigma^2} -\frac{\arccos^2(x/r)}{2\sigma_\phi^2}\right]\frac{I_0\left(\frac{rz}{\sigma^2}\right)}{z\sqrt{(r^2-x^2)}}dzdr \tag{7.33}$$

Equation (7.33) is valid in the case for which the receiver locks on the LOS signal. This requires the bandwidth of the carrier tracking loop to be much smaller than the fading bandwidth of the received signal. If the bandwidth of the tracking loop is larger than the fading bandwidth, then the phase ϕ is approximately zero because the receiver will lock on the phase of the total signal. In that case, the distribution function $p_X(x)$ is equal to the shadowed Rician distribution $p_\beta(x)$. However, with a larger tracking bandwidth the loop noise increases, so there is always a certain phase error. Therefore, in the bit error probability calculations one gets an upper bound by using (7.33) and a lower bound by using $p_X(x)=p_\beta(x)$.

Now assume that the desired term z_x has value x, then the conditional bit error probability is

$$P_e|x = P(x+Z_Y<0) = \frac{1}{2}\operatorname{erfc}\left(\frac{x}{\sigma_t\sqrt{2}}\right) \tag{7.34}$$

where erfc $(\cdot)$ is the complementary error function, given by $\left(2/\sqrt{\pi}\right)\int_x^{\infty}\exp\left(-t^2\right)dt$. Averaging over all possible values x of z_x using the expression for the pdf of z_x given in (7.34), we obtain the following expression for the bit error probability:

$$P_e = \int_{-\infty}^{\infty} (p_e|x) p_X(x) dx \tag{7.35}$$

Numerical Results Without Diversity

First, the bit error probability is shown for $T_C = 0.1\mu s$, $r_b = 2400$ bits/s, $N = 4095$, and $K = 1$. The fading parameters are adjusted according to (7.5) and (7.8). Table 7.1 shows the measured values for σ^2, m_S, and σ_S found in [7]. In Table 7.2, the modified values for σ^2 and σ_ϕ are shown. Also, the calculated values for σ_ϕ for narrowband operation are given, showing a considerable difference with the measured values [5-7] in the case of heavy shadowing. This difference may be due to the filtering of the received signal. It can be expected that σ_ϕ should be almost $\pi/2$ for heavy shadowing, because the distribution function approaches a Rayleigh pdf. This implies an almost uniformly distributed phase. However, in order to compare with [7], the measured value for heavy shadowing ($\sigma_\phi = 0.52$) has been used in all calculations. Assuming that for heavy shadowing σ_ϕ decreases by the same amount as for light and average shadowing, it is given by (0.52/2.9) in the presence of spread spectrum modulation.

Table 7.1
Channel model parameters

	Light	Average	Heavy
σ^2	0.158	0.126	0.0631
m_S	0.115	−0.115	−3.91
σ_S	0.115	0.161	0.806

Table 7.2
Modified channel model parameters

	Light	Average	Heavy
σ_ϕ narrowband	0.40	0.47	1.55
σ_ϕ spread spectrum	0.14	0.16	1.42
σ_1^2 spread spectrum	0.023	0.018	0.009

Figures (7.3) and (7.4) show that spread-spectrum modulation yields better performance than narrowband modulation for light and average shadowing. For heavy shadowing, the performance is worse at signal-to-noise ratios (SNRs) below 36 dB. The reason for this is that for light and average shadowing most of the signal power is received via the LOS, so if the multipath power is reduced by the use of spread-spectrum modulation, a less perturbed signal will be obtained. In the case of heavy shadowing, however, the direct-

LOS power is much smaller than the multipath power, resulting in an approximately Rayleigh faded signal.

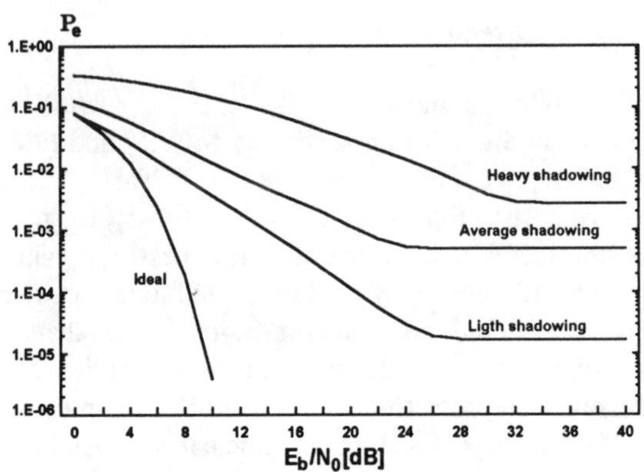

Figure 7.3 Bit error probability for ideal BPSK and for narrowband BPSK with light, average, and heavy shadowing.

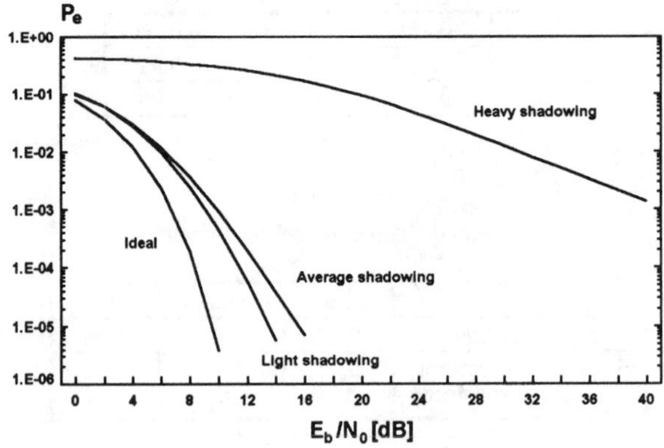

Figure 7.4 Bit error probability for spread-spectrum modulation with K=1 user, chip length T_c= 0.1 μs, Gold code length N=4095, and bit rate r_b=2.4 kbit/s.

The use of spread-spectrum modulation now decreases the total signal power considerably, with the result that the resulting bit error probability increases. In this case, diversity techniques can be used to improve the bit error probability.

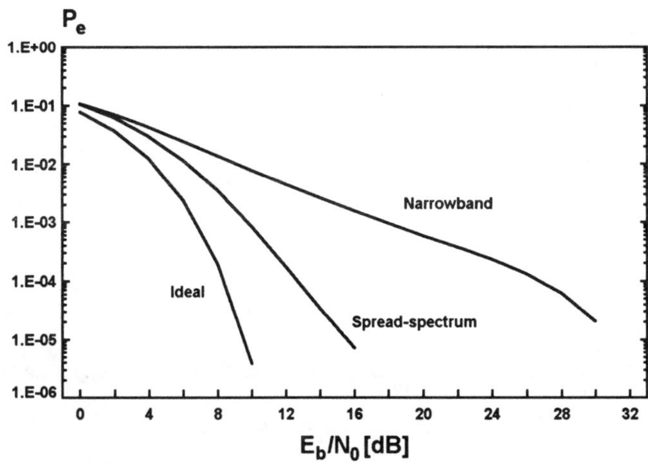

Figure 7.5 Comparison of the bit error probability with narrowband and spread-spectrum modulation with light shadowing and envelope fading only for K=1 user, chip length T_c= 0.1 μs, Gold code length N=4095, and bit rate r_b=2.4 kbit/s.

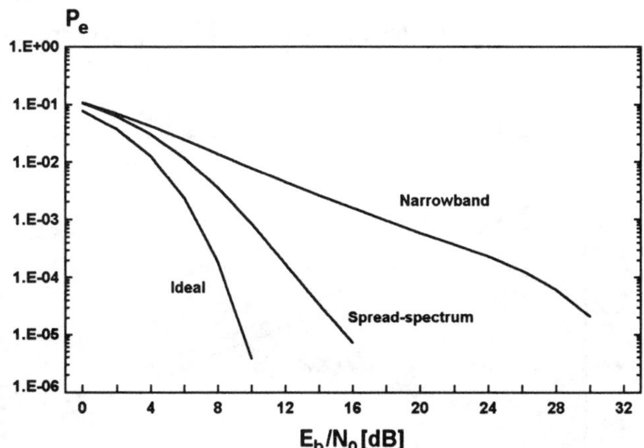

Figure 7.6 Comparison of the bit error probability with narrowband and spread-spectrum modulation with average shadowing and envelope fading only for K=1 user, chip length T_c= 0.1 μs, Gold code length N=4095, and bit rate r_b=2.4 kbit/s

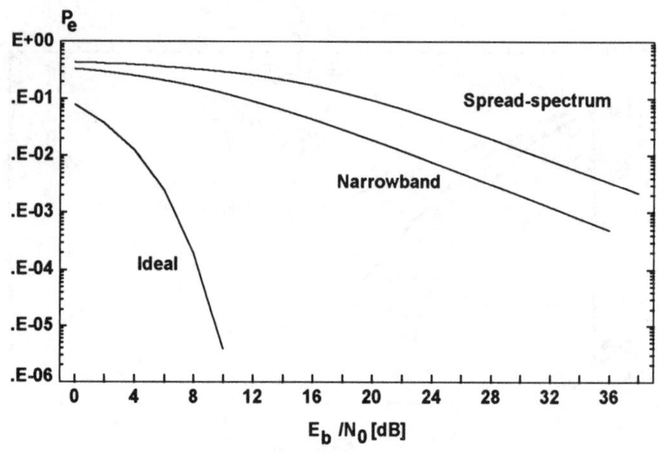

Figure 7.7 Comparison of the bit error probability with narrowband and spread-spectrum modulation with heavy shadowing and envelope fading only for K=1 user, chip length T_c= 0.1 μs, Gold code length N=4095, and bit rate r_b=2.4 kbit/s.

If the receiver is able to lock onto the total phase of the first path, then only the envelope fading has to be considered, which is investigated in Figures 7.5, 7.6, and 7.7, where the bit error probability is compared to the narrowband and spread-spectrum modulation.

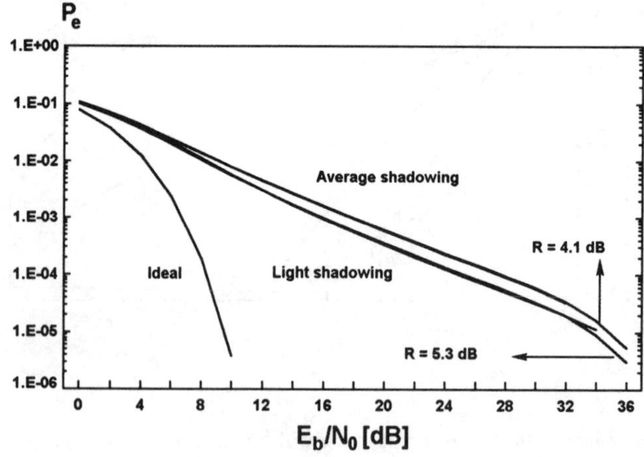

Figure 7.8 Comparison of the shadowed Rician distribution with the normal Rician distribution for light and average shadowing with narrowband transmission.

It is clear that the bit error probability for narrowband modulation changes considerably by removing the phase variation, while in the spread-spectrum modulation there is much less change. The phase variance, which is reduced by a factor of approximately 9 by the use of spread-spectrum modulation, exerts a nonnegligible influence only near the irreducible error probability level ($\leq 10^{-9}$ for spread-spectrum modulation).

We can see from Figure 7.8 that light and average shadowed Rician distributions can be represented by the normal Rician distribution with a Rician factor of 5.3 and 4.1 dB, respectively, with an error of less than 5% for bit error probability values between 10^{-1} and 10^{-5}. Outside this interval, the error tends to increase. It can be expected that matching will be possible only in a certain range, since the two distribution functions are different in nature because of the shadowing. Heavy shadowing, however, results in a negligible LOS power. Accordingly, we see a perfect match with the curve of Rayleigh fading in Figure 7.9, where the Rice factor is equal to zero. Figure 7.10 shows the results for spread-spectrum modulation with average shadowing and the number of users as a parameter. To maintain a bit error probability of 10^{-3} the SNR has to be increased by about 0.5 dB for $K=100$ users, 1 dB for $K=200$, and 2 dB for $K=400$ as compared to the SNR for a single user ($K=1$). Note that E_b/N_0 is the average signal-to-noise ratio, which does not include interference power. So increasing E_b/N_0 decreases the bit error probability, until the irreducible bit error probably caused by the interference power is reached. Figure 7.11 shows that as T_c decreases, the bit error probability decreases because of the decreased multipath power.

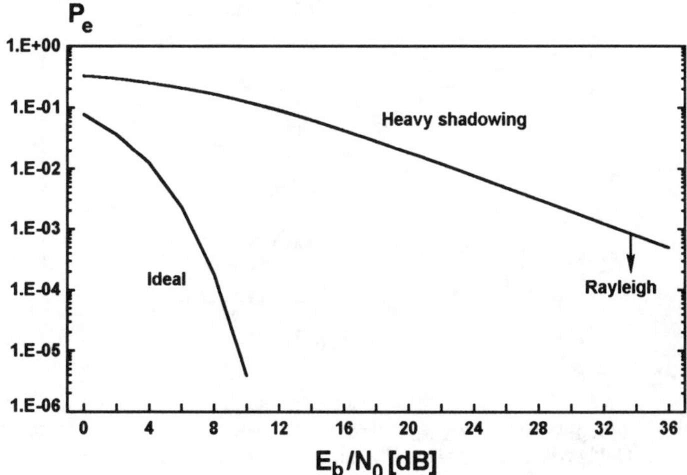

Figure 7.9 Comparison of the shadowed Rician distribution with the Rayleigh distribution for heavy shadowing with narrowband transmission.

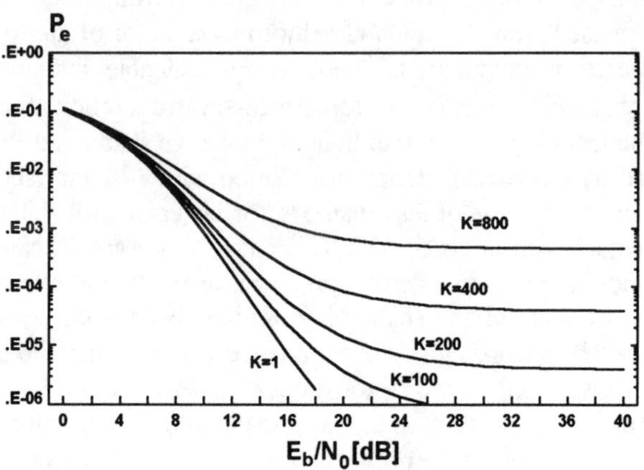

Figure 7.10 Bit error probability for spread-spectrum modulation with average shadowing for chip length T_c= 0.1 μs, Gold code length N=4095, bit rate r_b=2.4 kbit/s and K as parameter.

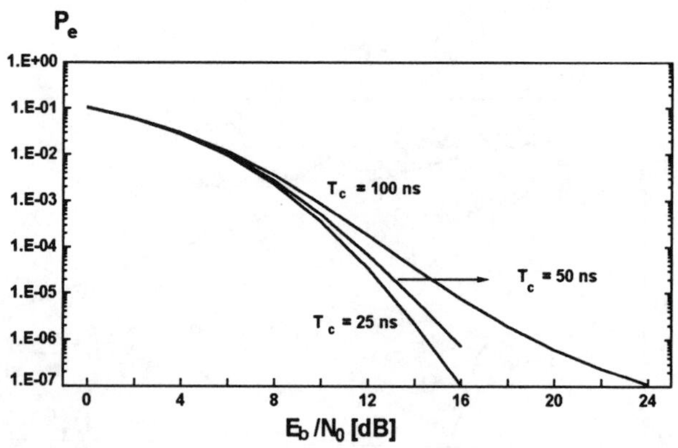

Figure 7.11 Bit error probability for spread-spectrum modulation with average shadowing for K=1, Gold code length N=4095, bit duration T_b=$N \cdot T_c$, and chip duration T_c as parameter.

7.3.2 BEP Analysis with Diversity

When spread-spectrum modulation is used with a chip time that is less than the delay spread of the channel, a number of resolvable paths L exist that can be used to improve the performance. Two types of diversity, maximal ratio combining and selection diversity, are considered here. For simplicity we now assume that code and carrier phase tracking errors are negligible. If the receiver tracks the code and carrier phase at path j from user i then a signal sample of the correlation output can be written:

$$z = A\beta_{11}T_b b_1^0 + \sum_{k=1}^{K} I_k + \eta \tag{7.36}$$

Here A is the amplitude of the received signal, β_{11} is the path gain of path 1 for user 1, T_b is the bit duration, b_1^0 is the instantaneous bit of user 1, I_k is the interference received from user k, and η is the Gaussian noise term.

Maximal Ratio Combining

When maximal ratio combining is used, the received signal is coherently correlated with a particular code for M different paths. Here M is the order of diversity. Each path is multiplied by the path gain β_{lk} and all correlation outputs are combined. The probability density function of the sum of the squared path gains β_{lk}^2 is the convolution of the M different squared path gain probability density functions. Since the amplitude of the first path has a shadowed Rician distribution $p_r(r)$ according to (7.1), the probability density function $p_1(x)$ of a squared shadowed Rician variable follows from a simple transformation:

$$p_1(x) = \frac{p_r(\sqrt{x})}{2\sqrt{x}} \tag{7.37}$$

All other paths are assumed to have Rayleigh distributed amplitudes. Their gains have a chi-square probability density function, so the sum of $M-1$ chi-square variables results in the following probability density function

$$p_2(x) = \frac{1}{2\sigma_2^2}\exp\left(-\frac{x}{2\sigma_2^2}\right) * \frac{1}{2\sigma_3^2}\exp\left(-\frac{x}{2\sigma_3^2}\right) * \ldots * \frac{1}{2\sigma_M^2}\exp\left(-\frac{x}{2\sigma_M^2}\right)$$

$$= \sum_{i=2}^{M} \frac{\left(2\sigma_i^2\right)^{M-3}}{\prod_{\substack{j=2 \\ j\neq i}}^{M}\left(2\sigma_i^2 - 2\sigma_j^2\right)} \exp\left(-\frac{x}{2\sigma_i^2}\right) \tag{7.38}$$

where * denotes convolution.

Now the probability density function of the sum of all M squared path gains can be obtained by a convolution of (7.37) and (7.38):

$$p_{mrc}(\alpha) = \int_0^\infty p_1(x)p_2(x-\alpha)dx$$

$$= \int_0^\alpha \frac{p_r\left(\sqrt{x}\right)}{2\sqrt{x}} \sum_{i=2}^{M} \frac{\left(2\sigma_i^2\right)^{M-3}}{\prod_{\substack{j=2 \\ j\neq i}}^{M}\left(2\sigma_i^2 - 2\sigma_j^2\right)} \exp\left[-\frac{(x-\alpha)}{2\sigma_i^2}\right]dx \tag{7.39}$$

where

$$\alpha = \sum_{m=1}^{M} \beta_{lk}^2 \tag{7.40}$$

The bit error probability P_e can be expressed as the probability that the decision variable is negative while the received data bit is positive, assuming that the data bits are equiprobable. Since we assume the noise plus interference to be Gaussian distributed, the bit error probability P_e can be written as:

$$P_e = \frac{1}{2}\int_0^\infty \operatorname{erfc}\left[\sqrt{\frac{\alpha A^2}{2\sigma_t^2}}\right] p_{mrc}(\alpha)d\alpha \tag{7.41}$$

where erfc(z) is the complementary error function, given by

$$\operatorname{erfc}(z) = \int_z^\infty \exp\left(-t^2\right)dt \tag{7.42}$$

Selection Diversity

With selection diversity, the strongest out of M different signals is selected. The cumulative distribution function (cdf) of the output signal is the multiplication of all M different path gain cdfs, so the resulting pdf is

$$p_{sd}(x) = \frac{d}{dx}\left[p_1(x)p_2(x)\ldots p_l(x)\ldots p_M(x)\right] \tag{7.43}$$

where $p_l(x)$ is a shadowed Rician cdf for $l=1$ and a Rayleigh cdf otherwise.

$$p_1(x) = \int_0^x p_r(r)dr \tag{7.44}$$

$$p_l(x) = 1 - \exp\left(-\frac{x^2}{2\sigma_i^2}\right), \qquad 2 < l \le M \tag{7.45}$$

The bit error probability is obtained by:

$$\begin{aligned} P_e &= \frac{1}{2}\int_0^\infty \operatorname{erfc}\left(\frac{xA}{\sigma\sqrt{2}}\right)p_{sd}(x)dx \\ &= \frac{1}{2}\int_0^\infty \operatorname{erfc}\left(\frac{xA}{\sigma\sqrt{2}}\right) \\ &\quad \frac{d}{dx}\left(\int_0^x p_r(r)\prod_{l=2}^{M}\left[1-\exp\left(-\frac{x^2}{2\sigma_l^2}\right)\right]dr\right)dx \end{aligned} \tag{7.46}$$

The evaluation of (7.46) involves a triple integral and a differentiation, which makes the numerical calculation extremely difficult. In the case of heavy shadowing, however, the shadowed Rician cdf $P_1(x)$ can be approximated by a Rayleigh cdf [10], which makes it possible to solve (7.43) mathematically.

Numerical Results with Diversity

The numerical results are computed for a bit rate of $1/T_b=2400$ bit/s, code length $N=4095$, chip time $T_c=0.1$ μs and $K=400$ users. Also the measurement results presented in Tables 7.1 and 7.2 are used. The SNR E_b/N_0 used in the figures is defined as:

$$\frac{E_b}{N_0} \triangleq \frac{A^2 T_b}{2N_0} \tag{7.47}$$

Figure 7.12 shows the bit error probability P_e for light shadowing, using maximal ratio combining. For the sake of comparison, P_e is also plotted for the ideal BPSK case, i.e., coherent BPSK without fading and interference.

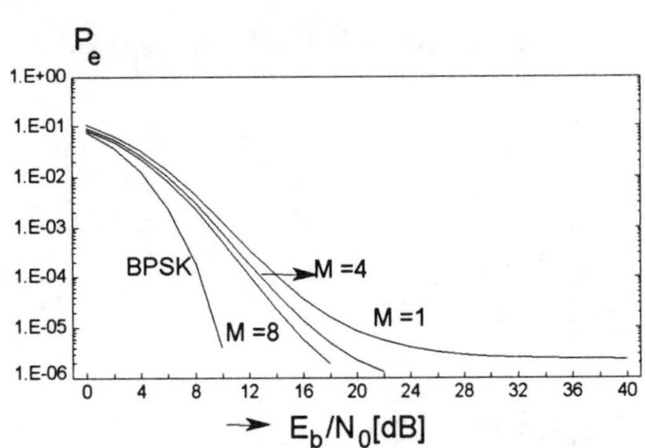

Figure 7.12 Bit error probability for light shadowing using maximal ratio combining.

With no diversity ($M=1$), the performance difference with the ideal BPSK plot is not much for small values of E_b/N_0, about 3.8 dB for $P_e=10^{-3}$. As a result, the gain of maximal ratio combining is limited. For $P_e=10^{-3}$, about 1.3 dB less power is required if $M=8$.

For average shadowing (Figure 7.13), the ratio of LOS power and multipath power is smaller and therefore the diversity gain is greater, about 2.2 dB for $P_e=10^{-3}$ and $M=8$. In the case of heavy shadowing (Figure 7.14), the irreducible bit error probability due to multiuser interference is very large. Diversity decreases this level from 5×10^{-2} to a minimum of 5×10^{-6} for MRC with $M=8$. The reason for the poor performance is that the multipath power is dominant, so the use of a spread spectrum without diversity decreases the total received power considerably and increases the bit error probability.

For selection diversity, the bit error probability is calculated for heavy shadowing only, because the pdf of the first path can be approximated by a Rayleigh pdf in this case, which removes most numerical difficulties in calculating (7.46). With light and heavy shadowing the performance of a system with selection diversity will be worse than with maximal ratio combining (MRC), since MRC is known to be a superior diversity technique as compared to selection diversity [16]. For heavy shadowing, Figure 7.14

shows that selection diversity gives an irreducible bit error probability that is 8 times greater than with MRC for $M=4$.

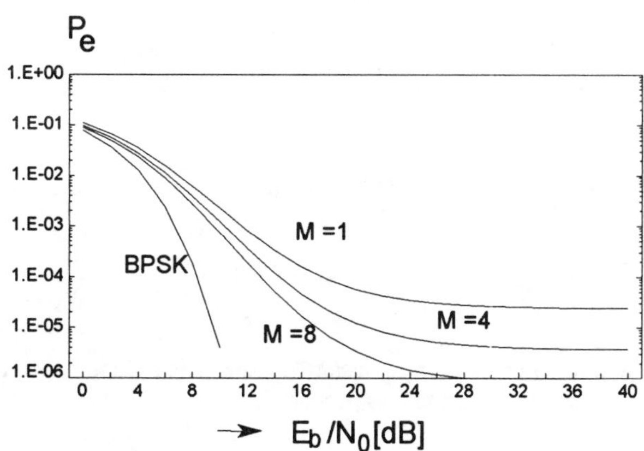

Figure 7.13 Bit error probability for average shadowing using maximal ratio combining.

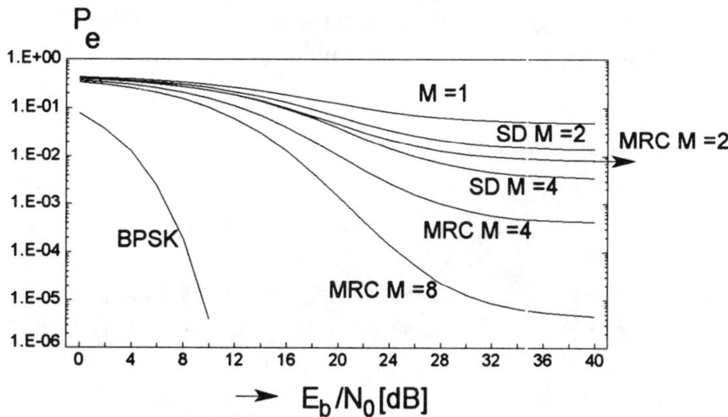

Figure 7.14 Bit error probability for heavy shadowing using selection diversity and combining maximal ratio combining and selection diversity.

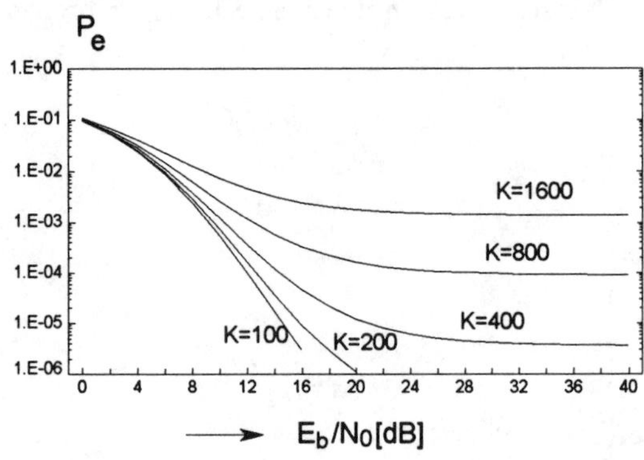

Figure 7.15 Bit error probability for average shadowing using maximal ratio combining with M=4, r_b=2400 bits/s, N=4095, and K as parameter.

To show the effect of changes in the parameters used in the previous results, Figures 7.15, 7.16, and 7.17 depict the bit error probability for different values of K, T_c, and N, respectively. These figures are calculated for average shadowing. The use of maximal ratio combining is assumed with an order of diversity M=4. If the number of users K is increased (Figure 7.15) the bit error probability increases because of the interference power, which is proportional to K. Decreasing the chip time T_c (Figure 7.15) while keeping the number of chips per bit constant leads to a decrease of the bit error probability because of increased multipath rejection.

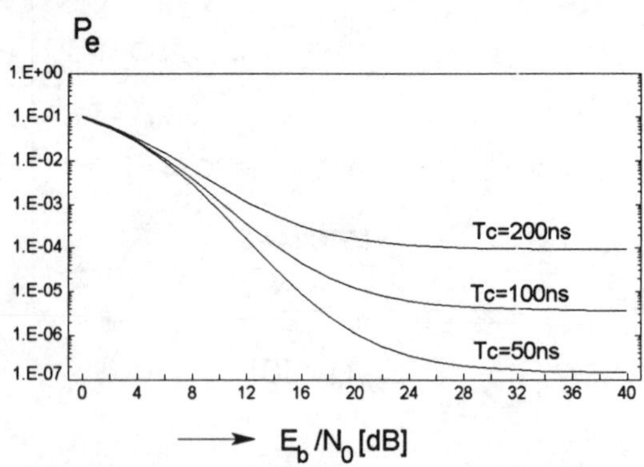

Figure 7.16 Bit error probability for average shadowing using maximal ratio combining with M=4, r_b=2400 bits/s, N=4095, K=400, and T_c as parameter.

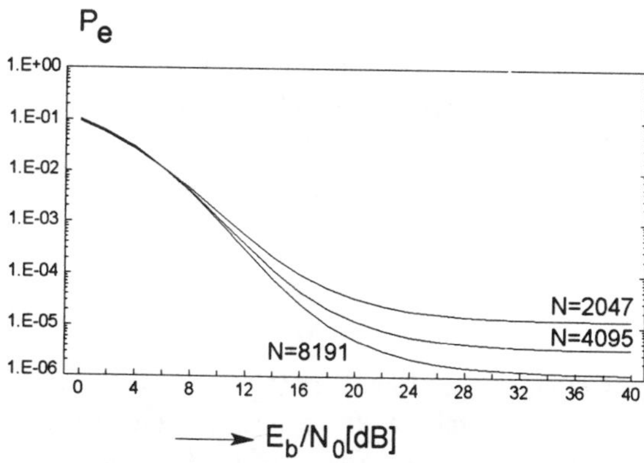

Figure 7.17 Bit error probability for average shadowing using maximal ratio combining with M=4, K=400, T_c=100 ns, and N as parameter.

A decrease of the code length N, while $T_c=T_b/N$ is fixed at 100 ns, causes an increase in the bit error probability (Figure 7.17), because the interference power is inversely proportional to N.

7.4 THROUGHPUT AND DELAY ANALYSIS

The previous analysis is particularly applicable to circuit switched communications, such as telephony. In these types of applications, bit error and outage probabilities are the main parameters of interest. For packet switched data transmission, the situation is quite different. In this case, the aim is to deliver certain packets of data to a receiver without errors (or with a very small error probability). This is usually accomplished by using some kind of retransmission protocol, whereby packets are retransmitted whenever the receiver detects the occurrence of errors. In such transmission systems, the main parameters of interest are the throughput (often expressed as the number of successfully delivered packets per time slot) and the packet delay. This section investigates the advantages and disadvantages of the use of spread-spectrum CDMA in packet switched communications.

A communication network is considered with such a bandwidth that in the case of perfect time division multiple access (TDMA) or FDMA, a total number of N users can be accommodated, each transmitting at a bit rate of $1/T_b$ with T_b as the bit duration. Instead of a fixed assignment scheme like TDMA and FDMA, consider instead a random access slotted CDMA scheme, where the data sequence is spread by a certain spreading code, consisting of N chips per bit. The code length and the total number of codes may be fixed to N, but it may also be larger than N. The only assumption about the codes that is made is that they can be approximated by random sequences [17]. Further, it is assumed that the total number of users is large enough to get a Poisson distribution function for the offered traffic. Therefore, the probability $p_{tr}(k)$ that k packets are generated during a certain time slot is given by:

$$p_{tr}(k) = \frac{G^k}{k!}\exp(-G) \tag{7.48}$$

Here, G is the average number of transmitted packets per time slot.

When a packet is transmitted, there is a certain probability $P_s(k)$ that it is received successfully. Erroneous packets have to be retransmitted after a certain random delay. The steady state throughput of this transmission system is defined as the average number of successfully received packets per time slot, and it is given by:

$$S = \sum_{k=1}^{K_{max}} kP_{tr}(k)P_s(k) \tag{7.49}$$

Here, K_{max} is the maximum number of users that can be simultaneously handled by the system, because the number of receivers or available code words is limited. If K_{max} is set to infinity, then the maximum available throughput is only limited by the packet success probability, which gradually decreases to zero for increasing numbers of users K.

It may be noted that slotted ALOHA is a special limiting example of slotted CDMA for K_{max}=N=1. In order to make a fair comparison between CDMA and slotted ALOHA, it is desirable to use the same bandwidth for both systems, which gives two options for the slotted ALOHA case: First, the data rate can be chosen equal to the CDMA chip rate. Second, the data rate can be chosen equal to the CDMA data rate, which makes it possible to divide the total bandwidth in N separate ALOHA channels [18]. In this way, a combination of FDMA and ALOHA is made where each user randomly selects a certain frequency band and a certain time slot. The second option can be expected to achieve higher throughput values, because in the first option, the large data rate will cause considerable intersymbol interference because of the relatively large multipath delay spread of the channel. Therefore, the bit error probability and hence the throughput for the first option will always be worse than in the multichannel ALOHA system, where the data rate is N times smaller.

Assuming that the total amount of traffic is randomly distributed over N channels, the throughput of the multichannel slotted ALOHA system can be simply calculated as N times the throughput of one narrowband slotted ALOHA channel, with an offered load that is equal to the total offered load divided by N. In fact, it is possible to normalize the obtained throughput values for both CDMA and ALOHA by dividing the total system throughput by *N*. Thus, the corresponding throughput per "channel" is found.

The corresponding average packet delay *D* is defined as the number of slot times it takes for a packet to be successfully received. Thus it is the average time duration (in slots) between the packet being offered to the transmitter and the packet being successfully received, and is given by:

$$D = 1.5 + T_d + \left[\frac{G}{S} - 1\right]\left[\frac{N_{AT}}{2} + 1 + 2T_d\right] \tag{7.50}$$

where $G/S-1$ is the average number of retransmissions, $N_{AT}/2$ is the mean retransmission delay, and T_d is the propagation delay. Now the throughput and delay can be evaluated for different conditions using (7.48)-(7.50) and the packet success probability, which can be obtained by the expressions derived as follows.

The packet success probability P_S can be evaluated for slow and fast fading. In the case of slow fading, the path gains are assumed to be constant during one packet time. For fast fading, it is assumed that the path gains are uncorrelated for two consecutive data bits. Further, the use of a forward error correcting code is included by the assumption that *t* bit errors can be corrected from a total number of N_p bits per packet.

The resulting expression for the packet success probability in the case of fast fading is:

$$P_s = \sum_{j=0}^{t} P_e^j (1 - P_e)^{N_p - j} \binom{N_p}{j} \tag{7.51}$$

In the case of slow fading, the packet success probability becomes:

$$P_s = \int_0^{\infty} \sum_{j=0}^{t} [P_e(x)]^j [1 - P_e(x)]^{N_p - j} \binom{N_p}{j} p(x)dx \tag{7.52}$$

where for narrowband:

$$P_e(x) = \frac{1}{2}\operatorname{erfc}\left[\frac{xA}{\sigma_r\sqrt{2}}\right], \; p(x) = p_r(x) \tag{7.53}$$

for spread-spectrum with no diversity:

$$P_e(x)=\frac{1}{2}\,\mathrm{erfc}\left[\frac{xA}{\sigma\sqrt{2}}\right],\ p(x)=p_{rss}(x) \tag{7.54}$$

for spread-spectrum with selection diversity:

$$P_e(x)=\frac{1}{2}\,\mathrm{erfc}\left[\frac{xA}{\sigma\sqrt{2}}\right],\ p(x)=p_{sd}(x) \tag{7.55}$$

for spread-spectrum with maximal ratio combining:

$$P_e(x)=\frac{1}{2}\,\mathrm{erfc}\left[\sqrt{\frac{xA^2}{2\sigma^2}}\right],\ p(x)=p_{mrc}(x) \tag{7.56}$$

Figures 7.18 and 7.19 show the normalized throughput (*S/N*) curves of narrowband slotted ALOHA without and with forward error correction, respectively.

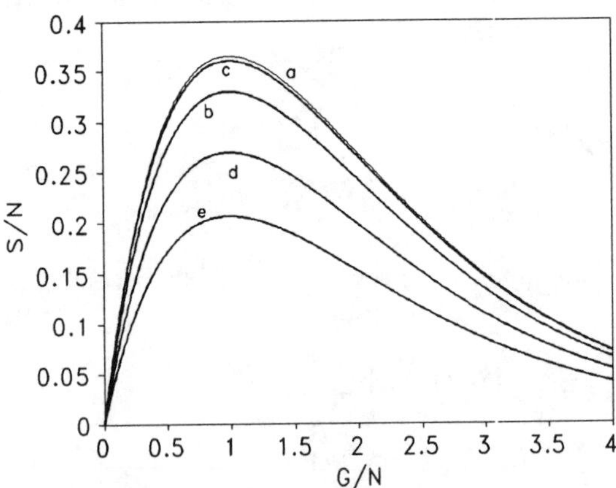

Figure 7.18 Normalized throughput of narrowband slotted ALOHA for (a) light shadowing, slow fading, (b) light shadowing, fast fading, (c) average shadowing, slow fading, (d) average shadowing, fast fading, (e) heavy shadowing, slow fading.

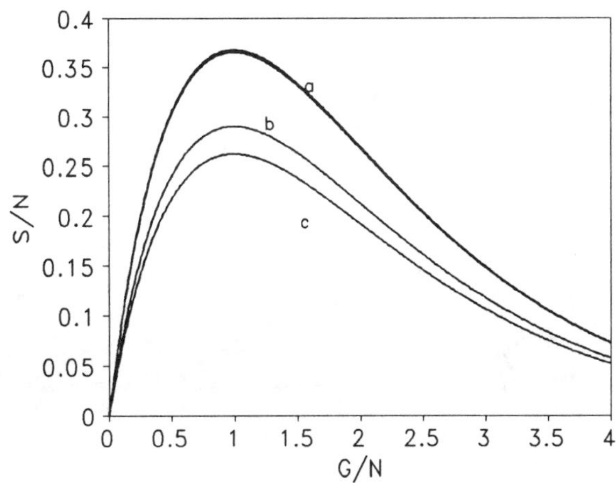

Figure 7.19 Normalized throughput of narrowband slotted ALOHA using FEC coding for (a) light and average shadowing, slow fading and fast fading (b) heavy shadowing, slow fading, (c) heavy shadowing, fast fading.

The results of these and the following figures were obtained for T_s/T_c=6.5, E_b/N_0=20 dB, N_p=256, and t=0 or t=10 in the absence or presence of forward error correction coding. The values of N and K_{max} used were 127 and $2N$, respectively. However, as explained earlier, the results are also valid for other values of N, as long as N is large compared to one.

Figure 7.20 shows that the normalized throughput of slotted CDMA without error correction is less than that of slotted ALOHA. However, when error correction is applied, like in Figures 7.19 and 7.21, then CDMA benefits far more than slotted ALOHA, which results in a larger normalized throughput than narrowband slotted ALOHA for light and average shadowing.

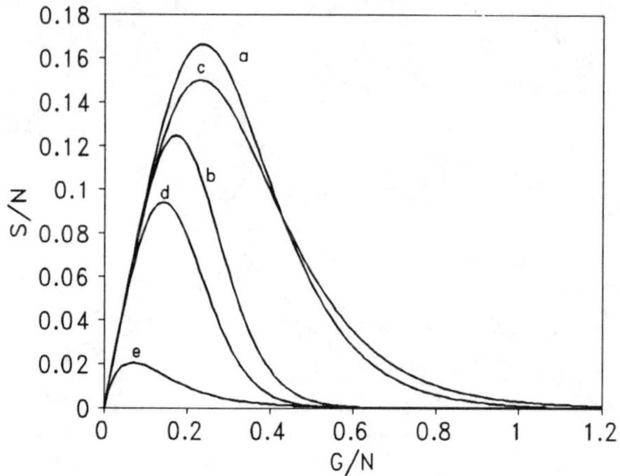

Figure 7.20 Normalized throughput of slotted CDMA for (a) light shadowing, slow fading, (b) light shadowing, fast fading, (c) average shadowing, slow fading, (d) average shadowing, fast fading, (e) heavy shadowing, slow fading.

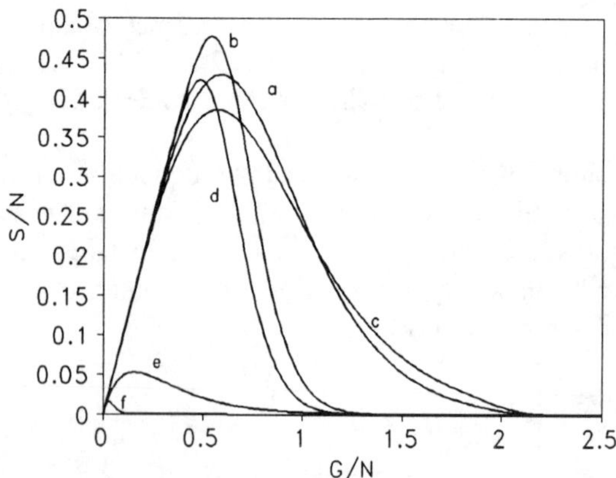

Figure 7.21 Normalized throughput of slotted CDMA, using FEC coding for (a) light shadowing, slow fading, (b) light shadowing, fast fading, (c) average shadowing, slow fading, (d) average shadowing, fast fading, (e) heavy shadowing, slow fading, (f) heavy shadowing, fast fading.

An interesting fact that can be seen in the previous figures is that fast fading yields a higher maximum throughput than slow fading when forward error correction coding is applied, while the performance of fast fading is worse if no error correction is used. This is because for fast fading, the bit errors are randomly spread over all packets, so each packet has the same success probability, which is relatively low when no error correction coding is used. In the case of slow fading, however, errors appear in bursts.

Due to the slow fading, the packet success probability varies, which means that compared to fast fading, a certain part of the packets has a higher packet success probability, resulting in a higher throughput when no forward error correction is applied. If error correction coding is used, then the packet success probability is greatly enhanced for fast fading as long as the SNR plus interference ratio, which is inversely proportional to the offered load is above some threshold value. Beneath that threshold, the packet success probability quickly drops to zero, as can be seen in the figures. For slow fading, error correction coding is less effective, because there are fewer packets with up to t bit errors that can be corrected. Therefore, the maximum throughput in the case of slow fading is less than for fast fading. However, for high values of the offered load, the throughput for slow fading decreases less fast than for fast fading, because there is always a certain part of the packets with a higher success probability than in the case of fast fading.

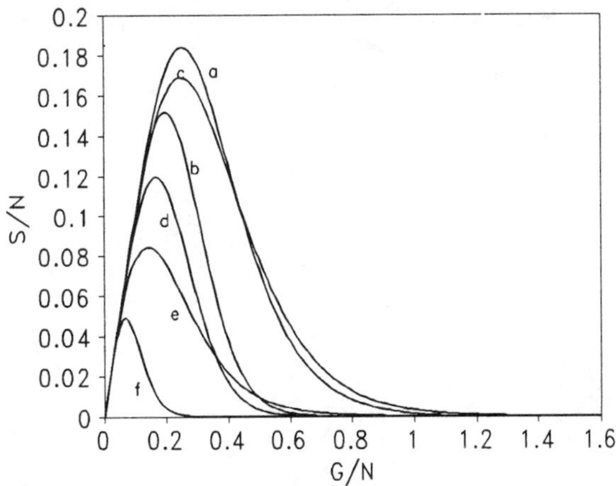

Figure 7.22 Normalized throughput of slotted CDMA, using MRC for (a) light shadowing, slow fading, (b) light shadowing, fast fading, (c) average shadowing, slow fading, (d) average shadowing, fast fading, (e) heavy shadowing, slow fading, (f) heavy shadowing, fast fading.

Figure 7.22 demonstrates that the use of maximal ratio combining provides a considerable improvement in throughput compared to Figure 7.20. However, if the throughput curves in the case of combined forward error correction and maximal ratio combing (Figure 7.23) are compared for the case of just forward error correction (Figure 7.21), then there is only a minor throughput improvement of about 10% due to maximal ratio combining in the cases of light and average shadowing.

For heavy shadowing, however, the use of maximal ratio combining increases the throughput by a factor of almost 4. This is because for heavy shadowing, the LOS signal power is negligible as compared to the multipath power. So if the receiver tries to demodulate only the first path, it uses only a small fraction of the total received power, assuming that the chip time is smaller than the multipath delay spread. In that case, maximal ratio combining is very useful to make use of all received signal power that is available.

In Figure 7.24, the normalized throughput curves for heavy shadowing using selection diversity are drawn. It is clear that selection diversity performs worse than maximal ratio combining by at least a factor of 2. However, it still considerably improves the throughput as compared to the case of no diversity.

Figure 7.25 shows the normalized throughput of slotted CDMA with the number of correctable bits as a parameter. As the error correcting capability increases, the maximum achievable throughput increases to high values. However, the user data throughput is of course decreased by the increasing number of bits used for error correction.

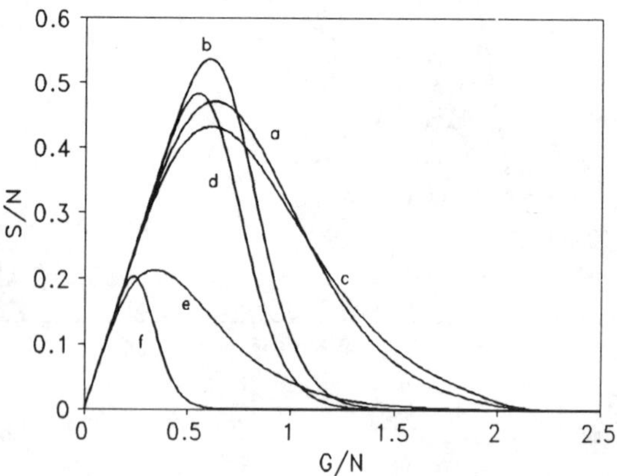

Figure 7.23 Normalized throughput of slotted CDMA, using MRC and FEC for (a) light shadowing, slow fading, (b) light shadowing, fast fading, (c) average shadowing, slow fading, (d) average shadowing, fast fading, (e) heavy shadowing, slow fading, (f) heavy shadowing, fast fading.

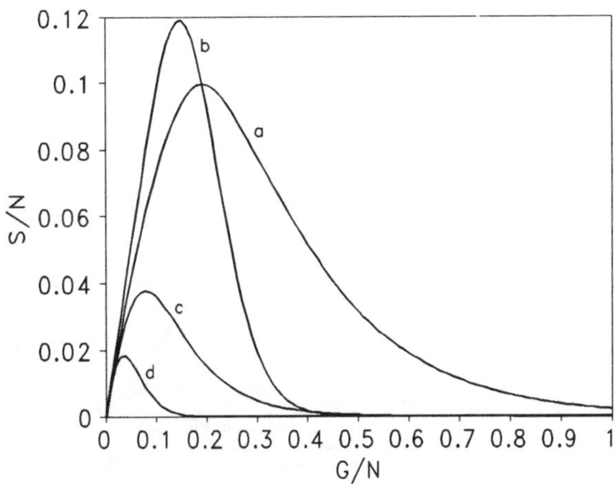

Figure 7.24 Normalized throughput of slotted CDMA for heavy shadowing, using selection diversity: (a) FEC, slow fading, (b) FEC, fast fading, (c) no FEC, slow fading, (d) no FEC, fast fading.

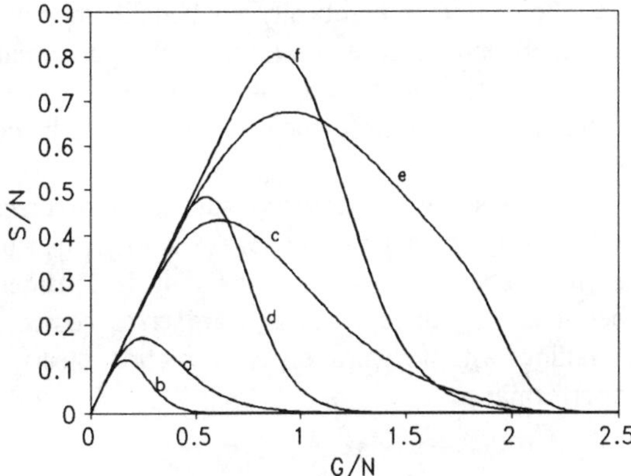

Figure 7.25 Normalized throughput of slotted CDMA for average shadowing, using MRC and FEC, with the number of correctable bits as a parameter: (a) t=0, slow fading, (b) t=0, fast fading, (c) t=10, slow fading, (d) t=10, fast fading, (e) t=20, slow fading, (f) t=20, fast fading.

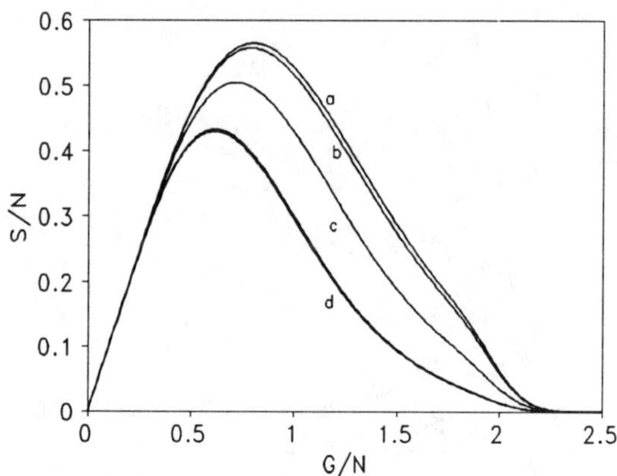

Figure 7.26 Normalized throughput of slotted CDMA for average shadowing, slow fading, using MRC and FEC, with T_S/T_C as a parameter:
(a) $T_S/T_C=10^4$, (b) $T_S/T_C=10^3$, (c) $T_S/T_C=10^2$, (d) $T_S/T_C=10$ and 1.

Figure 7.26 demonstrates the influence of the ratio T_S/T_C on the performance of CDMA. We can see that the best performance is obtained for a high T_S/T_C ratio. When T_S/T_C is of the order of 1 or less, then there is practically no benefit anymore of using spread-spectrum to reduce multipath interference. As a result, the performance converges to a certain lower bound. When T_S/T_C is increased, the performance increases up to a certain upper bound, where the multipath interference in the first path, containing the LOS signal, becomes negligible.

Figures 7.27 and 7.28 show the delay versus the normalized throughput. The results were obtained for values of N_{at}=3 and T_d=74 slots. It can be seen that for the narrowband, the difference between slow and fast fading is considerable without error correction, while it becomes negligible when forward error correction is used. In the case of CDMA, fast fading with forward error correction clearly provides the best throughput and delay performance.

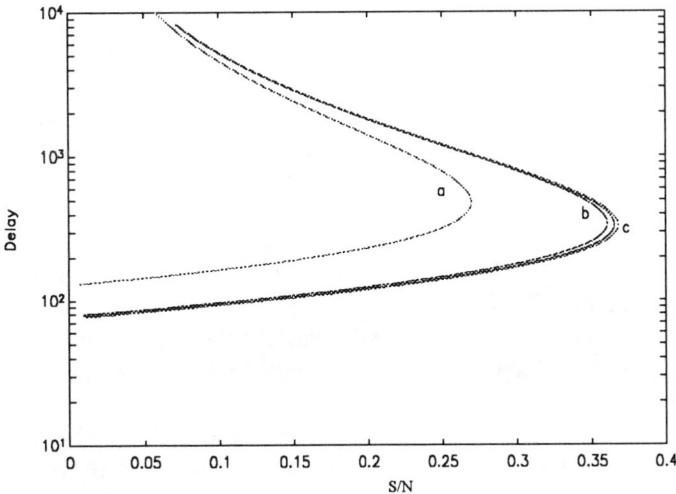

Figure 7.27 Normalized throughput-delay curves of narrowband slotted ALOHA for average shadowing: (a) slow fading, no FEC, (b) fast fading, no FEC, (c) slow and fast fading, FEC.

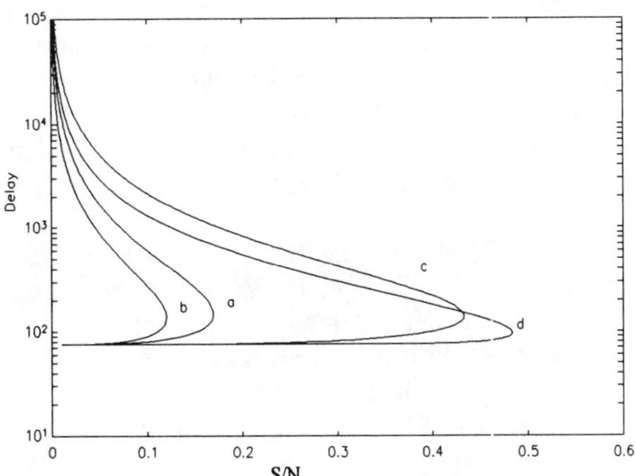

Figure 7.28 Normalized throughput-delay curves of slotted CDMA for average shadowing, using MRC: (a) slow fading, no FEC, (b) fast fading, no FEC, (c) slow fading, FEC, (d) fast fading, FEC.

7.5 CONCLUSIONS

In this chapter we investigated the performance of a land-mobile CDMA satellite system in terms of the bit error probability using coherent BPSK modulation. First a system without diversity was considered. A spread-spectrum system yields better results than narrowband transmission if the LOS path is dominant. In the case of heavy shadowing (i.e., where the LOS signal power is smaller than the multipath power), diversity techniques can be added to improve the performance.

Two diversity techniques, selection diversity and maximal ratio combining, were considered in the second part of the analysis presented in this chapter. For the case of light shadowing, path diversity improves the system performance by a small value. But there is a significant enhancement in the performance in the event of average and heavy shadowing. As expected, maximal ratio combining offers a higher performance than selection diversity.

Computational results further show that an increase in the number of users or a decrease of the code length for a fixed value of the chip time decrease the performance of the system. However, enhanced performance is obtained by decreasing the chip time for a fixed value of the code length. Thus there are two different ways to enhance the performance. First, the chip time can be decreased and, second, the order of diversity can be increased. The choice between these two depends on bandwidth limitations and hardware considerations. A decreased chip time requires a larger bandwidth and faster hardware. A higher order of diversity requires more hardware at the same speed. The path diversity techniques seem to be very useful to improve the performance but they cannot combat the shadowing.

The performance of slotted ALOHA and direct-sequence CDMA for packet switched communications was evaluated for the land-mobile satellite channel in terms of throughput and delay with BPSK modulation for light, average, and heavy shadowing. To evaluate the throughput and delay, the packet success probability was derived for the narrowband with FEC and for the spread spectrum with FEC and two types of path diversity (selection diversity and maximal ratio combining) considering slow and fast fading.

The performance of CDMA with and without diversity techniques can generally be expressed as a function of T_s/T_c, N/K, E_b/N_0 and the error correcting capability. Numerical results for the throughput and delay were presented for various combinations of the mentioned parameters. Without forward error correction coding, narrowband slotted ALOHA is found to give a better performance than CDMA. However, if error correction coding is applied, CDMA clearly outperforms narrowband slotted ALOHA. The use of forward error correction is far more beneficial for CDMA than the use of path diversity techniques. Maximal ratio combining is very beneficial in the case of heavy shadowing. Slow fading performs better than fast fading; however, with FEC fast fading has a higher maximum throughput than slow fading, except for heavy shadowing, where the LOS power is smaller than the multipath power. Further, the throughput of slow fading does not decrease as fast for higher values of the offered load than it does in the case of fast fading.

REFERENCES

[1] J.H. Lodge, "Mobile satellite communications systems: toward global personal communications," *IEEE Comm. Mag.*, Vol. 29, pp. 24-30, November 1991.

[2] J.D. Kiesling, "Land mobile sattelite systems," *Proc. IEEE*, Vol. 78, pp. 1107-115, July 1990.

[3] K.G. Johanssen, "Code division multiple access versus frequency division multiple access channel capacity in mobile satellite communication," *IEEE Trans. Vehicular Technol.*, Vol. 39, pp. 17-26, February 1990.

[4] K.S. Gilhousen, I.M. Jacobs, R. Padovani, A.J. Viterbi, L.A. Weaver and C.E. Wheatly, "Increased capacity using CDMA for mobile satellite communications," *IEEE J. Selected Areas Comm.*, Vol. 8, pp. 503-514, May 1990.

[5] C. Loo, "A statistical model for a land mobile satellite link," *IEEE Trans. Vehicular Technol.*, Vol. VT-34, pp. 122-127, August 1985.

[6] C. Loo, "Measurements and models of a land mobile satellite channel and their applications MSK signals," *IEEE Trans. on Vehicular Technol.*, Vol. VT-36, pp. 114-121, August 1987.

[7] C. Loo, "Digital transmission through a land mobile satellite channel," *IEEE Trans. Comm.*, Vol. 38, pp. 693-697, May 1990.

[8] J. van Rees, "Measurements of the wide band radio channel characteristics for rural, residential and suburban areas," *IEEE Trans. on Vehicular Technol.*, Vol. VT-36, pp. 2-6, February 1987.

[9] P. Beckman, "*Probability in Communication Engineering,*" Harcourt, Brace and World, New York, 1967.

[10] R.D.J. van Nee, H.S. Misser and R. Prasad, "Direct sequence spread spectrum in a shadowed Rician fading and land-mobile satellite channel," *IEEE J. Selected Areas Comm.*, Vol. 10, pp. 350-357, February 1992.

[11] R.D.J. van Nee and R. Prasad, "Spread sprectrum path diversity in a shadowed Rician fading land-mobile satellite channel," *IEEE Trans. Vehicular Technol.,* Vol. 42, pp. 131-135, May 1993.

[12] R. Prasad, R.D.J. van Nee and R.N. van Wolfswinkel, "Performance analysis of multiple access technique for land-mobile satellite communication," *Proceedings GLOBECOM'94*, San Francisco, pp. 740-744, November/December 1994.

[13] R.D.J. van Nee, R.N. van Wolfswinkel and R. Prasad, "Slotted ALOHA and code division multiple access techniques for land-mobile satellite personal communications," *IEEE J. Selected Areas Comm.*, Vol. 13, pp. 382-388, February 1995.

[14] L.W. Fenton, "The sum of log-normal probability distributions in scatter transmission systems," IRE *Trans. Comm. Syst.*, Vol. CS -8, pp. 57-67, March 1960.

[15] M.B. Pursley, "Performance evaluation for phase-coded spread spectrum multiple access communication part I: system analysis," *IEEE Trans. Comm.*, Vol. COM-25, pp. 795-799, August 1977.

[16] W.C.Y. Lee, "*Mobile Communications Design Fundamentals,*" H.W. Sams & Co, Indianapolis, 1986.

[17] R.C. Dixon, "*Spread Spectrum Systems,* " John Wiley & Sons, New York, 1976.

[18] W. Yue, "The effect of capture on performance of multichannel slotted ALOHA systems," *IEEE Trans. Comm.*, Vol. 39, pp. 818-822, June 1991.

Chapter 8
Hybrid Direct-Sequence/Slow Frequency Hopping CDMA Systems

8.1 INTRODUCTION

This chapter presents the derivation of the bit error probability of a hybrid direct-sequence (DS)/slow frequency hopping (SFH) CDMA system in an indoor Rician fading channel using BPSK, QPSK, and DPSK modulation.

The effect of selection diversity and maximal ratio combining on the performance is also investigated. The three modulation techniques are compared. A comparison between pure direct-sequence systems with hybrid systems is also presented by evaluating throughput and delay.

Hybrid systems [1-10] are attractive because they can combine the advantages of both direct-sequence and frequency hopping systems while avoiding some of their disadvantages. A hybrid system can combine the anti-multipath effectiveness of DS systems with the good antipartial-band-jamming and the good antinear-far problem features of FH systems. Hybrid systems may also use shorter signature sequences and hopping patterns, thus reducing the overall acquisition time. A disadvantage of hybrid systems is the increased complexity of their transmitters and receivers. There are two ways of looking at the hybrid DS/SFH systems:

1. From the viewpoint of direct-sequence, the hybrid system diminishes the near-far problem and increases the system capacity in terms of maximum number of users.

2. From the viewpoint of slow frequency hopping, the hybrid system creates a form of coded data redundancy in order to obtain better performance when confronted with multiuser interference.

This chapter is organized as follows. Section 8.2 describes the transmitter model for DPSK, BPSK, and QPSK modulation. The channel model is discussed in Section 8.3. Section 8.4 presents the DPSK, BPSK, and QPSK receiver model. The bit error probability for DPSK modulation is derived in Section 8.5. Computational results of the bit error probability and the outage probability are presented in this section. The effect of FEC coding using numerical results on the performance is discussed. Section 8.6 derives the bit error probability for BPSK and QPSK modulation. All three modulation techniques are compared using computational results. Computational results of the throughput and delay are discussed in Section 8.7 and conclusions are given in Section 8.8.

8.2 TRANSMITTER MODEL

Transmitter models of a hybrid DS/SFH system using DPSK, BPSK, and QPSK modulation schemes are discussed in this section.

8.2.1 DPSK Modulation

In Figure 8.1 the transmitter model of a hybrid DS/SFH system with DPSK modulation is shown.

In the case of channel coding, the channel encoder and the interleaver have to precede the DPSK modulator in the diagram of Figure 8.1. Each user produces a data waveform given by:

$$b_k(t) = \sum_j b_k^j P_T(t - jT) \tag{8.1}$$

where b_k^j is the jth data bit of user k and belongs to the set $\{0,1\}$. Furthermore $P_T(t)$ is a rectangular NRZ pulse of unit height and duration T. The signal $b_k(t)$ is first multiplied by a spreading sequence $a_k(t)$, which accomplishes DS:

$$a_k(t) = \sum_j a_k^j P_{T_c}(t - jT_c) \tag{8.2}$$

where the jth pulse of $a_k(t)$ belongs to the set of $\{-1,1\}$. Here T_c is the chip duration and $P_{tc}(t)$ a rectangle NRZ pulse of unit height and duration T_c. This sequence is periodical with period T.

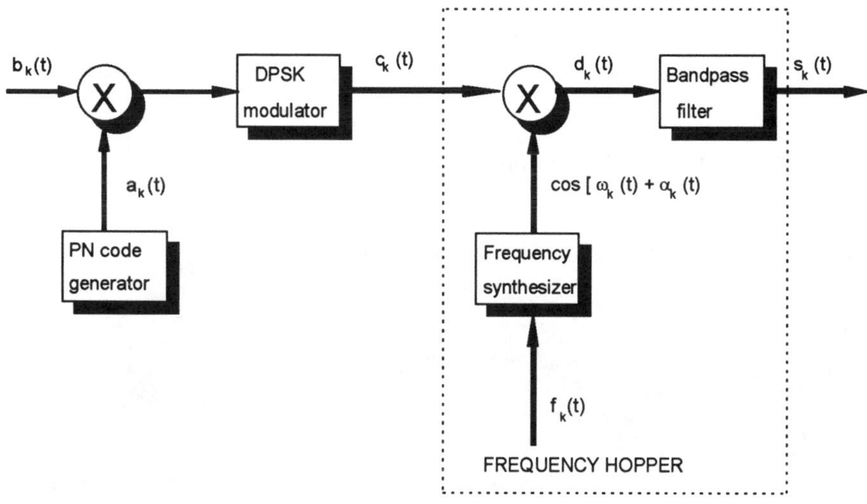

Figure 8.1 Transmitter model of a hybrid DS/SFH system.

This signal $a_k(t)b_k(t)$ is DPSK modulated and can be written as:

$$c_k(t) = a_k(t)b_k(t)\cos(\omega_c t + \theta_k) \tag{8.3}$$

where θ_k is a random phase associated with user k. As a consequence of the DPSK modulation b'_k now belongs to the set of $\{-1,1\}$:

$$d_k(t) = a_k(t)b_k(t)\cos(\omega_c t + \theta_k)\cos[\omega_k(t)t + \alpha_k(t)] \tag{8.4}$$

The frequency $f_k(t)$ and the phase $\alpha_k(t)$ are constant over a time interval of duration T_h. The bandpass filter in the frequency hopper, shown in Figure 8.1, removes unwanted frequency components present at the output of the multiplier. So the transmitted signal becomes:

$$s_k(t) = \sqrt{2P}\, a_k(t)b_k(t)\cos[\omega_c t + \omega_k(t)t + \theta_k + \alpha_k(t)] \tag{8.5}$$

where P is the transmitted power. The quantity $\alpha_k(t)$ represents the phase shift introduced by the frequency hopper when it switches from one frequency to another. We assume $\alpha_k(t)$ to be constant during the time intervals for which $f_k(t)$ is constant.

8.2.2 BPSK and QPSK Modulation

The sequences of data bits with a rate $2/T_b$ produced by user k is separated in two sequences of binary data, denoted by $b_{c,k}(t)$ and $b_{s,k}(t)$, respectively, each having a rate $1/T_b$.

$$
\begin{aligned}
b_{c,k} &= \sum_j b_{c,k}^j P_{T_b}\left(t - jT_b\right) \\
b_{s,k} &= \sum_j b_{s,k} P_{T_b}\left(t - jT_b\right)
\end{aligned}
\tag{8.6}
$$

where $b_{x,k}^j$ belongs to the set $\{-1,1\}$ and $P_{T_b}(t)$ is a rectangular pulse of duration T_b. These signals are first multiplied by the spreading sequences $a_{c,k}(t)$ and $a_{s,k}(t)$, which are defined the same way as in (8.3). These sequences are periodical with period T_b. The signals $a_{c,k}(t)b_{c,k}(t)$ and $a_{s,k}(t)b_{s,k}(t)$ are PSK modulated on two carriers in quadrature.

This modulated signal is then frequency hopped, according to the hopping pattern associated with user k. After appropriate filtering, this hopped signal for QPSK becomes:

$$
\begin{aligned}
s_k^Q(t) &= \sqrt{2P}a_{c,k}(t)b_{c,k}(t)\cos\left[\omega_c t + \omega_k(t)t + \alpha_k + \theta_k\right] \\
&+ \sqrt{2p}a_{s,k}(t)b_{s,k}(t)\sin\left[\omega_c t + \omega_k(t)t + \alpha_k + \theta_k\right]
\end{aligned}
\tag{8.7}
$$

where the phases are the same as those defined in section 8.2.1. The hopped signal for BPSK can be obtained by setting $b_{s,k}(t)$ in (8.7) to 0.

8.3 CHANNEL MODEL

The link between the kth user and the base station is characterized by a lowpass equivalent transfer function given by:

$$
h_k(t) = \sum_{l=1}^{L} \beta_{kl}\delta(t - \tau_{kl})\exp(j\gamma_{kl})
\tag{8.8}
$$

where kl refers to path l of user k. We assume there are L paths associated with each user. The lth path of the kth user is characterized by three random variables: the gain β_{kl}, delay τ_{kl}, and the phase γ_{kl}. In this study we make the following five assumptions concerning the channel:

1. Path gain β_{kl}, the phase γ_{kl}, and delays τ_{kl} of the different paths are statistically independent for different values of k and l;
2. The path gain β_{kl} is Rician distributed;
3. The phase factor γ is uniformly distributed over $[0,2\pi]$ and the path delay τ_{kl} is uniformly distributed in $[0,T_b]$;
4. The delay spread T_{max} is less than the bit duration T_b in order to avoid intersymbol interference;
5. The channel introduces additive white Gaussian noise $n(t)$, with two-sided power spectral density $N_0/2$.

The Rician probability density function is given by:

$$P_\beta(r) = \frac{r}{\sigma_r^2}\exp\left[-\frac{r^2+s^2}{2\sigma_r^2}\right] I_0\left(\frac{sr}{\sigma_r^2}\right) r \geq 0, s \geq 0 \qquad (8.8a)$$

Here $I_0(\cdot)$ is the modified Bessel function of the first kind and zeroth order, s is the peak value of the specular radio signal, and σ^2 is the average power of the scattered signal. The second central moment of the Rician distribution is given by $E(r^2) = 2\sigma_r^2 + s^2$. The average power of a bandpass signal with a Rician distributed envelope is given by $P = \frac{1}{2}E(r^2) = \frac{1}{2}s^2 + \sigma_r^2$, where $s^2/2$ represents the power of the dominant component and σ_r^2 is the power of the scattered signal. The ratio of the dominant received power to the scattered power is called the Rice factor, which is given by $R = s^2/2\sigma_r^2$.

8.4 RECEIVER MODEL

In this model we assume that the DPSK modulated signal is detected noncoherently and the BPSK and QPSK modulated signals are detected coherently. Because of the stable line of sight (LOS) component, phase recovery at the receiver can be achieved, which makes coherent detection possible.

8.4.1 DPSK Receiver

In the case of DPSK we consider a noncoherent receiver. This means that the receiver uses the phase information concealed in two consecutive bits. The receiver consists of a dehopper, a bandpass matched filter, a DPSK demodulator, and a hard-decision device. Besides, when channel coding is used, a deinterleaver and a hard-decision decoder are added to the receiver. Figure 8.2 shows the receiver structure.

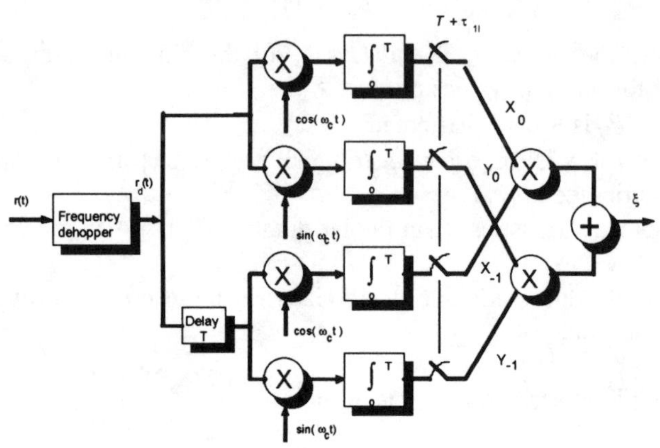

Figure 8.2 Frequency dehopper and DPSK demodulator.

The received signal can be written as:

$$r(t) = \sqrt{2P}\,\sum_{k}\sum_{l}\beta_{kl}a_k(t-\tau_{kl})b_k(t-\tau_{kl})\cos\left[\omega_c t + \omega_k(t-\tau_{kl})t + \phi_{kl}\right] + n(t) \tag{8.9}$$

with

$$\phi_{kl} = \alpha_{kl}(t-\tau_{kl}) + \theta_{kl} - \left[\omega_c + \omega_k(t-\tau_{kl})\right]\tau_{kl} + \gamma_{kl} \tag{8.10}$$

and n(t) is white Gaussian noise with two-sided power spectral density $N_0/2$ [W/Hz]. The received signal r(t) enters a bandpass filter, which removes out-of-band noise. The mixer of the dehopper performs the appropriate frequency translation.

We assume that the hopping pattern of the receiver is synchronized with the hopping pattern associated with the jth path of user 1 (which will be denoted as the reference path). The dehopper introduces a phase $\beta_l(t)$, which is constant over the hopping interval T_h. A bandpass filter following the mixer removes high-frequency terms. The resulting dehopper output signal is given by:

$$r_d(t) = \sqrt{P/2}\sum_k\sum_l \beta_{kl}\, a_k(t-\tau_{kl})b_k(t-\tau_{kl})\delta\left[f_l(t-\tau_{lj}), f_k(t-\tau_{kl})\right] \\ .\cos\left[\omega_c t + \omega_k(t-\tau_{kl})t + \phi_{kl}\right] + n_d(t) \tag{8.11}$$

where $n_d(t)$ can be considered as white Gaussian noise with two-sided power spectral density $N_0/8$. The delta function in this expression is defined as follows

$$\delta(u,v) = \begin{matrix} 1\ \text{for}\ u = v \\ 0\ \text{for}\ u \neq v \end{matrix} \tag{8.12}$$

in which both u and v are real. Note that the dehopper suppresses, at any instant t, all path signals whose frequency at instant t differs from $f_l(t-\tau_{lj})$. Evidently, the reference path signal is not suppressed; the other path signals from the reference user are suppressed only during a part of the first or last bit of a hop, depending on the relative delay of the considered path with respect to the reference path. Path signals from users different from the reference user contribute to the dehopper output only during those time intervals for which their frequency accidentally equals the frequency of the reference path signal.

We can express $r_d(t)$ by means of its quadrature components:

$$r_d(t) = x(t)\cos(\omega_c t) - y(t)\sin(\omega_c t) \tag{8.13}$$

with

$$x(t) = \sqrt{P/2}\sum_k\sum_l \beta_{kl} a_k(t-\tau_{kl})b_k(t-\tau_{kl})\delta\left[f_l(t-\tau_{lj}), f_k(t-\tau_{kl})\right] \\ .\cos\left[\phi_{kl} + \beta_l(t)\right] + n_c(t) \tag{8.14}$$

$$= \sqrt{P/2}\sum_k\sum_l \beta_{kl}\alpha_k(t-\tau_{kl})b_k(t-\tau_{kl})\delta\left[f_1(t-\tau_{lj}), f_k(t-\tau_{kl})\right]\cos(\varphi_{kl}) + n_c(t) \tag{8.15}$$

$$y(t) = \sqrt{P/2}\sum_k\sum_l \beta_{kl}\ a_k(t-\tau_{kl})b_k(t-\tau_{kl})\delta\left[f_l(t-\tau_{lj}), f_k(t-\tau_{kl})\right] \\ .\sin\left[\phi_{kl} + \beta_l(t)\right] + n_s(t) \tag{8.16}$$

$$= \sqrt{P/2}\sum_{k}\sum_{l}\beta_{kl}\ a_k(t-\tau_{kl})b_k(t-\tau_{kl})\delta\left[f_l(t-\tau_{lj}), f_k(t-\tau_{kl})\right]\sin(\psi_{kl}) + n_s(t) \quad (8.17)$$

The dehopper output signal $r_d(t)$ enters a DPSK demodulator, whose impulse response is matched to a T_b-seconds segment of $a_l(t)\cos(\omega_c t)$.

The bandpass signal at the matched filter output contains narrow peaks of width $2T_c$, at instants $\lambda T_b + \tau_{lj}$ ($\lambda = 0, \pm 1, \pm 2, \ldots$), which are caused by the reference path signal. The other $L-1$ path signal from the reference user gives rise to peaks at instants $\lambda T_b + \tau_{1l}$ ($l \neq j$). However, there is no peak during the first or last bit of a hop, because each of these $L-1$ path signals is partly suppressed by the dehopper. Path signals from the nonreference users also give rise to peaks; however, they are usually smaller than the peaks caused by the reference user. These peaks of the nonreference user are caused by the fact that the spreading codes are not ideal in the sense that they are correlated to some extent.

The DPSK demodulator output is a lowpass signal, which has peaks at instants $\lambda T_b + \tau_{1l}$ as we already have seen. A positive peak indicates that the corresponding channel encoder output bit is likely to be a logical one and a logical zero when the peak is negative.

We assume that the delay differences between path signals from the same user are larger than $2T_c$, so these peaks do not overlap; this assumption is justified when the PN sequence period N is relatively large. The DPSK demodulator output signal is sampled at the instants $\lambda T_b + \tau_{11}$, and this gives rise to the decision variables $\xi_l(\lambda)$.

8.4.2 DPSK Demodulator Output

In DPSK the message bits are coded with two consecutive code bits. If a binary one is to be sent during a bit interval, it is sent as a signal with the same phase as the previous bit. If a binary zero is sent, it is transmitted with the opposite phase of the previous bit. So the demodulator has to correlate two consecutive signal elements and create the decision variable ξ_l, which can be written as:

$$\xi = \mathrm{Re}\left[V_o V_{-1}^*\right] \quad (8.18)$$

with

$$\begin{aligned} V_o &= X_o + jY_o \\ V_{-1} &= X_{-1} + jY_{-1} \end{aligned} \quad (8.19)$$

in which V_o denotes the complex envelope at the current sampling instant at the output of the demodulator and the same for V_{-1} at the previous sampling instant. The real and imaginary parts of the complex envelope denote the output of the in-phase integrator and the output of the quadrature-branch integrator, respectively. The outputs of the in-phase and the quadrature branches are given by:

$$X_O = \sqrt{P/8}\sum_k\sum_l \beta_{kl}\cos(\psi_{kl})\int_{\lambda T_b}^{(\lambda+1)T_b} a_1(t)a_k(t-\tau_{kl})b_k(t-\tau_{kl}) \cdot\delta\left[f_1(t-\tau_{1j}),f_k(t-\tau_{kl})\right]dt+\eta \tag{8.20}$$

$$Y_o = \sqrt{P/8}\sum_k\sum_l \beta_{kl}\sin(\psi_{kl})\int_{\lambda T_b}^{(\lambda+1)T_b} a_1(t)a_k(t-\tau_{kl})b_k(t-\tau_{kl}) .\delta\left[f_1(t-\tau_{1j}),f_k(t-\tau_{kl})\right]dt+\nu \tag{8.21}$$

$$\eta = \int_{\lambda T_b}^{(\lambda+1)T_b} a_1(s)n_c ds \qquad \nu = \int_{\lambda T_b}^{(\lambda+1)T_b} a_1(s)n_s ds \tag{8.22}$$

where ν and η are zero-mean Gaussian random variables with equal variances given by $N_0T_b/16$. The bit under consideration is bit number $\lambda = jN_b + p$. Actually, the phase Ψ is time dependent, but it has a fixed value for the time interval under consideration, because the δ function will only be equal to one if $f_k(t-\tau_{kl}) = f_1(t-\tau_{1j})$, which means a fixed value for f and α.

Because of the limited interval for the delays, two bits will be under consideration: bits 0 and -1 of the sequence k (or equivalently λ and $\lambda-1$, respectively) as shown in figure 8.3.

So, in order to perform the cross-correlations of the bits of the nonreference user with the bits of the reference user, we have to split up the time interval $[0,T_b]$ in two parts, namely, $[0,\tau] \cup [\tau,T_b]$ as shown in Figure 8.3. We define:

$$R_{1k}(\tau_{kl}) = \int_0^{\tau_{kl}} a_1(t)a_k(t-\tau_{kl})dt \tag{8.23}$$

$$\hat{R}_{1k}(\tau_{kl}) = \int_{\tau_{kl}}^{T} a_1(t) a_k(t - \tau_{kl}) dt \tag{8.24}$$

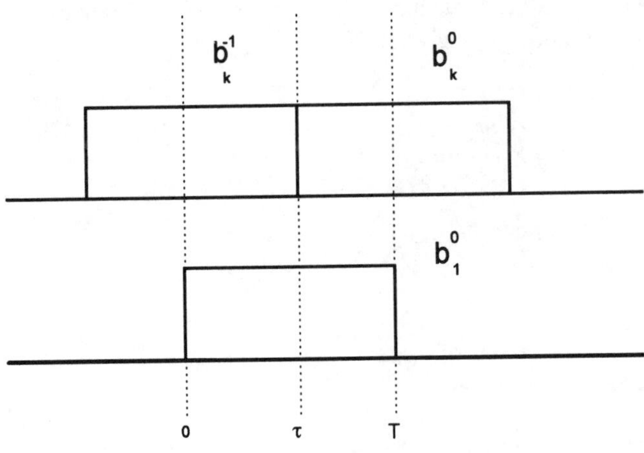

Figure 8.3 Time interval with the bits under consideration.

In these definitions, the periodicity of the spreading sequence has been taken into account. As a consequence, we can write:

$$X_o = \sqrt{P/8} \sum_k \sum_l \beta_{kl} \cos(\psi_{kl}) \left[d_1(b_k^{-1}) R_{1k}(\tau_{kl}) + d_2(b_k^0) b_k^0 \hat{R}_{1k}(\tau_{kl}) \right] + \nu \tag{8.25}$$

$$Y_o = \sqrt{P/8} \sum_k \sum_l \beta_{kl} \sin(\psi_{kl}) \left[d_1(b_k^{-1}) R_{1k}(\tau_{kl}) + d_2(b_k^0) b_k^0 \hat{R}_{1k}(\tau_{kl}) \right] + \eta \tag{8.26}$$

where b_k^{-1} and b_k^0 are the bits with numbers $\lambda - 1$ and λ of the kth user. The functions d_1 and d_2 account for the possible hits between the current frequency of user 1 and the frequencies associated with the two bits under consideration. X_{-1} and Y_{-1} are given by equations similar to (8.25) and (8.26) with λ becoming λ–1 (i.e., 0 becomes –1 and –1 becomes –2).

We define:

$$X_1 = \sum_{l, l \neq j} \beta_{1l} \cos(\psi_{1l}) R_{11}(\tau_{1l}) \tag{8.27}$$

$$X_k = \sum_l \beta_{kl} \cos(\psi_{kl}) \, R_{1k}(\tau_{kl}) \tag{8.28}$$

$$\hat{X}_1 = \sum_{l, l \neq j} \beta_{1l} \cos(\psi_{1l}) \hat{R}_{11}(\tau_{1l}) \tag{8.29}$$

$$\hat{X}_k = \sum_l \beta_{kl} \cos(\psi_{kl}) \hat{R}_{1k}(\tau_{kl}) \tag{8.30}$$

and similar expressions for Y. Without loss of generality, we can assume that $\tau_{lj} = 0$, $\Psi_{lj} = 0$, because we have considered the jth path between transmitter 1 and the receiver as the reference path. So, we can write:

$$X_o = \sqrt{P/8}\,\hat{R}_{11}(0)\beta_{1j} b_1^{\lambda} + \sqrt{P/8} \sum_k \left[d_1(b_k^{-1}) \, b_k^{-1} X_k + d_2(b_k^0) \, b_k^0 \hat{X}_k \right] + \nu \tag{8.31}$$

$$Y_o = \sqrt{P/8} \sum_k \left[d_1(b_k^{-1}) \, b_k^{-1} \, Y_k + d_2(b_k^0) \, b_k^0 \hat{Y}_k \right] + \eta \tag{8.32}$$

in which we define m as:

$$m = \sqrt{P/8}\, \hat{R}_{11}(0)\, \beta_{1j} \;\, b_1^{\lambda} \tag{8.33}$$

The complex envelope, which is considered to be a random variable, consists of the following four contributions:

1. A useful term, which is due to the jth path signal of the reference user;
2. A complex Gaussian noise term, of which the real and imaginary parts are statistically independent and have the same variance, $N_0T_b/16$;
3. A multipath interference term, which is due to the $L-1$ other path signals from the reference user. The contributions from the different paths are uncorrelated;
4. A multiple access interference term, which is due to the path signals of the $K-1$ nonreference users. Here also the contributions of the different paths are uncorrelated.

The equations derived in this section are the starting point for the derivation of the expression for the bit error probability for both selection diversity and maximal ratio combining.

8.4.3 BPSK and QPSK Receivers

We assume that a received signal is composed of the contributions of the different users, their different paths, and additive white Gaussian noise. Therefore, we can write the expression for the received signal for QPSK as:

$$\begin{aligned} r(t) = \sqrt{2P}\sum_k\sum_l \beta_{kl}a_{a,c}(t-\tau_{kl})b_{c,k}(t-\tau_{kl})\cos[\omega_c t+\omega_k(\tau)t+a_k+\theta_k+\tau_{kl}] \\ +\sqrt{2P}\sum_k\sum_l \beta_{kl}\alpha_{s,k}(t-\tau_{kl})b_{s,k}(t-\tau_{kl})\sin[\omega_c t+\omega_k(\tau)t+\alpha_k+\theta_k+\tau_{kl}]+n(t) \end{aligned} \tag{8.34}$$

where $n(t)$ is AWGN with single-sided power spectral density N_0 [W/Hz]. Since we consider user 1 as the reference user, the dehopping sequence is the one associated with user 1. After dehopping the in-phase and quadrature components, the signal is multiplied by the DS code associated with user 1, and multiplied by the carrier signal and integrated over the bit duration.

The bit under consideration is bit number λ. However, without loss of generality, we can assume that λ=0 and that the jth path between transmitter 1 and the receiver is the reference. Because of the dominant and stable component, the receiver can be time synchronous with this path. Besides, the receiver is able to acquire the phase of the stable component of this path, due to its tracking mechanism. Therefore, we say that τ_{1j}=0 and Ψ_{1j}=0, in which Ψ is the overall phase shift. All delays are then defined as relative delays according to this reference path.

8.4.4 BPSK and QPSK Demodulator Output

Let us focus on the in-phase signal. The output of the in-phase integrator is given by:

$$\begin{aligned} z_{o,c} = \sqrt{P/8}\sum_k\sum_l \beta_{kl}\cos(\psi_{kl})\int_{\lambda T_b}^{(\lambda+1)T_b} a_{c,l}(t)a_{c,k}(t-\tau_{kl})b_{c,k}(t-\tau_{kl})\delta[f_l(t),f_k(t-\tau_{kl})]dt \\ +\sqrt{P/8}\sum_k\sum_l \beta_{kl}\sin(\psi_{kl})\int_{\lambda T_b}^{(\lambda+1)T_b} a_{c,l}(t)a_{c,k}(t-\tau_{kl})b_{c,k}(t-\tau_{kl})\delta[f_l(t),f_k(t-\tau_{kl})]dt+v_c \end{aligned} \tag{8.35}$$

where v_c is a zero-mean Gaussian random variable with variance $N_oT_b/16$. The δ-function accounts for the possible hits: a hit occurs when the frequency $f_1(t)$ used by the reference user and the frequency $f_k(t-\tau_{kl})$ used by nonreference users are the same.

Because of the limited interval of the delays, two bits will be under consideration, namely, bits 0 and −1 of sequence k. We define the following correlation functions:

$$R_{1k}^{xy}(\tau_{kl}) = \int_0^{\tau_H} a_{x,1}(t)a_{y,k}(t-\tau_{kl})dt \tag{8.36}$$

$$\hat{R}_{1k}^{xy}(\tau_{kl}) = \int_{\tau_H}^{T} \alpha_{x,1}(t)a_{y,k}(t-\tau_{kl})dt \tag{8.37}$$

in which x and y may be both c or s. In these definitions, the periodicity has been taken into account. We define the following parameters:

$$X_k^{xy} = \sum_l \beta_{kl}\, f[\psi_{kl}]R_{1k}^{xy}(\tau_{kl}) \tag{8.38}$$

$$\hat{X}_k^{xy} = \sum_l \beta_{kl}\, f[\psi_{kl}]\hat{R}_{1k}^{xy}(\tau_{kl}) \tag{8.39}$$

with x and y defined as previously. When $x = y$, then the function f is the cosine function; otherwise it is the sine function. If $k = 1$, then $l \neq j$, so we have:

$$\begin{aligned} z_{o,c} &= \sqrt{P/8}T_b\beta_{1j}b_{c,1}^{o} \\ &+\sqrt{P/8}\sum_k\left[d_1(b_{c,k}^{-1})b_{bc,k}^{-1}X_k^{cc}+d_2(b_{c,k}^{o})b_{c,k}^{o}\hat{X}_k^{cc}+d_1(b_{s,k}^{-1})b_{s,k}^{-1}X_k^{cs}+d_2(b_{s,k}^{o})b_{s,k}^{0}\hat{X}_k^{cs}\right]+v \end{aligned} \tag{8.40}$$

where $b_{c,k}^{-1}$ $(b_{s,k}^{-1})$ and $b_{c,k}^{0}$ $(b_{s,k}^{o})$ are the current bit and the previous bit of in-phase (quadrature) data stream of the kth user. The functions d_1 and d_2 account for the possible hit between the current frequency of user 1 and the frequencies of the two bits under consideration.

We see that cross-rail interference occurs, in the sense that the bits associated with the in-phase (quadrature) component interfere with the quadrature (in-phase) component. In the case of BPSK we have:

$$z_o = \sqrt{P/8}T_b\,\beta_{1j}b_1^o + \sqrt{P/8}\,\sum_k\left[d_1(b_k^{-1})b_k^{-1}X_k + d_2(b_k^o)b_k^o\hat{X}_k\right] + \nu \tag{8.41}$$

in which X_k and $\hat{X}_k$ are defined as follows:

$$X_k = \sum_l \beta_{kl}\cos(\psi_{kl})\,R_{1k}(\tau_{kl}) \tag{8.42}$$

$$\hat{X}_k = \sum_l \beta_{kl}\cos(\psi_{kl})\hat{R}_{1k}(\tau_{kl}) \tag{8.43}$$

As in the case for QPSK, we have that $l \neq j$ if $k = 1$, according to (8.27) and (8.29). We see that the cross-rail term has disappeared in (8.41).

With the help of all these equations we are able to derive the expressions for the average bit error probability for both selection diversity and maximal ratio combining. This is done in the next section.

8.5 BIT ERROR PROBABILITY (BEP)

Three different main phenomena contribute to errors in the system under consideration.

First, even in the absence of noise and fading, errors may occur when a signal is hopped to a frequency slot that is occupied by another signal. Whenever two different signals occupy one frequency slot simultaneously, we say that a *hit* occurs.

Second, the different spread-spectrum codes assigned to the different users show mutual cross-correlation effects. This is because the codes are not ideal and of finite length. These cross-correlation effects result in side pulses in the detector, which might give rise to errors in the detection of the bits.

Third, even in the absence of the multiple access interference effects described above, errors may occur due to fading and additive white noise.

According to [1-4], we first decouple the effects of hits from other users due to frequency hopping from the multiple access interference due to the direct-sequence spread-spectrum signals. In [1-4] this is done by first evaluating the conditional error probability, given the number of hits, and then averaging with respect to the distribution of the hits.

In our model we evaluate the conditional probability not only given a number of hits, but also given the path gain β and the delay τ. Then we first average with respect to the number of hits and next with respect to the path gains and the delays. We can actually say that given a number of hits from other users, the hybrid system is almost equivalent to a DS-SSMA system with noncoherent reception.

The conditional bit error probability, which is computed as the mean of several situations, corresponding to the possible hit situations produced by the multiple access interferers, is then given by:

$$P_e(\beta,\tau_{kl}) = \sum_{n_i=0}^{K-1} P_e(\,|n_i,\beta,\tau_{kl})P(n_i) \tag{8.44}$$

where $P_e(n_i,\beta,\tau_{kl})$ is the conditional bit error probability in the absence of coding, assuming there are n_i active interferers out of $K - 1$ users. An expression for this probability is derived in the next section. Probability $P(n_i)$ is the probability of having n_i active interferers out of $K - 1$ users. For random hopping patterns with a number of frequencies equal to q, the probability that any two users use the same frequency is given by $1/q$. So, $P(n_i)$ is given by:

$$P(n_i) = \binom{K-1}{n_i}\left(\frac{1}{q}\right)^{n_i}\left(1-\frac{1}{q}\right)^{K-1-n_i} \tag{8.45}$$

which is a binomial distribution. To use this approximation we have to meet the following requirements:

1. All users yield the same average power at the receiver (a symmetrical system);
2. All $K - 1$ nonreference users have the same path power, so that the multi-access interference power only depends on the number of hits;
3. All path signals of the reference user have the same path power.

When these key requirements are met and the PN sequence period N is sufficiently large, the multipath and multiple access interference terms consist of a large number of statistically independent contributions with the same distribution. This means that each interference term can be well approximated by a Gaussian random variable.

The influence of the multipath term is actually dependent on the bit position in the hopping interval. For the first bit in a block, there is no hit with certitude with the previous bit, because they do not necessarily have the same frequency. For all other bits in the blocks, there will be a hit with certitude with the previous and current bit conveyed by interfering paths. We neglect the effect of this first bit, which is a reasonable assumption, because N_b is relatively large in the case of slow frequency hopping.

8.5.1 DPSK Modulation with Selection Diversity

For selection diversity of order M, the decision variable $\xi_{SD}(\lambda)$ is the maximum of M random variables $\xi_l(\lambda)$. Since the data bits are equiprobable, the bit error probability can be written as:

$$P\left[\xi_{max}^{M} < 0 | b_l^0\, b_l^{-1} = 1\right] \tag{8.46}$$

If we assume multi-channel reception in a time-invariant Rician fading channel and DPSK modulation for fixed delays, phase angles, bits and assuming the maximum of the envelope has been found (all other signals are seen as noise), we can use the bit error probability given by [11]:

$$P_e(n_i, \beta_{max}, \tau_{kl}) = Q(a,b) - \frac{1}{2}\left(1 + \frac{\mu}{\sqrt{\mu_o \mu_{-1}}}\right)\exp\left(-\frac{a^2 + b^2}{2}\right) \cdot I_0(ab) \tag{8.47}$$

where Q is the Marcum Q-function, which is defined as:

$$Q(a,b) = \int_b^\infty x \exp\left(-\frac{a^2 + x^2}{2}\right) I_o(ax) \tag{8.48}$$

and where I_0 is the modified Bessel function of the first kind and zeroth order, which is defined as:

$$I_0(ab) = \frac{1}{2\pi} \int_o^{2\pi} \exp(a\, b \cos\theta) d\theta \tag{8.49}$$

The parameters a, b, μ, μ_{-1}, and μ_o are given next:

$$a = \frac{m}{\sqrt{2}} \left| \frac{1}{\sqrt{\mu_o}} - \frac{1}{\sqrt{\mu_{-1}}} \right| \tag{8.50}$$

$$b = \frac{m}{\sqrt{2}} \left| \frac{1}{\sqrt{\mu_o}} + \frac{1}{\sqrt{\mu_{-1}}} \right| \tag{8.51}$$

$$\mu_o = \mathrm{var}\left(V_o \middle| L, \tau_{kl}, n_i, b\right) \tag{8.52}$$

$$\mu_{-1} = \mathrm{var}\left(V_{-1} \middle| L, \tau_{kl}, n_i, b\right) \tag{8.53}$$

$$\mu = \mathrm{cov}\left(V_o, V_{-1} \middle| L, \tau_{kl}, n_i, b\right) \tag{8.54}$$

After this, the bit error probability given in (8.44) has to be averaged for all possible values of β_{max} and for the delays.

We now have to weight this result by considering β_{max}, the Rician fading statistics, i.e., the pdf of the maximal path gain. For an order of diversity equal to M and assuming, of course, that all path gains are equally distributed independent Rician variables, this pdf becomes:

$$f_{\beta_{max}}(\beta_{max}) = M \left[\int_0^{\beta_{max}} \frac{z}{\sigma_r^2} \exp\left(-\frac{s^2+z^2}{2\sigma_r^2}\right) I_0\left(\frac{sz}{\sigma_r^2}\right) dz \right]^{(M-1)} \frac{\beta_{max}}{\sigma_r^2} \exp\left(-\frac{s^2+\beta_{max}^2}{2\sigma_r^2}\right) I_0\left(\frac{s\beta_{max}}{\sigma_r^2}\right) \tag{8.55}$$

It is through this pdf that the influence of the order of diversity can be understood. In Figure 8.4 the influence of the order of diversity M on the pdf of the maximal path gain is shown.

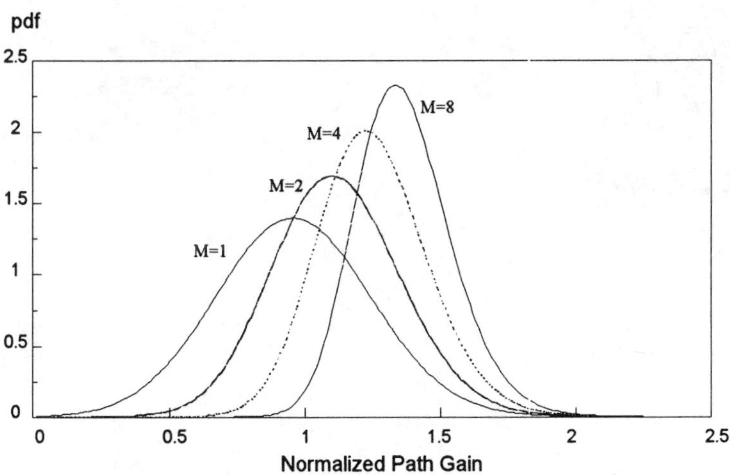

Figure 8.4 Influence of the order of diversity on the pdf of the maximal path gain.

The increase of the order of diversity M results in a narrower pdf, which results in a higher probability of determination and realization of the maximal path gain.

The bit error probability only conditioned on the delays τ_{kl} with selection diversity is given by:

$$P_e = \int_{\mu}\int_{\mu-1}\int_{\mu_o}\int_{\beta_{max}} P_e(\beta_{max}, \mu_o, \mu_{-1}, \mu) f_{\mu-1} f(\mu_{-1})_{\mu_o} f_\mu(\mu) d\beta_{max} d\mu_o d\mu_{-1} d\mu \tag{8.56}$$

It has been mentioned by many authors that integration over all path delays τ_{kl} is very time consuming. Therefore an approximation has to be made. We know that if the number of users K and the number of paths L is sufficiently large, the central limit theorem can be used. As a result of this, the conditional variances μ_o , μ_{-1}, and μ are assumed to have a Gaussian distribution. The parameters belonging to the Gaussian distribution are the mean and the variance of the stochastic variable under consideration. The conditioning on the delays can now be removed by weighting the conditional bit error probability by the appropriate Gaussian probability density functions. Therefore, we finally get an expression for the average bit error probability:

$$P_e = \int_{\beta_{max}} \int_{\tau_{kl}} P_e(\beta_{max}, \tau_{kl}) f_{\beta_{max}}(\beta_{max}) f_{\tau_k}(\tau_k) d\beta_{max} d\tau_{kl} \tag{8.57}$$

in which $f_{\beta max}(\beta_{max})$ is the pdf given in (8.55) and $f_{\mu o}(\mu_o)$ is the Gaussian pdf. This integral can be simplified by considering that μ appears explicitly in the function P_e [see equation (8.47)]. We also know that:

$$\int_{-\infty}^{\infty} \mu f(\mu) d\mu = E(\mu) \tag{8.58}$$

In the next section we discuss the performance of the hybrid DS/SFH system with maximal ratio combining.

8.5.2 DPSK Modulation with Maximal Ratio Combining

As we noted earlier, the decision variable in maximal ratio combining of order M is given by $\xi_{mrc}(\lambda)$, which is the average of M random variables ξ_l. To be able to give an expression for the bit error probability for the case of maximal ratio combining, we have to assume that the multiple access and the multipath interference are Gaussian. This means that the term

$$\sqrt{P/8}\left[\sum_k \left(d_1(b_k^{-1})b_k^{-1}X_k + d_2(b_k^0)b_k^0\hat{X}_k\right) + j\left(d_1(b_k^{-1})b_k^{-1}Y_k + d_2(b_k^0)b_k^0\hat{Y}_k\right)\right] \tag{8.59}$$

in V_0 is complex Gaussian and also for the similar parts of V_{-1}. [See (8.31) and (8.32)].

We have to keep in mind that this assumption is a simplification of the mathematical description of the bit error probability in the case of maximal ratio combining. However, this assumption enables us to derive some closed expressions for the bit error probability and to begin to understand the behavior of the system. We consider channels in which the path gains have identical average power and in which the path delays are uniformly distributed in $[0, T_b]$. The decision variable is given by:

$$\xi_{mrc} = \frac{1}{2} \sum_{i=1}^{M} \left(V_{0,i} V_{-1,i}^{*} + V_{0,i}^{*} V_{-1,i}\right) \tag{8.60}$$

If the interference is assumed Gaussian, the sum of the interference and the AWGN is also Gaussian. These total noise terms will be denoted N_1 for V_o and N_2 for V_{-1}, respectively. So the decision variable then becomes:

$$\xi_{mrc} = \mathrm{R_e}\left[\sum_{i=1}^{M} \left(AT\beta_i b_1^o + N_{1i} \right)\left(AT\beta_i b_1^{-1} + N_{2i} \right)^* \right] \tag{8.61}$$

where β_i denotes the gain and N_{1i} and N_{2i} the Gaussian random variables, associated with the ith path. We now have to make two assumptions about the noise variables.

First, we assume that N_{1i} and N_{2i} are independent for each i. In our case this is justified, because calculations have shown that μ is small in comparison with μ_o and μ_{-1}. This implies that the cov[N_{1i},N_i] is negligible compared with var[N_{1i}] and var[N_{2i}].

Second, we assume that the pair (N_{1i},N_{2i}) is independent of the pair (N_{1j},N_{2j}) for $i \neq j$. Mathematically, this is not correct, because the delays $\{\tau_{kl}\}_i$ and $\{\tau_{kl}\}_j$ are not independent. However, there are two reasons to make the assumption physically reasonable.

First, each set of delays is taken with reference to a different time origin (corresponding to the arrival time of the signal on the corresponding combined path). Second, we know that any two resolved paths (i,j) are separated by at least one chip time period.

The computation of the total noise term can be obtained from the computation of μ_o by taking the expectation over the path delays. We denote this expectation by $E_\tau(\mu_o)$.

For path k, we have a signal power given by $P\beta_k^2$ and a total noise power given by $N_T = E\tau(\mu_o)$. Therefore, the signal-to-noise ratio (SNR) of resoluted path k is given by:

$$(\mathrm{SNR})_K = \frac{P\beta_k^2}{E_\tau(\mu_o)} \tag{8.62}$$

With this information we are able to give an expression for the bit error probability for the case of MRC.

If we assume that the values of the path gains are known, we can use the bit error probability given in [10] for maximal ratio combining and diversity of order M:

$$P_{e,mrc}(\gamma_b, n_i) = \frac{1}{2^{(2M-1)}} \exp(-\gamma_b) \sum_{k=0}^{M-1} p_k \gamma_b^k \tag{8.63}$$

with

$$p_k = \frac{1}{k!} \sum_{n=0}^{M-1-k} \binom{2M-1}{n} \tag{8.64}$$

$$\gamma_b = \frac{E}{N_T} \sum_{k=1}^{M} \beta_K^2 \tag{8.65}$$

in which N_T represents the total noise power. To remove the conditioning on the interferers we have to substitute (8.63) into (8.44). We then obtain the probability, which is only conditioned on γ_b. To find the average bit error probability, we only have to weight this conditional probability with the probability density function of γ_b, which is given in [11] as:

$$f_{\gamma_b}(\gamma_b) = \frac{1}{2E/N_T} \left[\frac{\gamma_b}{(E/N_T)s_M^2} \right]^{\frac{(M-1)}{2}} \exp\left[-\frac{\left(s_M^2 + \gamma_b N_T/E\right)}{2} \right] I_{M-1}\left(\frac{s_M}{\sigma_r^2} \sqrt{\frac{\gamma_b}{E/N_T}} \right) \tag{8.66}$$

with $s_M^2 = Ms^2$ and where I_{M-1} is the $(M-1)^{\text{th}}$-order modified Bessel function. We finally compute the bit error probability by:

$$P_{e,mrc} = \int_{\gamma_b} P_{e,mrc}(\gamma_b)\, f_{\gamma_b}(\gamma_b) d\gamma_b \tag{8.67}$$

Another measure for the performance is the outage probability. In the next section we discuss the outage probability for the hybrid DS/SFH system with DPSK modulation.

In Figure 8.5 the average BEP is presented for selection diversity (SD) and maximal ratio combining (MRC) with the order of diversity M as parameter. From Figure 8.5 we can see that the performance of MRC is better than the performance of SD; however, for a low order of diversity the difference in performance is less than for higher orders of diversity. The asymptotes of the curves are due to the multiuser interference; this limitation would disappear if the multipath interference were zero, the SNR were infinite, and if the number of frequencies in the hopping pattern were infinite.

In Figure 8.6 the effect of different parameter settings on the average BEP for the case of MRC is presented with the constraint of a fixed bandwidth. This means that NqR_b is kept constant.

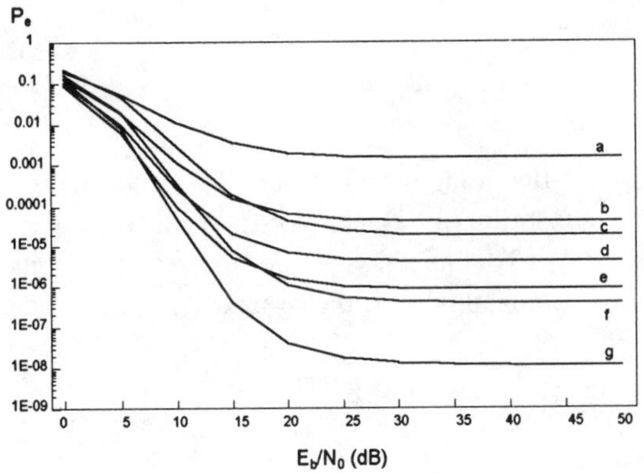

Figure 8.5 BEP for DS/SFH with SD and MRC with M as parameter; q=10; L=5; R_b= 64 kbit/s; K=15; N=255; R=6.8 dB:
(a) nondiversity (b) M=2 with SD (c) M=2 with MRC
(d) M=3 with SD (e) M=4 with SD (f) M=3 with MRC
(g) M=4 with MRC.

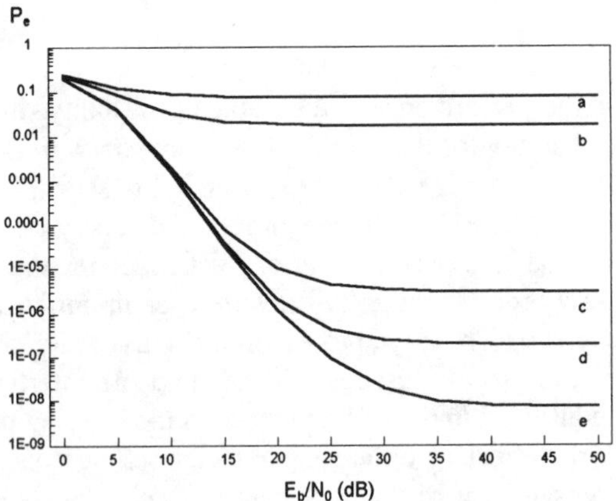

Figure 8.6 BEP of DS/SFH with MRC for different parameter settings given a fixed bandwidth; M=2; T_m=250 ns; R=6.8 dB:
(a) N=127; q=21; R_b=144 kbit/s; L=5 (b) N=255; q=10; R_b=144 kbit/s; L=10
(c) N=255; q=24; R_b=64 kbit/s; L=5 (d) N=127; q=98; R_b=32 kbit/s; L=2
(e) N=255; q=49; R_b=32 kbit/s; L=3

We see that a bit rate of 144 kbit/s yields a relatively poor performance. It is also obvious that code period of N=127 yields a worse performance than with the code period of N=255. We can explain these results because the number of resolvable paths is influenced by the rms delay spread T_m, the bit rate R_b, and the period of the spreading codes N. An increase of the data rate increases the number of resolvable paths for the same spreading code period. The increase of the number of resolvable paths yields a higher level of multipath interference and this causes a relatively high level of multipath interference.

Secondly, from the different combinations of constant N and q at fixed bit rate, we see that a doubling of the spreading code period yields a better performance than a doubling in the number of frequencies.

Third, the value of the maximum delay spread is of importance as well. A decrease of this parameter decreased the number of resolvable paths. Accordingly, this results in an increase in performance, because of the diminishing level of multipath interference.

In Table 8.1 we present a comparison of hybrid DS/SFH and DS for selection diversity (SD) for two bit rates and for two values of the order of diversity M. The transmission bandwidth B_T, which is proportional to $R_b N_q$, is taken as a parameter. The results for DS are obtained from Chapter 5.

Table 8.1

BEP comparison of DS and hybrid DS/SFH with SD for two bit rates, two values of M, and the bandwidth as parameter; E_b/N_0=40 dB; T_{max}=250 ns; R=6.8 dB; K=15

M	R_b (kbit/s)	DS N=255	Hybr. DS/SFH q=2 N=127	Hybr. DS/SFH q=10 N=127	Hybr. S/SFH q=5 N=255
4	32	3.2×10^{-5}	2.0×10^{-5}	4.8×10^{-8}	2.9×10^{-11}
4	64	6.0×10^{-4}	3.1×10^{-2}	1.0×10^{-3}	1.1×10^{-5}
8	32	6.0×10^{-6}	9.0×10^{-7}	2.2×10^{-10}	1.6×10^{-12}
8	64	3.0×10^{-5}	1.8×10^{-2}	3.4×10^{-4}	8.2×10^{-7}

We see that under the constraint of the same transmission bandwidth the performance of hybrid DS/SFH is slightly better than DS for the bit rate of R_b=32 kbit/s and for both orders of diversity. For the bit rate of R_b=64 kbit/s and under the same constraint, the performance favors DS, for both orders of diversity. A higher bit rate means a higher number of resolvable paths, and this means a higher level of multipath interference. Apparently, for a low number of resolvable paths, which is the case for the lowest bit rate, hybrid DS/SFH with a number of frequencies q=2 is able to deal with the worse correlation properties of spreading codes with period N=127. For a higher number of resolvable paths, the number of frequencies of q=2 is insufficient to overcome the effect of increased multipath interference and the effect of the low spreading code period, for both orders of diversity.

The situation for which the transmission bandwidth is increased by a factor of 5 for the hybrid system, through the combinations N=127 and q=10 or N=255 and q=5, shows something different. In case of N=127 and q=10 the performance of the hybrid system for the highest bit rate is better than DS, and for the highest bit rate it is worse than DS for both orders of diversity. It is obvious that even a number of frequencies as high as q=10 is not able to overcome the effect of the spreading codes and the multipath effects, due to the increased number of resolvable paths in the case of the highest bit rate.

The combination of N=255 and q=5 shows that the performance for both bit rates favors the hybrid system for both orders of diversity. Besides, this combination yields a better performance than the previous combination of N and q, due to the better correlation properties of the spreading code with period N=255.

8.5.3 Outage Probability for DPSK Modulation

The outage probability is defined as the probability that the instantaneous bit error probability exceeds a preset threshold. We denote the threshold value as ber_0. The instantaneous value of the bit error probability can be obtained from (8.57). The averaging over β_{max} should be removed and a fixed value for β_{max}, β, should be substituted. Equation (8.57) then becomes

$$P_e(\beta)=\int_{-\infty}^{\infty}\int_{-\infty}^{\infty} P_e(\beta|\mu_o,\mu_{-1})f_{\mu_o}(\mu_o)f_{\mu_{-1}}(\mu_{-1})d\mu_o d\mu_{-1}=\text{ber}(\beta) \tag{8.68}$$

The outage probability (in case of selection diversity) can then be calculated as follows:

$$P_{out}=P(0\le\beta_{\max}\le\beta_0)=P[\text{ber}(\beta)\ge\text{ber}_0]=\int_0^{\beta_0} f_{\beta_{\max}}(\beta_{\max})d\beta_{\max} \tag{8.69}$$

in which β_o is the value of β at which the instantaneous bit error probability is equal to ber_0. The integrand is just the PDF of β_{max} given in (8.55).

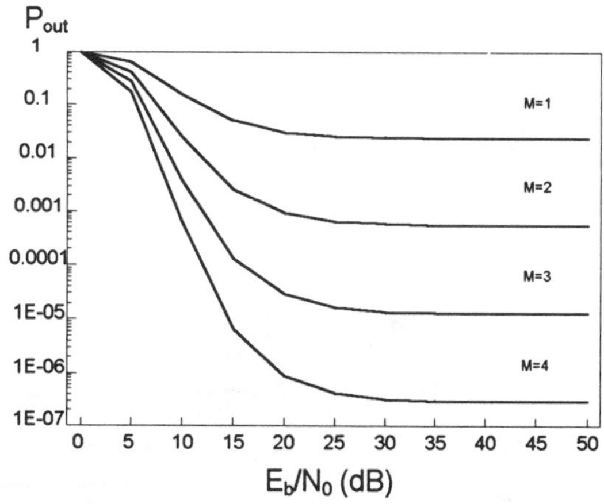

Figure 8.7 Outage probability for the SD case for a fixed bandwidth with M as parameter; N=255; q=10; R=6.8 dB; L=5; R_c=64 kbit/s; T_{max}=250 ns; K=15:
(a) M=1 (b) M=2 (c) M=3 (d) M=4.

We consider the outage probability as a function of the signal to white noise ratio (ratio of the energy per bit and the density of the AWGN). The threshold value is taken to be 0.01 in our case. In Figure 8.7 the effect of the order of diversity on the outage probability for the SD case is presented.

It is obvious that an increase in the order of diversity decreases the outage probability considerably. This is due to the fact that we have considered Figure 8.5, which shows that a higher order of diversity yields less spread in the PDF of the maximum path gain β_{max}.

In Figure 8.8 we present the effect of different parameter settings on the outage probability, with the constraint of a fixed bandwidth. We see from these plots that relative high bit rates yield very poor performance in terms of outage probability. As with the bit error probability, this is due either to the worse correlation properties of codes with N=127 or to the relatively large number of resolvable paths at high bit rates. As discussed before, an increase in the number of resolvable paths causes an increase in the multipath interference level.

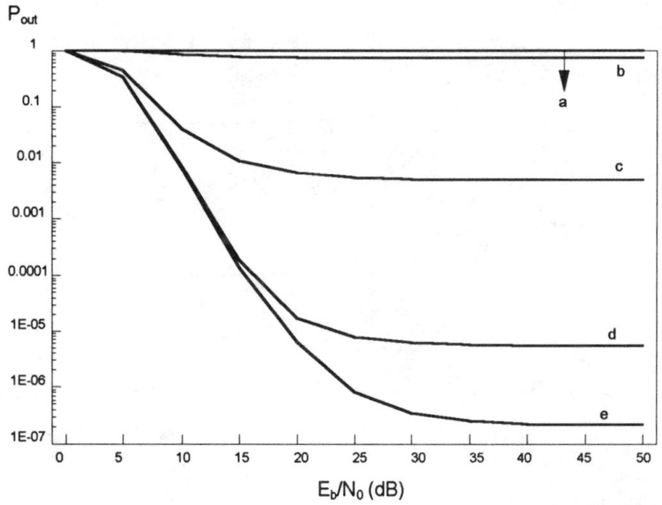

Figure 8.8 Outage probability for the SD case with different parameter settings given a fixed bandwidth; M=2; R=6.8 dB:
(a) N=127; q=21; R_b=144 kbit/s; L=5
(b) N=255; q=10; R_b=144 kbit/s; L=10
(c) N=255; q=24; R_b=64 kbit/s; L=5
(d) N=127; q=98; R_b=32 kbit/s; L=2
(e) N=255; q=49; R_b=32 kbit/s; L=3.

Under the constraint of a fixed bandwidth, only the combination of relatively low bit rates and a sufficiently large spreading code period yields a reasonable performance. If the constraint of the fixed bandwidth is left, performance can be enhanced by using a sufficiently large number of frequencies in combination with large spreading code periods.

8.5.4 FEC Coding

In Figure 8.9 the effect of FEC coding on the performance of the hybrid system is presented, again with the constraint of a fixed bandwidth and as a function of the ratio of source bit energy and the spectral density of the AWGN.

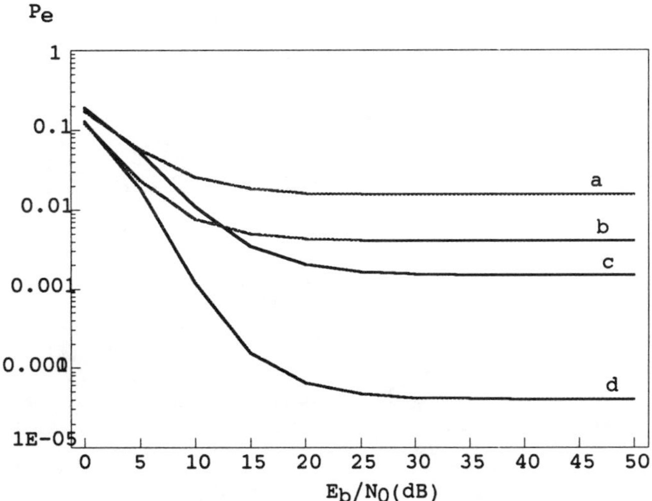

Figure 8.9 BER of hybrid DS/SFH in case of SD with and without FEC coding at fixed bandwidth; q=10, R=6.8 dB; K=15; T_{max}=250 ns; R_c=64 kbit/s:
(a) M=1, Golay code, N=127, L=6 (b) M=2, Golay code, N=127, L=6
(c) M=1, no coding, N=255, L=5 (d) M=2, no coding, N=255, L=5.

The code used is the (23,12) Golay code, which transforms a block of 12 bits into a block of 23 bits. This code has a code rate r_c equal to 12/23 (=0,52). This means that for the constraint of a fixed bandwidth, we can compare the following two cases:

A: no coding, N=255 and q=10;
coding, N=127 and q=10;

B: no coding, N=255 and q=20;
coding, N=255 and q=10

We see that the performance with coding worsens performance. This is due to the value of the spreading code period of N=127, which has worse correlation properties than N=255, together with a larger number of resolvable paths, which increases the multipath interference and, hence, too many errors occur. And since the error correction code is able to correct a limited number of errors, performance is not enhanced. Even a selection diversity of order 2 together with FEC coding does not give a better performance than no diversity without coding.

In Figure 8.10 we consider the bandwidth fixed, but now we change the number of frequencies in the hopping pattern instead of the spreading code sequence.

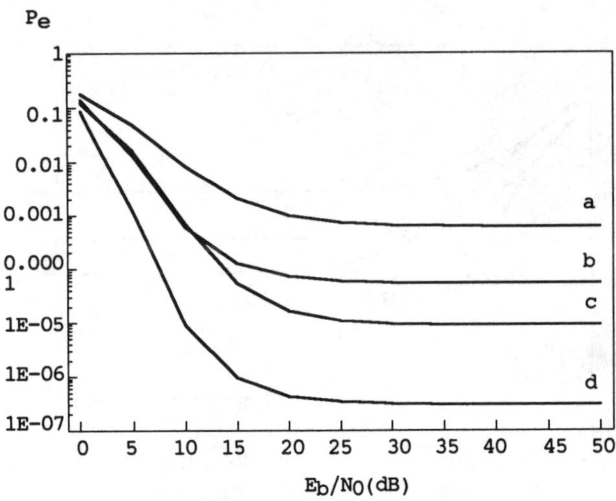

Figure 8.10 BER of hybrid DS/SFH in case of SD with and without FEC coding at fixed bandwidth; N=255, R=6.8 dB; K=15; T_{max}=250 ns and R_c=64 kbit/s:
(a) M=1, no coding, q=20, L=5; (b) M=1, Golay code, q=10, L=10
(c) M=2; no coding, q=20, L=5 (d) M=2, Golay code, q=10, L=10.

From Figure 8.10 we see that the use of FEC coding enhances the performance. Although the bandwidth is twice that of Figure 8.9, it is obvious that with a constant spreading code period N=255, a number of frequencies q=10, and Golay coding the performance is better than for the same situation without coding.

Despite the relatively high number of resolvable paths in the case of coding, the sufficiently long spreading code, the large number of frequencies, and the error capabilities of the FEC code make it possible to enhance performance.

8.6 BEP FOR BPSK AND QPSK MODULATION

The multipath term of the in-phase signal of the QPSK receiver, denoted as $I_{c,mp}$, is given by:

$$I_{c,mp} = \sqrt{P/8}\left[b_{c,1}^{-1} X_1^{cc} + b_{c,1}^{o} \hat{X}_1^{cc} + b_{s,1}^{-1} X_1^{cs} + b_{s,1} \hat{X}_1^{cs}\right] \tag{8.70}$$

For BPSK the multipath term, denoted as I_{mp}, is given by:

$$I_{mp} = \sqrt{P/8}\left[b_1^{-1} X_1 + b_1^o \hat{X}_1\right] \tag{8.71}$$

The multiple access noise, denoted as $I_{c,ma}$ corresponding to the assumption of n_i active interferers for QPSK, is given by:

$$I_{c,ma} = \sqrt{P/8}\sum_{k=2}^{n_i+1}\left[b_{c,k}^{-1} X_k^{cc} + b_{c,k}^{o} \hat{X}_k^{cs} + b_{s,k}^{-1} X_k^{cs} + b_{s,k} \hat{X}_k^{cs}\right] \tag{8.72}$$

and for BPSK this term is given by:

$$I_{ma} = \sqrt{P/8}\sum_{k=2}^{n_i+1}\left[b_1^{-1} X_1 + b_1^o \hat{X}_1\right] \tag{8.73}$$

The bit error probability, assuming n_i interferers, is defined as the probability that the decision variable is lower than 0 assuming a 1 was transmitted (or conversely). For QPSK we then get:

$$P_e^Q = P\left[z_{o,c} < 0 \mid b_{c,1}^o = 1\right] \tag{8.74}$$

and for BPSK we have:

$$P_e^B = P\left[z_o < 0 \mid b_1^o = 1\right] \tag{8.75}$$

Here $z_{o,c}$ and z_o are given by (8.40) and (8.41), respectively. To compute the power of the multipath and multiple access terms for BPSK and QPSK exactly, we have to make some comments.

We assume that we deal with a given channel, which means that the absolute value of the delay range is known. The bit duration of BPSK was assumed to be T_b and the delay was assumed to be uniformly distributed over this bit duration. Accordingly, for the case of QPSK the bit duration is twice that of BPSK. Therefore, the delay for QPSK has to be considered as being distributed uniformly over half of the QPSK bit duration. Considering this and according to an approach proposed in [3], the powers of the multipath and multiple access terms are very well approximated by:

$$P_{MP} = \frac{P}{8}(L-1)\left(\sigma_r^2 + \frac{s^2}{2}\right)\frac{2T_b^2}{3N} \tag{8.76}$$

and

$$P_{MA} = \frac{P}{8} n_i \, L \left(\sigma_r^2 + \frac{s^2}{2} \right) \frac{2T_b^2}{3N} \tag{8.77}$$

for both QPSK and BPSK.

The bit error probability can be computed as a function of the bit energy to total noise power spectral density assuming the value of β_{1j}. The bit error probability has to be averaged over the distribution of β_{1j}. If we assume that the total noise is Gaussian with zero mean, we can use the bit error probability given by:

$$P_e = \frac{1}{2} \text{erfc}\left(\sqrt{\frac{E}{N_T}} \right) \tag{8.78}$$

We define the following normalized variable:

$$\alpha_{1j} = \frac{\beta_{1j}}{\left(2\sigma_r^2 + s^2 \right)^{0,5}} \tag{8.79}$$

We can finally write the bit error probability, given a number of active interferers, as:

$$P_e(n_i) = \frac{1}{2} \text{erfc}\left\{ \left[\left(\frac{E_b(2\sigma_r^2 + s^2)}{N_o} \right)^{-1} \frac{1}{\alpha_{1j}^2} + \frac{2L}{3N\alpha_{1j}^2}(1 + n_i) \right]^{-0.5} \right\} \tag{8.80}$$

The bit error probabilities are given as functions of the mean received bit energy to noise ratio. It is assumed that the multipath interference associated with the reference user is due to L paths instead of $L - 1$.

8.6.1 Selection Diversity

Selection diversity means that the largest of a group of M signals, carrying the same information, is selected. The order of diversity is equal to the number of resolvable paths times the number of antennas. The decision variable is denoted by z_{max}, which is the largest among the set of M values. To obtain the average bit error probability, we have to average the probability in (8.80) with the pdf of the maximum path gain β_{max}, instead of β. A change of variables gives:

$$\alpha = \frac{\beta}{\left(s^2 + 2\sigma_r^2\right)^{0,5}} \tag{8.81}$$

and

$$y = \frac{z}{\left(s^2 + 2\sigma_r^2\right)^{0,5}} \tag{8.82}$$

For the pdf of the maximum path gain we then find:

$$f_{a_{max}}(a) = M\left\{\int_0^a 2y(1+R)\exp\left[-R-(1+R)y^2\right]I_0\left[2\sqrt{R(1+R)}y\right]dy\right\}^{(M-1)} \\ 2a(1+R)\exp\left[-a^2(1+R)-R\right]I_0\left[2\sqrt{R(1+R)}a\right] \tag{8.83}$$

The average bit error probability is given as a function of the mean bit energy to white noise ratio at the receiver.

8.6.2 Maximal Ratio Combining

For maximal ratio combining, the contributions of several resolved paths are added together. The combiner that achieves the best performance is the one in which each matched filter output is multiplied by the corresponding complex-valued channel gain. This complex gain compensates for the phase shift and introduces a weighting associated with the signal strength. The realization of such a receiver is based on the assumption that the channel attenuations and phase shifts are perfectly known.

Considering the decision variable z_o for MRC, we have:

$$z_o = \mathrm{R_e}\left[\sqrt{P/8}\; T_b \sum_{i=1}^{M} \beta_{1i}^2 + \sum_{i=1}^{M} \beta_{1i} N_{1i}\right] \tag{8.84}$$

Each term involved in the combination process is corrupted by AWGN and the multiuser and the multipath interference. We make the assumption that the noise due to the multiuser and the multipath interference is also Gaussian and that all noise terms can be added. We also assume that the noise terms affecting two different paths are independent. On close examination this is not correct; however, for the same reason as discussed in Section 8.5.2, this assumption can be used.

Using the bit error probability given in [11], we have:

$$P_e = \frac{1}{2}\,\text{erfc}\left(\sqrt{\gamma_b}\right) \tag{8.85}$$

where the SNR per bit is given by:

$$\gamma_b = \frac{T_b^2\,E_b}{16\,N_T}\sum_{m=1}^{M}\beta_{1m}^2 \tag{8.86}$$

We need to average on the pdf of $t = \sum_{m=1}^{M}\beta_{1m}^2$. The average of this random variable is given by $M\left(2\sigma_r + s^2\right)$, which also represents the average gain at the receiver. We will perform a normalization of t with respect to the average value. The normalized variable v is then given by:

$$v = \sum_{m=1}^{M}\frac{\beta_{1m}^2}{M\left(2\sigma_r^2 + s^2\right)} \tag{8.87}$$

The pdf of v is then given by:

$$f(v) = M\left[\,1+R\right]\left[\,v\left(1+\frac{1}{R}\right)\right]^{(M-1)/2}\exp\left[-MR - Mv(1+R)\right]I_{m-1}\left(\,2M\sqrt{(vR(1+R)}\,\right) \tag{8.88}$$

The following bit error probability, averaged over this pdf, provides the average bit error probability as a function of the mean received bit energy to noise ratio:

$$P_e(n_i) = \frac{1}{2}\,\text{erfc}\left\{\left[\left(\frac{E_b(2\sigma_r^2 + S^2)}{N_o}\right)^{-1}\frac{1}{Mv} + \frac{2L}{3NMv}\,(1+n_i)\right]^{-0,5}\right\} \tag{8.89}$$

8.6.3 Comparison of QPSK, BPSK, and DPSK Modulation

In Section 8.5 we presented the numerical results of the performance of the hybrid DS/SFH system with DPSK as the modulation scheme. To get a good idea about the performance of DPSK compared to the performance of other modulation schemes, we have compared the BEP of DPSK with the BEP of BPSK and QPSK.

As far as the comparison of QPSK and BPSK is concerned, some constraints have to be taken into account. If we require the bit rate in both systems to be the same, we will have $T_q = 2T_b$, where T_q and T_b are the bit durations for QPSK and BPSK, respectively. In addition, we require the bandwidths to be the same, which means that the ratio N_q/T should be the same for both systems. This leads to the following requirement:

$$\frac{N_b\, q_b}{T_b} = \frac{N_q\, q_q}{T_q} \tag{8.90}$$

Considering the bit rate constraints, we have $N_q\, q_q = 2N_b\, q_b$. If we finally assume the same number of frequencies for both systems, we get $N_q = 2N_b$.

This has consequences for multipath and multiple access noise. Considering the expressions for these types of noise, given in (8.76) and (8.77), we see that the multi-access and multipath noise for BPSK is twice of that of QPSK. This means that, under the condition of constant bandwidth and bit rate, the performance of a QPSK system with a code length of 128 chips is equivalent to the performance of a BPSK system with a code length of 256. In this comparison, we have taken N_q=255 for QPSK, N_b=127 for BPSK, and N=255 for DPSK.

In Table 8.2 the BEP is presented for the three modulation techniques. We have compared the performance for q=10 and q=50, and M=1 and M=3 under the constraints of a fixed bandwidth and bit rate. Both SD and MRC have been considered.

Table 8.2

BEP comparison of three modulation schemes; E_b/N_o=25 dB; L=8; K=15

Diversity	M	Q	QPSK	BPSK	DPSK
SD	1	10	2.10^{-3}	6.10^{-3}	4.3×10^{-2}
SD	3	10	7.10^{-6}	2.10^{-4}	1.2×10^{-2}
SD	1	50	1.10^{-3}	3.10^{-3}	6.7×10^{-3}
SD	3	50	3.10^{-7}	8.10^{-6}	7.5×10^{-4}
MRC	1	10	2.10^{-3}	6.10^{-3}	4.3×10^{-2}
MRC	3	10	2.10^{-7}	6.10^{-6}	4.0×10^{-3}
MRC	1	50	1.10^{-3}	3.10^{-3}	6.9×10^{-4}
MRC	3	50	5.10^{-9}	3.10^{-7}	1.7×10^{-4}

From Table 8.2 we see that systems using DPSK perform poorer than systems using BPSK or QPSK. This is due to the fact that DPSK is assumed to be detected noncoherently at the receiver, as opposed to BPSK and QPSK, which are assumed to be detected coherently at the receiver. In the DPSK demodulator the noise variance is twice as large as for BPSK or QPSK. Besides, under the constraint of fixed bandwidth and data rate, the system with QPSK yields better performance than the one with BPSK. This is due to the fact that the multiuser and multipath interference of the BPSK system is twice the interference of the QPSK system, according to (8.76) and (8.77). This is the consequence of the condition of the signature sequences $N_q = 2N_b$, in order to have the same bandwidth. Without the bandwidth and bit rate constraint, BPSK and QPSK should give approximately the same performance.

8.6.4 Comparison of Two Hybrid DS/SFH Models

A comparison of the model used in this performance analysis with other models described in the literature is valuable. Therefore, we compare the performance of our model, denoted model 1, with the performance of another model described in [2], denoted model 2.

The model given in [2] also describes the performance of a hybrid DS/SFH system with DPSK modulation over Rician fading channels. However, in order to simplify the calculations in model 2, random spreading code sequences were used. This means that for each user k, the set of spreading codes (a) consists of a sequence of mutually independent random variables, taking values $\{+1, -1\}$ with equal probability. This means that the signature sequences, assigned to the different users, are mutually independent [3]. Particularly for the necessary averaging over the path delays, this provides some extreme simplifications.

The model presented here takes model 2 as a starting point, but does not use random signature sequences. Instead, deterministic spreading sequences are used, which requires some extended calculations with regard to the averaging over the path delays.

In Table 8.3 the bit error rates (BERs) for the two models are presented. The only difference in the comparison is that model 1 uses a spreading code period of 127 and model 2 uses a code period of 63.

Table 8.3

Comparison of the BERs of two models of hybrid DS/SFH with MRC: L=4; K=10; q=30; R=3 dB; E_b/N_0= 30 dB

Order of Diversity	Model 1 with N=127	Model 2 with N=63
M=1	1.1×10^{-2}	1×10^{-2}
M=2	1.4×10^{-3}	8×10^{-4}
M=3	2.3×10^{-4}	6×10^{-5}
M=4	4.4×10^{-5}	6×10^{-6}

From this table, we see that model 2 yields a lower BER than model 1, despite the shorter signature sequence. The difference between the bit error probability becomes smaller for higher orders of diversity. At first sight we would expect that the system with code length N=127 would yield a better result than the system with code length N=63. However, the lower BER obtained with model 2 is due the use of random spreading code sequences. The expectation of the correlation functions of these random sequences with respect to delay τ yields a different result than the expectation of the correlation functions of the deterministic signature sequences.

8.7 THROUGHPUT AND DELAY ANALYSIS

In this section we present the numerical results of the throughput and delay analysis of the CDMA network based on hybrid DS/SFH with selection diversity. The normalized throughput is given as a function of the offered traffic G. General expressions for the throughput and delay are given in Chapters 5 and 6. We have investigated the performance at a signal to white noise ratio of 20 dB. Further, it is assumed that the CDMA threshold is 30 active users. The indoor radio channel is assumed to be of the slow Rician fading type.

In Section 8.7.1 we consider the throughput for different parameter settings without FEC coding. Section 8.7.2 gives the effect of FEC coding on the throughput S_n, and Sections 8.7.3 and 8.7.4 give the delay results for a system without and a system with FEC coding, respectively. The results are evaluated only for DPSK modulation.

8.7.1 Throughput Performance Without Coding

In Figure 8.11 the influence of the order of diversity M for a packet size of 128 bits is presented. We see that the throughput S of the system increases with an increase in the order of diversity. This is due to the fact that with an increase of M the BER decreases. This results in a larger packet success probability.

Further, it is obvious that the maximum of the throughput shifts to the right as the order of diversity increases.

The effect of the order of diversity on the throughput with a packet size of 1024 bits is given in Figure 8.12. We see that the throughput still increases with an increase of the order of diversity, however, the maximal throughput for M=4 is almost half the maximal throughput of a packet size of 128 bits. This is due to the fact that the increased packet size decreases the packet success probability.

However, an increase in the order of diversity yields a smaller BER. So, to obtain a reasonable throughput of large packets, one should use a receiver with a sufficient large order of diversity.

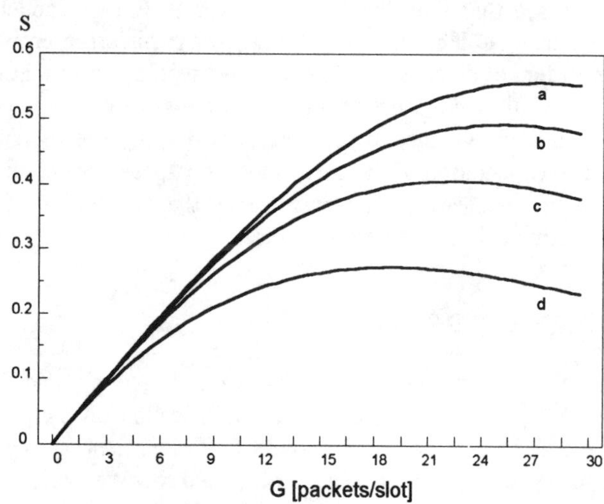

Figure 8.11 Effect of the order of diversity on the throughput with a packet size of 128 for the SD; q=10; N=255; T_{max}=100 ns; L=4; R=6.8 dB: (a) M=4; (b) M=3; (c) M=2; (d) M=1.

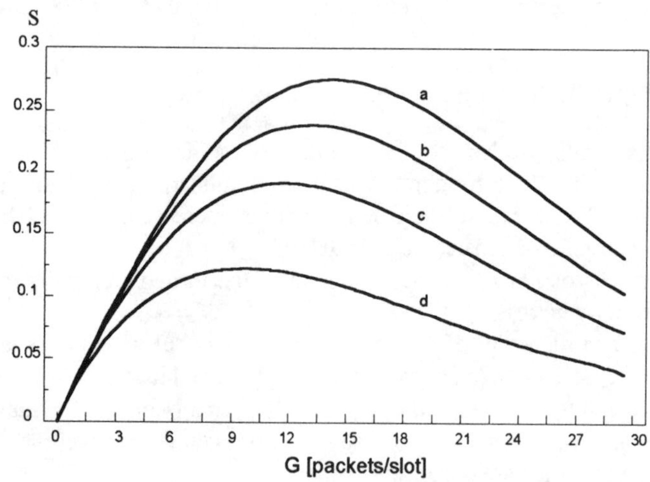

Figure 8.12 Effect of the order of diversity on throughput with a packet size of 1024 for the SD case; E_b/N_o=20 dB; q=10; L=4; T_{max}=100 ns; N=255; R=6.8 dB: (a) M=4; (b) M=3; (c) M=2; (d) M=1.

In Figure 8.13 we have considered the throughput performance for different parameter settings with the constraint of a fixed bandwidth and a fixed data rate. The plots show a large range in which the maximum throughput varies. We see that in general the throughput decreases with an increase in the number of resolvable paths in combination with a small spreading code period N. Even the relatively small packet size of 42 bits does not give a reasonable value for throughput when L is relatively large.

The cause for the poor performance is the increased level of multipath interference when L increases. This causes the BER to increase, which yields a smaller packet success probability. When the period of the spreading codes decreases, the multiuser interference increases, which also causes a higher BER. The increase in the number of frequencies in the hopping pattern is not capable of diminishing the effect of increasing the number of resolvable paths and decreasing the spreading code period. It is obvious that with an appropriate choice of the parameters, a reasonable value for throughput can be obtained.

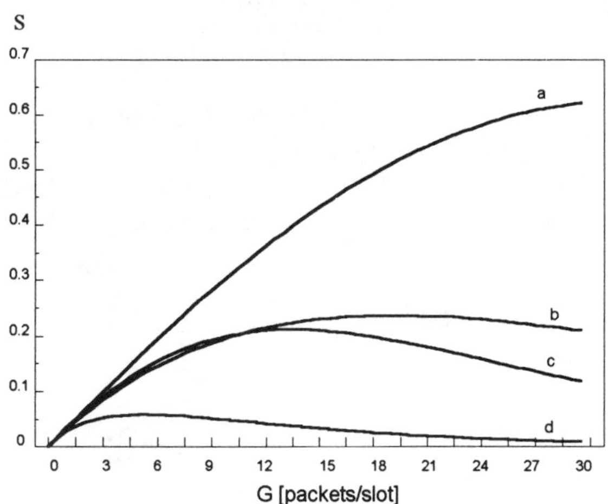

Figure 8.13 Effect of the different parameter settings on the throughput for a packet size of 42 bits given a fixed bandwidth; M=2; R=6.8 dB:
(a) N=255; q=10; L=4; T_{max}=100 ns (b) N=127; q=20; L=2; T_{max}=100 ns
(c) N=255; q=10; L=10; T_{max}=250 ns (d) N=127;q=20; L=5; T_{max}=250 ns.

8.7.2 Influence of FEC Coding on Throughput

Figure 8.14 shows the comparison of the throughput with a packet size of 128 bits for systems with and without coding. We see that nondiversity without coding yields the poorest result, with respect to the throughput. The best result is obtained with a high order of diversity and Golay coding.

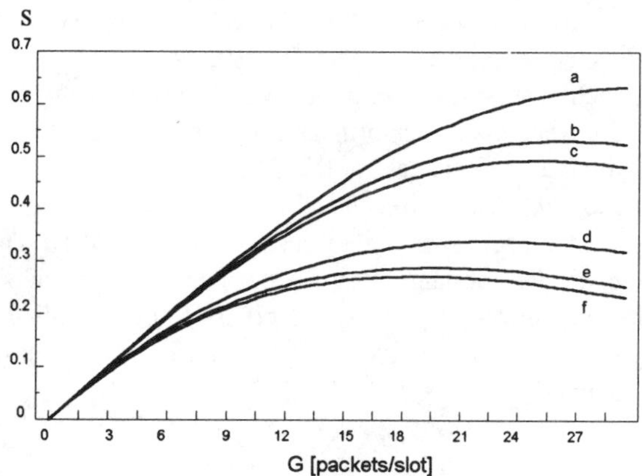

Figure 8.14 Comparison of the throughput with and without FEC coding for a packet size of 128 bits; T_{max}=100 ns; N=255; q=10; R=6.8 dB:
(a) M=3, Golay code, L=8 (b) M=3, Hamming code, L=8 (c) M=3, no coding, L=4
(d) M=1, Golay code, L=8 (e) M=1, Hamming code, L=8 (f) M=1 no coding, L=4.

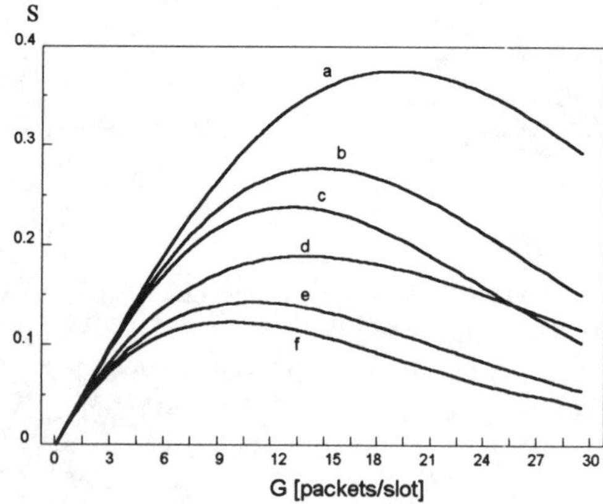

Figure 8.15 Comparison of the throughput with and without FEC coding for a packet size of 1024 bits; E_b/N_o=20 dB; N=255; q=10; R =6.8 dB:
(a) M=3, Golay code, L=8 (b) M=3, Hamming code, L=8 (c) M=3 no coding, L=4
(d) M=1, Golay code, L=8 (e) M=1, Hamming code, L=8 (f) M=1, no coding, L=4.

Figure 8.15 shows the effect of FEC coding on the throughput for a packet size of 1024 bits. We see that nondiversity with and without coding yields a relatively poor result. Only diversity of order 3 with Golay coding yields a reasonable throughput.

In general, we can say that in order to obtain acceptable performance for large packet sizes, we have to use a sufficiently large order of diversity and FEC coding.

8.7.3 Delay Performance Without Coding

We consider the delay as a function of the offered traffic. Further, we take the retransmission delay T_r equal to one. We also assume the same parameter settings as with the corresponding plots of the throughput.

Figure 8.16 presents the effect of the order of diversity on delay for a packet size with the same parameters as in Figure 8.11. We see that a high order of diversity causes a relative low delay in the system. Systems without diversity (M=1), however, yield relatively large delays in the system.

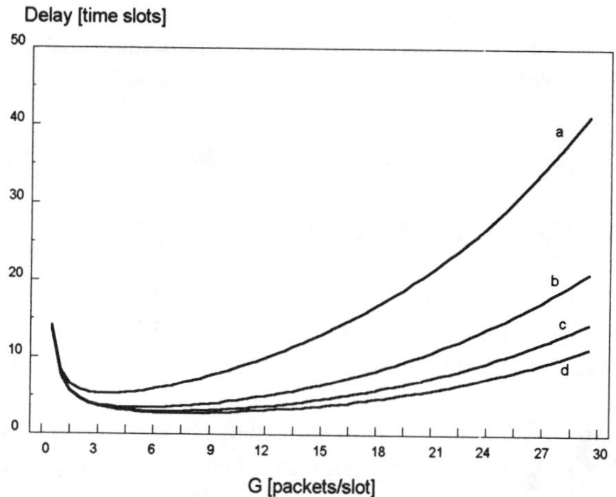

Figure 8.16 Effect of the order of diversity on the delay with a packet size of 128 bits: (a) M=1; (b) M=2; (c) M=3; (d) M=4.

In Figure 8.17 the effect of the order of diversity on the delay is shown, but now for a packet size equal to 1024 bits. The parameter setting is equal to the one in Figure 8.12. It is obvious that large packet sizes in combination with a low order of diversity yield a considerable delay in the system. Only a high order of diversity gives a reasonable value of the delay.

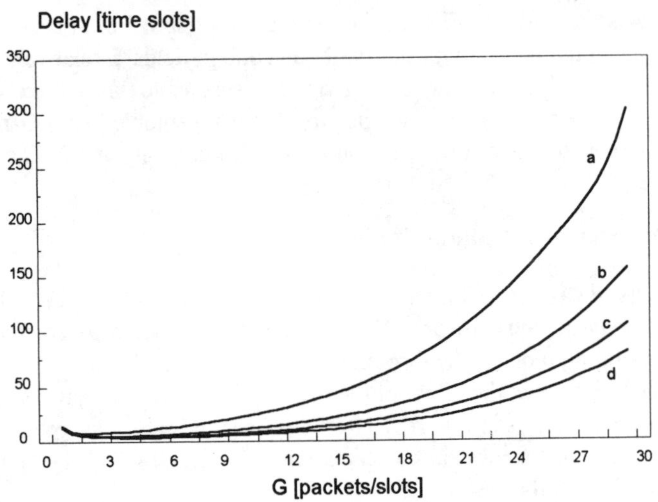

Figure 8.17 Effect of the order of diversity on the delay with a packet size of 1024 bits: (a) M=1; (b) M=2; (c) M=3; (d) M=4.

8.7.4 Effect of FEC Coding on Delay Performance

Figure 8.18 presents the comparison of the delay of systems using FEC coding and the delay of systems without coding for a packet size of 128 bits. The parameters are the same as those of Figure 8.14. We see that FEC coding decreases the delay in the system considerably.

Golay coding in combination with an order of diversity of 3 yields the lowest delay for this packet size. Nondiversity without coding yields the poorest result for this relatively small packet size.

Figure 8.19 presents the effect of FEC coding for a packet size of 1024 bits, with the same parameter setting as Figure 8.15. We see that FEC coding with diversity yields a relatively lower delay than FEC coding without diversity. However, the Golay code with M=1 yields almost the same delay as a system with M=3 without coding.

In general, we see that for large packet sizes FEC coding and diversity are required to obtain sufficiently low delays in the system.

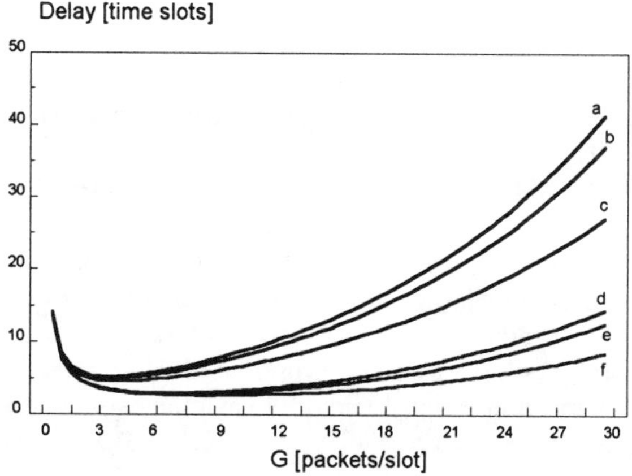

Figure 8.18 Effect of FEC coding on the delay for a packet size of 128 bits: (a) M=1 no coding; (b) M=1 Hamming code; (c) M=1 Golay code; (d) M=3 no coding; (e) M=3 Hamming code; (f) M=3 Golay code.

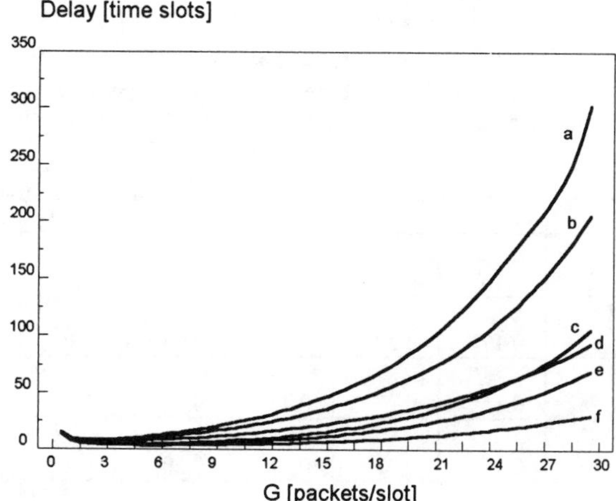

Figure 8.19 Effect of FEC coding on the delay for a packet size of 1024: (a) M=1 no coding; (b) M=1 Hamming code; (c) M=1 Golay code; (d) M=3 no coding; (e) M=3 Hamming code; (f) M=3 Golay code.

8.7.5 Comparison Between Hybrid DS/SFH and DS

In Chapter 6 the results of the throughput of a slotted CDMA network using DPSK modulation in a microcellular mobile radio environment were presented. The network consists of transmitter-receiver combinations, which operate with a direct-sequence spread-spectrum technique.

For microcells, which have a size of 0.4 to 2 km in diameter, the radio channel can be very well characterized by a Rician fading channel. The larger size of microcells, when compared with picocells, results in a larger value for the RMS delay spread (e.g., 2 μs for microcells). As a consequence, this yields a larger number of resolvable paths.

In Table 8.4 we present a comparison of the throughput of a CDMA system based on pure DS and a system based on hybrid DS/SFH, using selection diversity. We have done the comparison in a single microcell environment; the single cell means that we did not consider the interference of adjacent cells. The results for pure DS are obtained from Chapter 6.

The radio channel is considered to be of the fast Rician fading type, with a Rice factor of R=12 dB. This relatively high value for the Rice factor means a relatively strong LOS component. Further, we consider the RMS delay spread and the data rate to be constant for both systems, which means that the number of resolvable paths is only determined by the spreading code period N, according to the equation $L = \lfloor T_{max} / T_c \rfloor + 1$. The bandwidth, which is proportional with $R_c N q$, is used as the parameter.

Table 8.4

Comparison of the throughput of a CDMA system with selection diversity based on pure DS and on hybrid DS/SFH with the bandwidth as parameter; M=2, R=12 dB, N_p=42, C=30, R_b=32 kbit/s, T_{max}=1961 ns

	G=5	G=10	G=15	G=20	G=25	S_{max}	$G(S_{max})$
DS N=255 L=16	0.16	0.23	0.11	0.025	0.005	0.24	9
Hybr. DS/SFH N=127 q=2 L=8	0.17	0.27	0.14	0.030	0.020	0.27	10
Hybr. DS/SFH N=127 q=10 L=8	0.15	0.29	0.43	0.58	0.74	0.74	25
Hybr. DS/SFH N=255 q=2 L=16	0.17	0.32	0.31	0.20	0.17	0.34	12
Hybr. DS/SFH N=255 q=5 L=16	0.17	0.33	0.49	0.65	0.81	0.81	25

For the combination N=255 for DS and N=127 with q=2 for hybrid DS/SFH, which means the same bandwidth, we see that the performance of the hybrid system is slightly better than that of pure DS.

The number of resolvable paths for the pure DS case is twice that of hybrid)S/SFH, which means that the multiuser interference in the DS case is higher than for he hybrid case. Despite the larger code length in the DS case, the relatively low number f frequencies in the hybrid system is able to overcome the poorer correlation properties f N=127.

The combination N=255 with q=2 for the hybrid system gives a performance of wice the bandwidth of pure DS. We see from Table 8.4 that the maximum throughput S_{max} for the hybrid DS/SFH system is less than twice the maximum throughput of the DS ystem. An increase in bandwidth by a factor of 5, through the combination N=255 with =5 or N=127 with q=10, only gives an improvement in maximum throughput by a actor 3. The first combination yields the best performance.

The throughput in fast fading channels is lower than in slow fading channels. Iowever, a relatively large Rice factor improves throughput performance in fast fading hannels. To obtain reasonable throughput performance (e.g., S_{max} > 0.40), a hybrid)S/SFH with a sufficient number of frequencies is preferred over a pure DS system. Iowever, this requires a larger bandwidth for the hybrid system than pure DS.

.8 CONCLUSIONS

Ve have theoretically and numerically assessed the performance of a hybrid DS\SFH ommunication system with DPSK modulation for selection diversity and maximal ratio ombining and a CDMA network based on hybrid DS/SFH transceivers with DPSK iodulation. The numerical results have been obtained with help of a theoretical iathematical model.

With the BEP results for the hybrid DS/SFH system with DPSK, we compared vith the performance of a hybrid system based on BPSK and QPSK modulation. We lso compared the BEP performance of a hybrid model with DPSK and a model escribed in the literature.

Furthermore, the throughput of a CDMA network based on hybrid DS/SFH was ompared with the throughput of a CDMA system based on pure DS, in a fast Rician ading radio channel. From these results we draw the following conclusions.

. The modeling of multipath and multiuser interference is very complicated and an approximation by random Gaussian variables is allowed only when some key requirements are met.

. A comparison of two models of the hybrid DS/SFH system shows that the model described in the literature yields a lower bit error probability than the model described in this chapter. This is due to the use of random signature sequences in the model described in the literature, in contrast with the model of this chapter, which uses deterministic signature sequences.

3. The average BEP performance of systems employing MRC is better than the BE performance of those employing SD for the same order of diversity. The advantag of MRC over SD increases with an increasing order of diversity.

4. Under the constraint of a fixed bandwidth (qN is constant) and a given bit rate fc the hybrid DS/SFH with DPSK and without coding, the combination of a relative large signature sequence N and a certain number of frequencies q generally yielc better performance, both for the BEP and for the outage probability, than a sma value of N and a large value of q. We also know that performance degrades with a increase in bit rate, due to the high level of multipath interference.

5. A comparison of the BEP of the hybrid DS/SFH system with the DS system for tl same transmission bandwidths shows that for low bit rates the hybrid system yields slightly better performance. For higher bit rates, performance favors DS. With a increase in bandwidth by a factor of 5 for the hybrid system, performance at low b rates favors of the hybrid system; however, at high bit rates the performance of tl hybrid system, for a low code length N and relatively large q, is worse than tl performance of pure DS. The combination of large N with a small q for the san higher bit rate yields better performance for the hybrid system.

6. The use of FEC coding, under the constraint of a fixed bandwidth, only improv the performance for the combination of large signature sequences in combinatic with a relatively small number of frequencies, especially for $M > 1$.

7. A comparison of the modulation schemes DPSK, BPSK, and QPSK shows th DPSK generally yields the poorest result, due to the poorer noise resistance at tl DPSK demodulator. Under the condition of a fixed bandwidth and data rate, QPS yields better performance than BPSK, due to the reduction of the multiuser a multipath interference in case of QPSK. Since the indoor channel provides relatively strong LOS component, which can be used for synchronization coherent receivers, QPSK or BPSK is preferred over DPSK when relatively low b error probabilities are necessary.

8. The model that describes the throughput and the delay in this chapter is only val under special conditions, concerning the statistical properties of the different use and the receivers at the base station. The model is a simplification of the ve complex behavior of the CDMA network, but it provides a first impression of tl throughput and delay performance of the CDMA system.

9. The throughput of a slotted CDMA network based on hybrid DS/SF communication systems increases for an increasing order of diversity. In additio the corresponding delay in the system then decreases. The higher order of diversi

provides a lower bit error probability and, accordingly, a higher packet success probability.

0. Relatively small packet sizes yield a high throughput and accordingly a low delay in the system. To obtain a reasonable value for the throughput and delay in case of large packet sizes, the combination of the Golay code and a sufficiently high order of diversity is necessary.

1. The throughput decreases with an increase in the number of resolvable paths given a spreading code period N under the condition of a fixed bandwidth. So the performance gets worse when either the delay spread or the data rate is increased.

2. A comparison of the throughput of a hybrid DS/SFH system with a CDMA system based on DS for a fast Rician fading channel shows that the hybrid system yields a slightly higher maximum throughput than pure DS under the condition of a fixed bandwidth. The maximal throughput increases when the transmission bandwidth of the hybrid system is increased; however, the throughput does not increase proportionally. Further, at larger bandwidths a higher offered traffic can be put through the channel.

3. Although a fast fading channel generally provides a worse throughput than slow fading channels, the reasonable throughput performance over the fast fading channel is mainly due to the relatively high Rice factor.

REFERENCES

1] J. Wang and M. Moeneclaey, "Hybrid DS/SFH spread-spectrum multiple access with predetection diversity and coding for indoor radio," *IEEE J. Selected Areas Comm.*, Vol.10, No. 4, pp. 705-713, May 1992.

2] J. Wang and M. Moeneclaey, "Hybrid DS/SFH-SSMA with predetection diversity and coding over indoor radio multipath Rician-fading channels," *IEEE Trans. Comm.*, Vol. 40, No. 10, pp. 1654-1662, October 1992.

3] E.A. Geraniotis, "Noncoherent hybrid DS/SFH spread-spectrum multiple access communications," *IEEE Trans. Comm.*, Vol. COM-34, pp. 862-872, September 1986.

4] E.A. Geraniotis and M.B. Pursley, "Error probabilities for slow-frequency-hopped spread-spectrum multiple access communications over fading channels," *IEEE Trans. Comm.*, Vol. COM-30, pp. 996-1009, May 1982.

5] R. Prasad, E. Walther and R. Ponson, "Performance analysis of hybrid SFH/DS CDMA networks for personal communication systems," *Proceedings PIMRC'92*, Boston, USA, pp. 362-366, October 1992.

6] L.Vandendorpe, R.Prasad and R.G.A. Rooimans, "Hybrid slow frequency hopping/direct-sequence spread-spectrum communications systems with B- and QPSK modulation in an indoor wireless environment," *Proceedings of the Fourth International Symposium on Personal, Indoo and Mobile Radio Communications*, Yokohama, Japan, 9-11 September 1993.

[7] L. Vandendorpe and R. Prasad, "Hybrid slow frequency hopping/direct-sequence sprea spectrum indoor communication systems with QPSK modulation," *Ann. Télécomm.*, Vol. 49, pp 518-526, September/November 1994.

[8] F. Çakmak, R.G.A. Rooimans and R. Prasad, "Performance comparison of CDMA network based on hybrid DS/SFH, DS and SFH with B- and Q-PSK modulation in an indoor Ricia fading environment,", *Proceedings PIMRC'94*, The Hague, The Netherlands, pp. 1045-1049 September 1994.

[9] R. Prasad, R.G.A. Rooimans, L. Vandendorpe and F Çakmak, "Packet switched hybrid DS/SFH CDMA networks with B-, Q-, and D-PSK modulation in an indoor Rician fading environment," *Proceedings GLOBECOM'94*, San Francisco, USA, pp. 69-73, November/December 1994.

[10] R. Prasad, F. Çakmak and R.G.A. Rooimans, "Throughput and delay analysis of hybrid DS/SFH CDMA using a measured delay profile in a pico cellular PCN," *Proceedings GLOBECOM'95* Singapore, pp. 1147-1151, November 1995.

[11] J.G. Proakis, "*Digital Communications*," McGraw-Hill International Editions, Computer Scienc Series, New York, 2nd edition, 1989.

Chapter 9
Slotted CDMA Protocol Using the Markov Chain Model

9.1 INTRODUCTION

This chapter presents a performance analysis for a slotted CDMA protocol using the direct-sequence spread-spectrum method for indoor wireless data communications with differential phase shift keying (DPSK) modulation. The traffic load is described by a discrete time Markov chain model for a finite number of simultaneous users. The throughput and steady-state delay are the performance parameters. Further, stability aspects of the CDMA protocol, such as the expected drift and state occupancy probabilities, are investigated.

The throughput and average delay of a CDMA protocol for indoor data communications were evaluated in Chapter 5 assuming the binomial arrival model. This chapter presents a detailed investigation of throughput, delay, and stability of a slotted DS-CDMA protocol with a slow indoor Rician fading channel using Markov chain model. Poisson and binomial traffic models are the particular cases of the generalized Markov chain model [1,2].

The packet flow in a random access CDMA protocol is shown in Figure 9.1. The channel input consists of new packets and retransmitted packets. Some of these packets are received successfully, while those which are received in error are retransmitted after a random delay.

The probabilities of new packets being transmitted in the next slot in the origination mode is p_0 and retransmission of the packet in the backlog mode in any given slot is p_r. When $p_0=p_r=p$ and C is the maximum number of users, the Markov chain (general arrival) model simplifies to:

- the Poisson model when $p \rightarrow 0$ and $C \rightarrow \infty$
- the binomial model when C is finite.

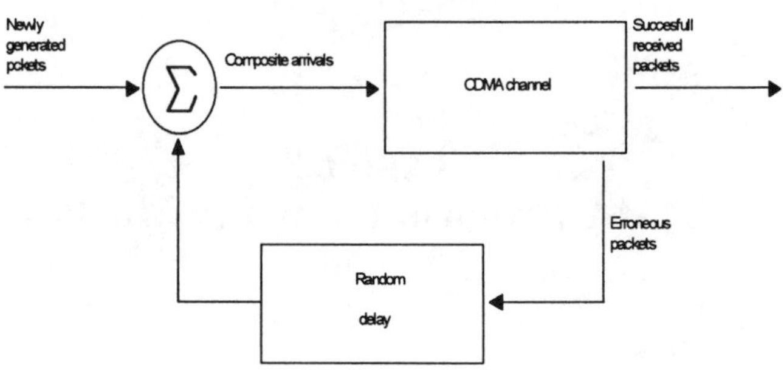

Figure 9.1 Packet flow in a random access CDMA protocol.

A system that consists of C users can be described at the beginning of every slot by the number of terminals that are in the backlog mode, which is a integer from 0 to C, called the system state (C+1 states). When $p_r > p_o$ and C is finite, a simple closed-form expression for the composite arrival distribution cannot be obtained as in the Poisson and binomial cases.

This chapter is organized as follows. Section 9.2 presents the performance analysis using the Markov chain model. Computational results are presented for the throughput, delay, and stability of the CDMA system considering DPSK modulation and maximal ratio combining. The effect of forward error correcting codes on the performance of a CDMA system is discussed in Section 9.3. Conclusions are given in Section 9.4.

9.2 PERFORMANCE ANALYSIS

This section presents the throughput, delay and stability of a CDMA protocol by describing the traffic load using a discrete time Markov chain model for a finite number of simultaneous users. To calculate the throughput, delay, and stability, we first have to evaluate the packet success probability in a Rician fading channel. The receiver may employ either selection diversity or maximal ratio combining methods. We have discussed the packet success probability for slow and fast (only for the purpose of comparison) fading channels, considering both selection diversity and maximal ratio combining. The packet success probability is obtained with the help of the bit error probability (BEP). A detailed derivation of the BEP was given in Chapter 5.

.2.1 Packet Success Probability with Selection Diversity

n selection diversity with DPSK, the decision variable used to select the path with the argest gain is the output of the M demodulated signals where M is the order of diversity. he order of diversity is less than or equal to the number of antennas times the number f resolvable paths L.

Channel characteristics are constant for a (very) long period compared to a ignaling interval for the slow fading transmission case. When we send data in packets, his definition is interpreted as that in a slow fading channel; all bits of a packet are eceived with the same average power.

Assuming a slow Rician fading channel with DPSK modulation, we can write acket success probability $P_{rk}(\mathrm{TX})$ for systems using selection diversity as given in hapter 5:

$$P_{rk}(\mathrm{TX}) = \int_{-\infty}^{\infty}\int_{-\infty}^{\infty}\int_{0}^{\infty} [1 - P_{sd}(\beta_{max}, \mu, \mu_0)]^{N_d} \cdot p_{\beta_{max}}(\beta_{max})\, p_{\mu}(\mu)\, p_{\mu_o}(\mu_o) d\beta_{max} d\mu\, d\mu_o \tag{9.1}$$

vhere

$$P_{sd}(\beta_{max}, \mu, \mu_0)] = \frac{1}{2}\left[1 - \frac{\mu}{\mu_0}\right] \exp\left[-\frac{A^2 \beta_{max}^2 T_b^{\,2}}{\mu_0}\right] \tag{9.2}$$

vhere N_d is the number of data bits per packet; $\mu=E[(z_0\_m)(z_{-1}\_m)^*|\{t_{lk}\},L]$; $\mu_0=\mathrm{var}(z_0|\{t_{lk}\},L)$; $m=E[z_0|\beta_{max},b_1^{\,0}]=E[z_{-1}|\beta_{max},b_1^{\,-1}]$; $E[\cdot]$ denotes statistical average and ar$[\cdot]$ denotes variance; $\{\tau_{lk}\}$ is delay; z_0 and z_{-1} are the envelope of the signal at the urrent sampling instant and the previous sampling instant, respectively; L is the naximum number of resolvable paths given by $L=\lfloor T_{max} / T_c \rfloor+1$ with $\lfloor x \rfloor$ as the largest nteger smaller then or equal to x; T_{max} is the delay spread; and T_c is the chip duration.

Here μ_0 and μ are approximated by Gaussian random variables and p_{μ} and p_{μ_0} aussian pdfs. Furthermore, μ can be removed and (9.1) and (9.2) can be simplified nalytically as follows:

$$\int_{\infty}^{\infty} \left[1 - \frac{\mu}{\mu_0}\right] p_{\mu}\, d\mu = 1 - \frac{E_{\tau}(\mu)}{\mu_0} \tag{9.3}$$

eaving a double integral to evaluate. The $p_{\beta_{max}}(r)$ term is the pdf of choosing the trongest path, given by:

$$p_{\beta_{max}}(r) = M\,[1 - Q(\frac{r}{\sigma},\frac{s}{\sigma})\,]^{M-1}\frac{r}{\sigma^2}\exp\left[-\frac{r^2+s^2}{2\sigma^2}\right] I_0(\frac{rs}{\sigma^2}) \qquad (9.4$$

where M is the order of diversity and $Q(x, y)$ is the Marcum Q-function [3].

For fast Rician fading, the packet success probability is given by:

$$P_{rk}(\text{TX}) = [1 - P_{erS}(\text{TX})\,]^{N_d} \qquad (9.5$$

where P_{erS} is the bit error probability, given by:

$$P_{erS}(\text{TX}) = \int_{-\infty}^{\infty}\int_{-\infty}^{\infty}\int_{0}^{\infty} P_{sd}(\beta_{max},\mu,\mu_0)\, p_{\beta_{max}}(\beta_{max})p_{\mu}(\mu)p_{\mu_o}(\mu_o)d\beta_{max}d\mu d\mu_0 \qquad (9.6$$

For fast fading, the channel variations are fast relative to the signaling interval. Thus each signaling symbol undergoes fading independent of other symbols. Therefore, when we send data in packets in a fast fading channel, all data bits undergo fading independently.

9.2.2 Packet Success Probability with Maximal Ratio Combining

In the case of maximal ratio combining with DPSK, the contributions of several resolved paths are added. The receiver, which is set for a reference user, combines all the spread-spectrum correlation peaks of the demodulated signals noncoherently and forms the decision variable. Maximal ratio combining with DPSK is also known as prediction combining (PDC) [4].

The packet success probability in the case of maximal ratio combining for the slow fading channels is given as follows:

$$P_{rk}(\text{TX}) = \int_{0}^{\infty} [1 - P_2(\gamma_b)\,]^{N_d}\, p(\gamma_b)\, d\gamma_b \qquad (9.7)$$

where

$$P_2(\gamma_b) = \frac{1}{2^{2M-1}}\exp(-\gamma_b)\sum_{k=0}^{M-1} p_k\gamma_b^{\,k} \qquad (9.8)$$

$$p_k = \frac{1}{k!} \sum_{n=0}^{M-1-k} \binom{2M-1}{n} \tag{9.9}$$

$$\gamma_b = \frac{E_b}{N} \sum_{k=1}^{M} \beta_k^{\,2} \tag{9.10}$$

and,

$$p(\gamma_b) = \frac{1}{2\frac{E_b}{N}} \left[\frac{\gamma_b}{\frac{E_b}{N} s_M^{\,2}} \right]^{\frac{M-1}{2}} \exp\left[-\frac{s_M^{\,2} + \gamma_b \frac{N}{E_b}}{2} \right] I_{M-1}\left(\sqrt{\gamma_b \frac{N}{E_b}}\, s_M\right) \tag{9.11}$$

Here, M is the order of diversity, N is the sum of the Gaussian noise and Gaussian multiuser interference, and β_k is the path gain of the kth combined path. Further, $s_M^2 = Ms^2 / \sigma^2$, and γ_b is the sum of the signal-to-noise ratios (SNRs) of the M combined paths, and $I_{M-1}(\cdot)$ is a modified Bessel function of the first kind and order M-1.

The packet success probability for fast fading is found to be:

$$P_{rk}(\text{TX}) = [1 - \int_0^{\infty} P_2(\gamma_b)\, p(\gamma_b)\, d\gamma_b]^{N_d} \tag{9.12}$$

9.2.3 Markov Chain Model

A well-known model to investigate the performance analysis of a multiple access system is the Markov chain model. The behavior of the terminals and the slotted CDMA protocol is described by a discrete time Markov chain. The terminal model uses a Markov chain with three operational modes for each of C indentical, independently operating terminals: the origination (O), the transmission (T), and the backlog mode (B). A terminal can only be in one mode at a time, thus a terminal in the backlog mode cannot transmit a new packet until the backlog packet is received correctly (terminal is blocked). The terminal model is illustrated in Figure 9.2 [1]. In the origination mode, new packets are generated and transmitted in the next slot with probability p_o. Terminals enter the backlog mode when a transmitted packet is not received successfully. In the backlog mode, packets are retransmitted with probability p_r.

The terminals can only be one mode at a time. Thus a terminal in the backlog mode cannot transmit a new packet until the backlogged packet is received correctly (terminal is blocked). The state of the entire system, consisting of C users, can be characterized at the start of a given time slot by the number of terminals that are in the backlogged mode. This can be any integer between 0 and C, called the system state. Thus, there are $C+1$ possible states corresponding to 0, 1, 2, ..., C backlogged terminals.

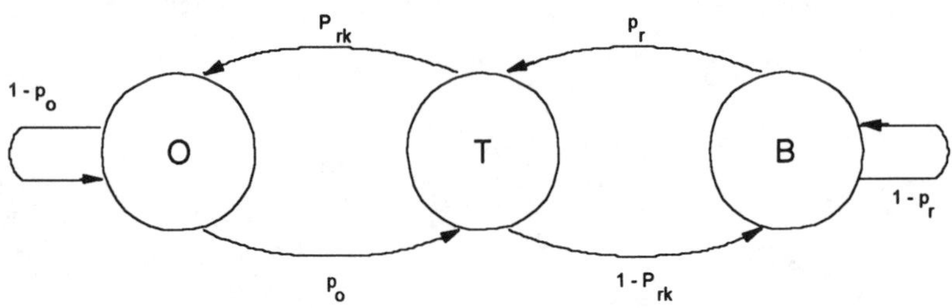

Figure 9.2 Terminal states in a Markov chain model.

The sequence of successive states (from slot to slot) forms a discrete-time Markov chain and is characterized by the state transition matrix P. The square matrix P (dimension $C+1$), consists of one-step state transition probabilities p_{nm}, from state n to state m.

For the random access CDMA, transitions can take place in a number of ways since there can be more than one successful transmission per time slot. Considering the system state n, the movement from one state to another is determined by the difference between the number of unsuccessful new transmissions (UNTX) and successful retransmissions (SRTX), since the number of successful new transmissions (SNTX) and the number of unsuccessful retransmissions (URTX) have no effect on the system state. Denoting the binomial probability function as:

$$\text{bin}(a,b,p) \triangleq \binom{b}{a} p^a (1-p)^{b-a} \tag{9.13}$$

let NTX be the number of new transmissions and RTX the number of retransmissions, and the total number of transmissions (new and retransmissions) in a particular slot as TX = NTX + RTX. Then we write:

$$\text{Prob}\{\text{SRTX} = k, \text{UNTX} = l \mid n\} = \sum_{\text{NTX}=l}^{C-n} \sum_{\text{RTX}=k}^{n} \text{bin}(l, \text{NTX}, P_{kr}(\text{TX}))\text{bin}(k, \text{RTX}, P_{rk}(\text{TX})) \cdot \text{bin}(\text{NTX}, C-n, p_o)\text{bin}(\text{RTX}, n, p_r) \tag{9.14}$$

where

$0 \le l \le C\text{-}n\,, \quad 0 \le k \le n$

P_{rk} is the packet success probability, and P_{kr} is the packet error probability with $P_{rk}+P_{kr}=1$.

A transition from state n to state m occurs whenever UNTX exceeds SRTX by $(m\text{-}n)$ for $m \ge n$ or SRTX exceeds UNTX by $(n\text{-}m)$ for $n \ge m$. Thus the state transition probabilities can now be calculated from:

$$p_{nm} = \begin{cases} \displaystyle\sum_{j=0}^{\min(n,C-m)} \text{Prob}\{\text{SRTX}=j,\ \text{UNTX}=m-n+j \mid n\},\ m \ge n \\ \displaystyle\sum_{j=0}^{\min(m,C-n)} \text{Prob}\{\text{SRTX}=n-m+j,\ \text{UNTX}=j \mid n\},\ m \le n \end{cases} \tag{9.15}$$

Equations (9.14) and (9.15) together give the state transition matrix P, which is irreducible, aperiodic, and positive recurrent unless all $P_{rk}(k)=1$, for all $1 \le k \le C$.

To obtain the steady-state composite arrival distribution, the long-term state occupancy probabilities $\mu(n)$ are given by the solution of

$$\mu^T = \mu^T P \quad \wedge \quad \sum_{n=0}^{C} \mu(n) = 1 \tag{9.16}$$

where

$$\mu^T = [\mu(0),\ldots,\mu(C)] \quad \wedge \quad P = \begin{bmatrix} P_{00}\cdots\cdots\cdots P_{0c} \\ P_{10}\cdots\cdots\cdots P_{1c} \\ \\ \cdots\cdots\cdots\cdots\cdots \\ P_{c0}\cdots\cdots\cdots P_{cc} \end{bmatrix} \tag{9.17}$$

The solution of the C+1 equations can be found with the help of [5] and given values of C, p_o, p_r and $P_{rk}(k)$, k=0, 1, 2, ..., C. In contrast to [1], we do need $P_{rk}(0)$ for the calculation of the state transition probabilities. In our calculations we assumed $P_{rk}(0)=P_{rk}(1)$. To solve (9.16), we rewrite this equation to:

$$0 = \mu^T (P - I) \tag{9.18}$$

The method used to solve (9.18) is to replace the last column of $P–I$ by a column of ones. This is allowed because the sum of entries in each row in the matrix $P–I$ equals 0 [5]. It means that no information is lost when we leave out one column. Using the last part of (9.16) we replace the last column by a column with entries equal to one, and call the new matrix M. Then we define a vector E^T=[0, 0, ..., 0, 1], and combine the two terms in (9.16), which results in $\mu^T M=E^T$. The state occupancy probabilities can now be found from $\mu^T=E^T M^{-1}$, which equals the last row of matrix M^{-1}.

9.2.4 Throughput, Delay, Drift, and Stability

The steady-state composite arrival distribution is given by [1]:

$$f_G(m|n) = \sum_{j=\max(m-n,0)}^{\min(m,C-n)} \text{bin}(j, C-n, p_o)\, \text{bin}(m-j, n, p_r) \tag{9.19}$$

The throughput β is defined as the average number of successfully received packets per time slot. Now, the normalized steady-state throughput S can be calculated:

$$S = \frac{\beta}{C} \tag{9.20}$$

where β is the throughput in packets per slot, given by:

$$\beta = \sum_{m=1}^{C} m\, P_{rk}(m) \left[\sum_{n=0}^{C} f_G(m|n)\, \mu(n) \right] \tag{9.21}$$

With the Markov chain model, we cannot evaluate the throughput directly for a given value of the offered traffic. This is because the offered traffic depends on p_o, p_r, C, and the expected backlog. However, the offered traffic can be estimated by:

$$G = (C - \bar{n})\, p_o + \bar{n}\, p_r \tag{9.22}$$

If $p_o = p_r$, the offered traffic is $G = p_o C$. In (9.22), $\bar{n}$ is the expected backlog or average system state, which can be written as:

$$\bar{n} = \sum_{n=0}^{C} n\, \mu(n) \tag{9.23}$$

The transmission delay is defined as the average number of slot times it takes for a packet to be received successfully. That is the average time duration between the packet being offered to the transmitter and the packet being received successfully.

The steady-state delay is now given by

$$D \triangleq \frac{\bar{n}}{\beta} \tag{9.24}$$

Another performance parameter to measure a system's stability/dynamics is called the expected drift. The expected drift in state n is given as:

$$d(n) = \sum_{m=0}^{C} (m-n)\, p_{nm} \tag{9.25}$$

The system is said to be stable for values of n with zero drift and negative slope, because at such a point the system has no tendency to move from the current state and, if it does, it will move to a lower system state (i.e., lower backlog). Unstable points are points with zero drift and positive slope.

9.2.5 Computational Results

Unless otherwise specified, the calculations here have been done using Gold codes with code sequence lengths N = 127 or 255 chips, N_d= 1024 data bits per packet, a system capacity of C=30 users, Rice factors R=6.8 or 11 dB, data bit rate R_b=144 kbit/s, and E_b/N_0=20 dB.

In Tables 9.1 and 9.2 the relationship between the delay spread T_{max}, the various bit rates, different code lengths, and the maximum number of resolvable paths L is given.

As shown in Tables 9.1 and 9.2, the number of resolvable paths is related to the delay spread and the signaling rate, which depend on the data bit rate and the code sequence length. Namely, for N=127, R_b=144 kbit/s, and a delay spread T_{max} between 55 and 110 ns, the number of resolvable paths is L=2.

Table 9.1
Relation between the delay spread (in ns) and the number of resolvable paths, for different data rates and code sequence length N=127

	32 kbit/s	64 kbit/s	144 kbit/s
$L = 1$	$T_{max} < 246$	$T_{max} < 123$	$T_{max} < 55$
$L = 2$	$246 \le T_{max} < 492$	$123 \le T_{max} < 246$	$55 \le T_{max} < 110$
$L = 3$	—	$246 \le T_{max} < 370$	$110 \le T_{max} < 165$
$L = 4$	—	—	$165 \le T_{max} < 220$
$L = 5$	—	—	$220 \le T_{max} < 275$

Table 9.2
Relation between the delay spread (in ns) and the number of resolvable paths, for different data rates and code sequence length N=255

	R_b = 32 kbit/s	R_b = 64 kbit/s	R_b = 144kbit/s
$L = 1$	$T_{max} < 123$	$T_{max} < 62$	$T_{max} < 27$
$L = 2$	$123 \le T_{max} < 246$	$62 \le T_{max} < 124$	$27 \le T_{max} < 54$
$L = 3$	$246 \le T_{max} < 370$	$124 \le T_{max} < 186$	$54 \le T_{max} < 81$
$L = 4$	—	$186 \le T_{max} < 246$	$81 \le T_{max} < 108$
$L = 5$	—	$246 \le T_{max} < 310$	$108 \le T_{max} < 135$
$L = 6$	—	—	$135 \le T_{max} < 162$
$L = 7$	—	—	$162 \le T_{max} < 189$
$L = 8$	—	—	$189 \le T_{max} < 216$

We compare the performance of selection diversity and maximal ratio combining in Figure 9.3 by presenting the throughput S versus the offered traffic G.

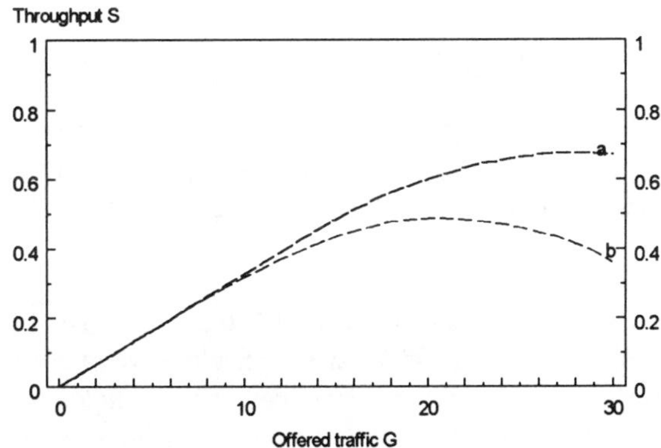

Figure 9.3 Throughput curves for slow Rice fading, for R=6.8 dB, N=127, C=30, p_o=p_r, M=4, L=4, N_d=42: (a) maximal ratio combining and (b) selection diversity.

As expected, we see from Figure 9.3 that maximal ratio combining (MRC) yields a higher throughput than the throughput offered by selection diversity.

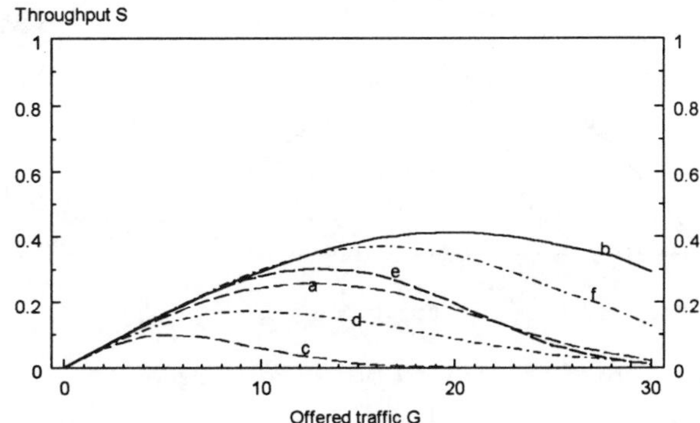

Figure 9.4 Throughput curves for fast and slow Rice fading with MRC, for R=6.8 dB, p_o=p_r, N_d=128, and
M=2, L=2, (a) fast (b) slow
M=2, L=4, (c) fast (d) slow
M=4, L=4, (e) fast (f) slow

In Figure 9.4, we compare the performance of slow and fast fading channels for a Rice factor of R=6.8 dB, N_d=127, and C=30. From this figure, we see that for each case, slow fading gives a higher throughput than the throughput due to fast fading. Curves (c) and (e) for fast fading and also curves (d) and (f) for slow fading show that the throughput increases with the higher order of diversity. For M=4, L=4, and slow fading (curve f), the improvement in throughput is lower than for M=2, L=2 (curve b). For fast fading, M=4 and L=4 (curve e) has a better performance than M=2 and L=2 (curve a), while the reverse is true for slow fading channels.

Further, we see from Tables 9.1 and 9.2 that a change in the value of L corresponds to a change in delay spread T_m at a particular data rate, or a change in data rate R_b at a particular delay spread. From curves a and c and b and d (for a fixed value of M), the throughput decreases with the increasing value of delay spread or data bit rate (if the value of L is to be higher). Note that the degradation in throughput happens only if the increased value of data rate or delay spread causes L to be larger. Otherwise, the throughput does not change. A higher order of diversity, with fixed L, enhances the throughput.

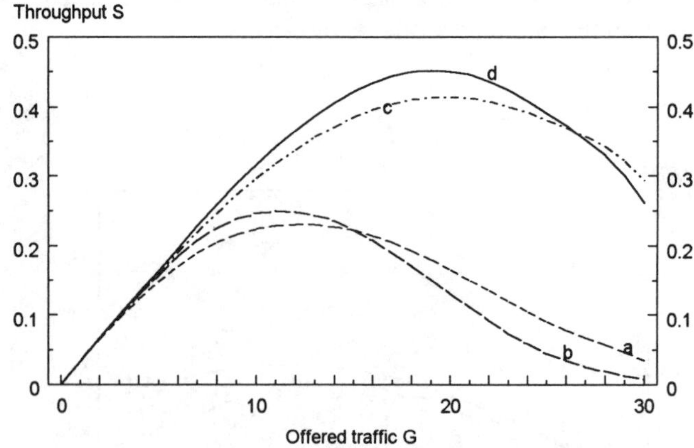

Figure 9.5 Throughput curves for different Rice factors, slow fading with MRC, N=127, C=30, p_o=p_r, M=2, L=2; (a) R=6.8 dB, N_d=2048 (b)R=1dB,N_d=2048, (c) R=6.8 dB, N_d=128 (d) R=11 dB, N_d=128.

To study the influence of the Rice factor on the performance, the throughput curves are plotted in Figure 9.5 for slow Rice fading with maximal ratio combining, N=127, C=30, p_o=p_r, M=2, and L=2. We see that the throughput curves with a Rice factor of R=11 dB yield higher maximum throughput than the throughput for R=6.8 dB. We see that R=11 dB enhances the throughput. But for increased offered traffic the throughput is smaller for R=11 dB than the throughput for R=6.8 dB.

The effect of packet length on the throughput is also investigated. Figure 9.6 gives the throughput curves of various packet lengths for slow Rice fading with MRC. As expected, smaller packet lengths cause an increase in throughput. We see from Figure 9.6 that for smaller packet lengths throughput is always higher.

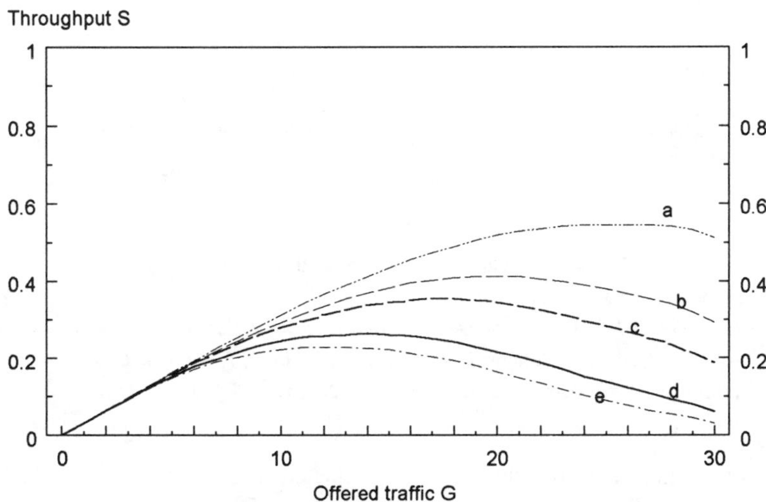

Figure 9.6 Effect of different N_d on the throughput curves for slow Rice fading with MRC, R=6.8 dB, N=127, C=30, p_o=p_r, M=2, L=2:
(a) N_d=42 (b) N_d=128 (c) N_d=256 (d) N_d=1024 (e) N_d=2048.

In Figure 9.7 throughput results are plotted for MRC with different code sequence lengths N. A doubled code sequence length N causes the chip duration to be half, and therefore the number of resolvable paths also increases (double) for a particular value of RMS delay spread (see Tables 9.1 and 9.2).

From Figure 9.7, we see that longer code sequence lengths and a higher number of resolvable paths with fixed values of T_m and R_b improve the performance. The change in the number of resolvable paths is not due to changes in either data rate or delay spread, but due to a change in the code sequence length.

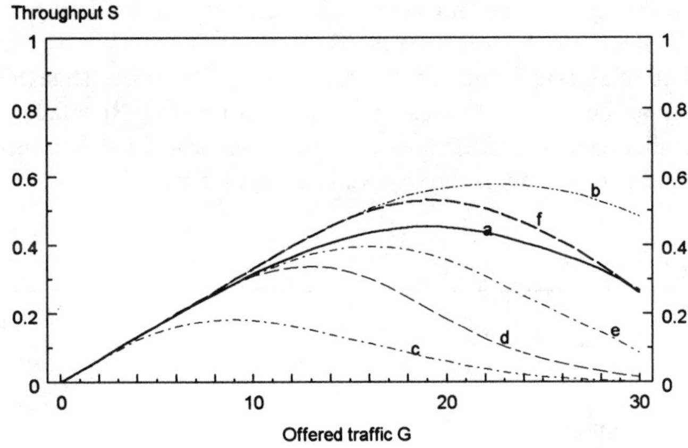

Figure 9.7 Effect of different code sequence length on the throughput curves. Slow Rice fading with R=6.8 dB, C=30, p_o=p_r, N_d=128:
(a) N=127, M=2, T_m=100 ns, L=2, (b) N=255, M=2, T_m=100 ns, L=4
(c) N=127, M=2, T_m=200 ns, L=4, (d) N=255, M=2, T_m=200 ns, L=8
(e) N=127, M=4, T_m=200 ns, L=4, (f) N=255, M=4, T_m=200 ns, L=8.

The effect of p_o and p_r on the throughput is depicted in Figures 9.8 and 9.9. In both figures, the curve with p_o=p_r is included for the purpose of comparison.

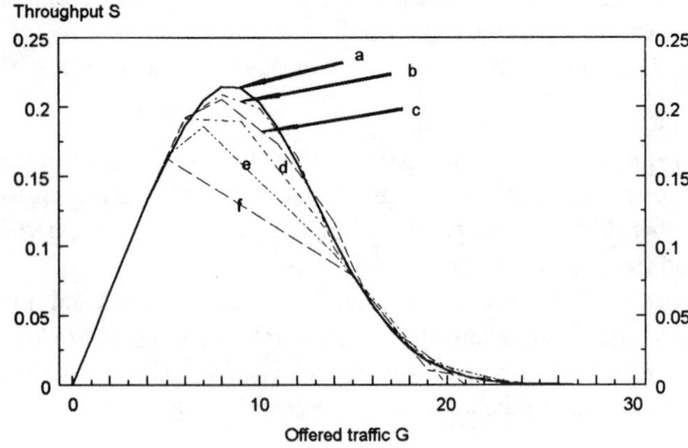

Figure 9.8 Throughput curves for MRC using p_r as a parameter, N=255, R=11 dB, M=2, L=16, $N d$=1024: (a) p_r=p_o, (b) p_r=0.5, (c) p_r=0.55, (d) p_r=0.6, (e) p_r=0.65, (f) p_r=0.7.

From Figure 9.8, we can see that the throughput curve with p_r=p_o yields its highest maximum throughput.

We see that an increase in p_r results in a degradation of the throughput. This can be explained because with high probability of retransmission p_r the channel gets overloaded sooner than p_r=p_o.

From Figure 9.9, we see that for p_o ≤0.3 the performance gets worse.

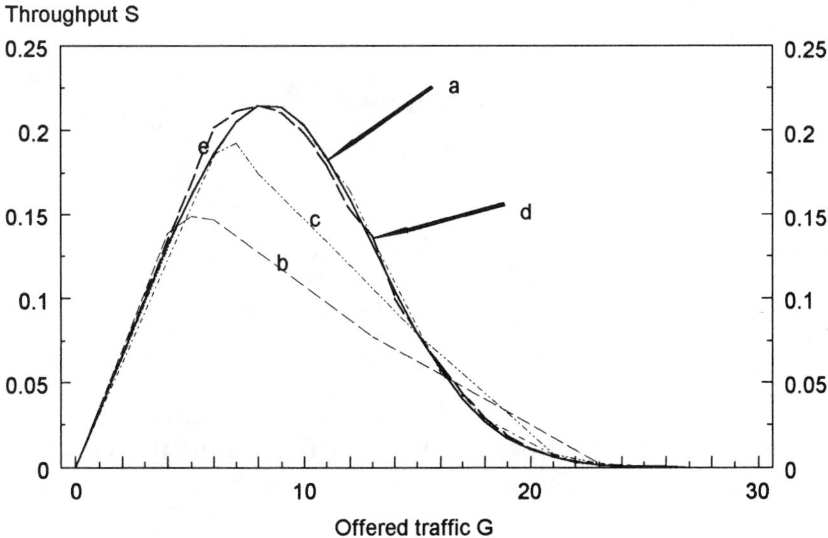

Figure 9.9 Throughput curves for MRC using fixed p_o as a parameter, N=255, R=11 dB, M=2, L=16, N_d=1024: (a) p_o=p_r (b) p_o=0.15 (c) p_o=0.2 (d) p_o=0.3 (e) p_o=0.5.

The steady-state delay is plotted in Figure 9.10 for N=127, N_d=128, p_o=p_r. Figure 9.10 shows that the worst performance is reached by M=2 and L=4 (curves b and e). The delay for R=11 dB has generally been found to be higher beyond the maximal throughput than for R=6.8 dB.

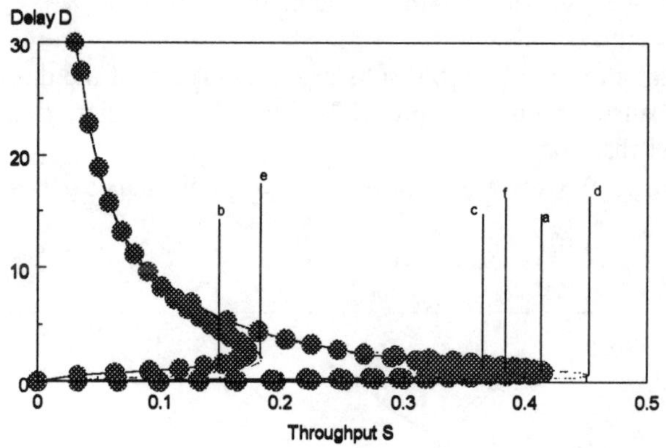

Figure 9.10 Delay versus throughput for MRC, with N=127, N_d=128, p_o=p_t:
for R=6.8 dB (a) M=2, L=2 (b) M=2, L=4 (c) M=4 L=4
for R=11 dB (d) M=2, L=2 (e) M=2, L=4 (f) M=4 L=4.

9.3 PERFORMANCE ANALYSIS USING FORWARD ERROR CORRECTING (FEC) CODES

As discussed in Chapter 5, error correcting codes improve the performance of the system. If the received bit error probability (BEP) does not meet the error probability requirements, FEC codes can be often used to reduce the error probability to an acceptable level. FEC codes involve systematic addition of extra bits to the transmitted message. These extra bits convey no information by themselves, but they make it possible to correct errors in the regenerated message bits [6-8]. Some interesting results are presented in this section.

9.3.1 Selection Diversity with FEC Codes

In the case of selection diversity, for the slow fading channels, the packet success probability is given by:

$$P_{rk}(\text{TX}) = \int_{-\infty}^{\infty}\int_{-\infty}^{\infty}\int_{0}^{\infty}\left[1 - P_{ec_s}(\text{TX})\right]^{N_d} \cdot p_{\beta_{\max}}(\beta_{\max})p_{\mu}(\mu)p_{\mu_o}(\mu_o)d\beta_{\max}d\mu\, d\mu_o \tag{9.26}$$

where

$$P_{ec_s}(\text{TX}) = \frac{1}{n}\sum_{m=t+1}^{n} m\binom{n_c}{m}[P_{sd}(\beta_{max}, \mu, \mu_0)]^m [1 - P_{sd}(\beta_{max}, \mu, \mu_0)]^{n_c - m} \tag{9.27}$$

with n_c as the code length, m as the number of errors, and t is the number of correctable bits. The packet success probability for the fast fading with selection diversity is found to be:

$$P_{rk}(\text{TX}) = [1 - P'_{ecl}(\text{TX})]^{N_d} \tag{9.28}$$

where

$$P'_{ec_l}(\text{TX}) = \frac{1}{n_c}\sum_{m=t+1}^{n_c} m\binom{n_c}{m}[P_{er_s}(\text{TX})]^m [1 - P_{er_s}(\text{TX})]^{n_c - m} \tag{9.29}$$

9.3.2 Maximal Ratio Combining with FEC Codes

Similarly, for maximal ratio combining, and slow fading, the packet success probability is given by:

$$P_{rk}(\text{TX}) = \int_{0}^{\infty} [1 - P_{ec_M}(\text{TX})]^{N_d}\, p(\gamma_b)\, d\gamma_b \tag{9.30}$$

where

$$P_{ec_M}(\text{TX}) = \frac{1}{n_c}\sum_{m=t+1}^{n_c} m\binom{n_c}{m}[P_2(\gamma_b)]^m [1 - P_2(\gamma_b)]^{n_c - m} \tag{9.31}$$

For the purpose of comparison, the packet success probability for fast fading with maximal ratio combining is as follows:

$$P_{rk}(\mathrm{TX}) = [1 - P''_{ecl}(\mathrm{TX})]^{N_d} \tag{9.32}$$

where

$$P''_{ecl}(\mathrm{TX}) = \frac{1}{n_c}\sum_{m=t+1}^{n_c} m\binom{n_c}{m}[P_{erM}(\mathrm{TX})]^m\,[1 - P_{erM}(\mathrm{TX})]^{n_c-m} \tag{9.33}$$

and

$$P_{erM}(\mathrm{TX}) = \int_0^{\infty} P_2(\gamma_b)\,p(\gamma_b)\,d\gamma_b \tag{9.34}$$

9.3.3 Computational Results

We consider a maximum number of users C=30, unless stated otherwise, E_b/N_0=20 dB, and data bit rate R_b=144 kbit/s.

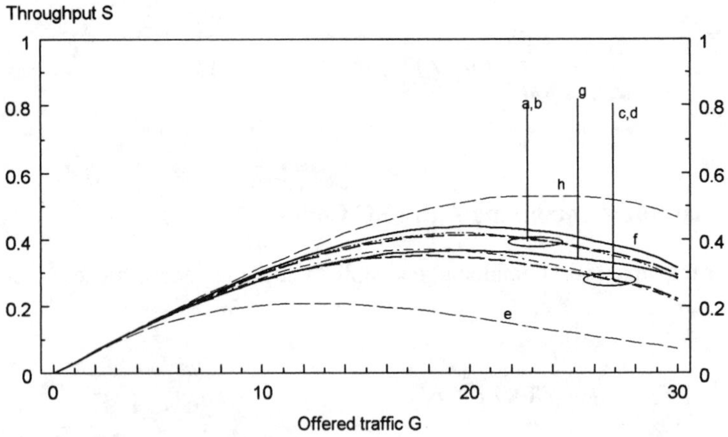

Figure 9.11 Comparison of the throughput with and without FEC codes for a Rice factor of R=6.8 dB, T_m=100 ns, p_o=p_r, N=127, N_d=128, M=2:
(a) no coding, L=2 (b) HC(15,11), L=3 (c) HC(7,4), L=4
(d) HC(3,1), L=6 (e) BCH(7,1), L=13 (f) BCH(15,7), L=4
(g) RS(7,1), L=13 (h) RS(15,9), L=4.

In Figure 9.11, the performance, is compared with system and without coding, for slow fading with maximal ratio combining with a Rice factor of R=6.8 dB, N=127, M=2, N_d=128, and T_m=100 ns.

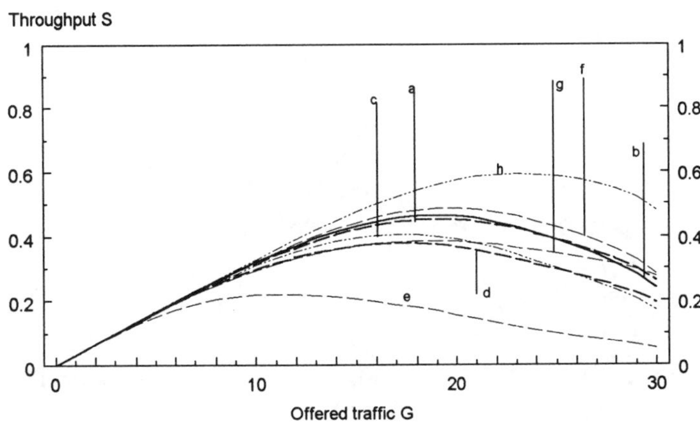

Figure 9.12 Comparison of the throughput with and without FEC codes for a Rice factor of R=11 dB, T_m=100 ns, p_o=p_r, N=127, N_d=128, M=2:
(a) no coding, L=2 (b) HC(15,11), L=3 (c) HC(7,4), L=4
(d) HC(3,1), L=6 (e) BCH(7,1), L=13 (f) BCH(15,7), L=4
(g) RS(7,1), L=13 (h) RS(15,9), L=4.

We see that FEC codes improve the system throughput. BCH (7,1) code yields the poorest performance (curve e). The best result is obtained with RS(15,9) code (curve h).

Figure 9.12 shows the effect of FEC codes on the throughput for a Rice factor of 11 dB. In general, the throughput curves yield a higher maximum result than in Figure 9.11. We see that except for RS(15,9) and BCH(15,7) codes, other FEC codes yield a relatively poor result. The RS(15,9) code curve in Figure 9.12 gives the best throughput, and BCH(7,1) yield the poorest result.

The steady-state MRC delays (with slow fading) with and without FEC codes are compared in Figure 9.13. We have considered the same parameter settings as Figure 9.11 with the corresponding plots of the throughput. We see that the lowest delay is reached by using RS(15,9) code, and a system using BCH(7,1) code yields the largest delay.

From Figures 9.11, 9.12, and 9.13, we conclude that the best improvement is given by RS(15,9) code. Since the performance using RS(15,9) is the best of the cases considered, hereafter computational results are presented only for RS(15,9) code.

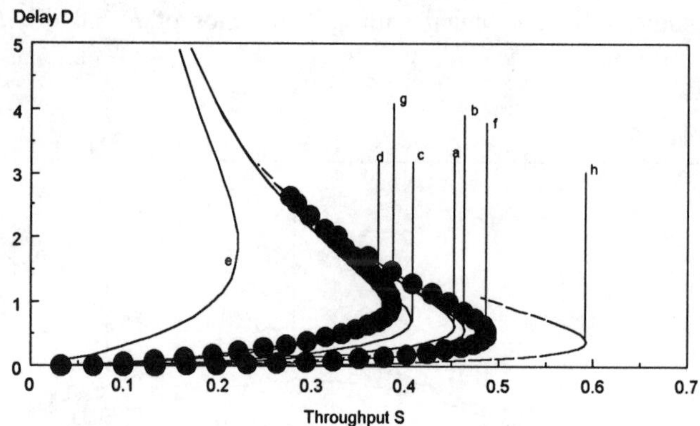

Figure 9.13 Delay versus throughput, with and without FEC codes for a Rice factor of R=11 dB, T_m=100 ns, p_o=p_r, N=127, N_d=128, M=2.
(a) no coding, L=2 (b) HC(15,11), L=3 (c) HC(7,4), L=4
(d) HC(3,1), L=6 (e) BCH(7,1), L=13 (f) BCH(15,7), L=4
(g) RS(7,1), L=13 (h) RS(15,9), L=4.

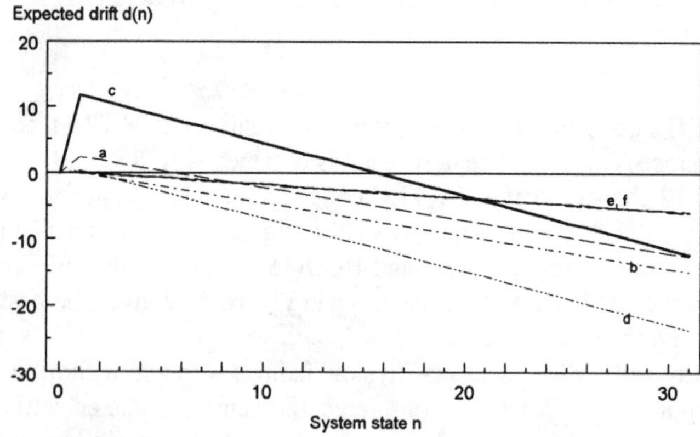

Figure 9.14 Expected drift $d(n)$ in system state n for MRC, with and without FEC codes for a Rice factor of R=11 dB, T_m=100 ns, N=127, N_d=128, M=2, and p_o=p_r.
for no coding, L=2, (a) p_r=0.5 (b) p_r=0.2 (c) p_r=0.8
for RS(15,9), L=4, (d) p_r =0.5 (e) p_r=0.2 (f) p_r=0.8.

In Figure 9.14, the expected drift $d(n)$ in system state n is plotted for N=127, R=11 dB, N_d=128, and p_o=p_r. This system is always stable for RS(15,9) code. For systems without coding, when p_o=p_r=0.2, the system is stable, but when p_r and p_o get larger, the system becomes unstable (curves a and c), because at first the system has a positive slope and then cross zero with negative slope.

The state occupancy probabilities of the system with the same parameters used in Figure 9.14 are depicted in Figure 9.15, where we can see seen that, for a system without coding and when p_r is increased, an average system state is increased. For RS(15,9) code, there is no slow change in system state, and the system has only low system states.

Figure 9.16 presents the expected drift in state n with fixed p_o=0.2 and different p_r.. Figure 9.16 shows, that the system is stable for all values of p_r, because the curves have a negative slope.

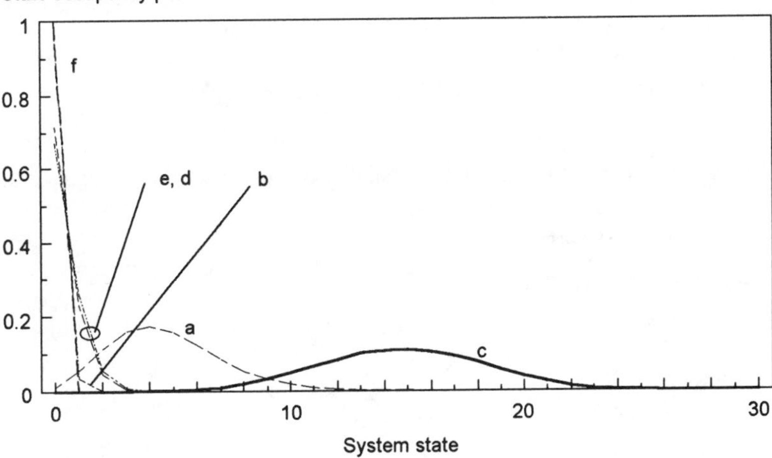

Figure 9.15 The state occupancy probability $m(n)$ in system state n for MRC, with and without FEC codes for a Rice factor of R=11 dB, T_m=100 ns, N=127, N_d=128, M=2, p_o=p_r.:
for no coding, L=2, (a) p_r=0.5 (b) p_r=0.2 (c) p_r=0.8
for RS(15,9), L=4, (d) p_r=0.5 (e) p_r=0.2 (f) p_r=0.8.

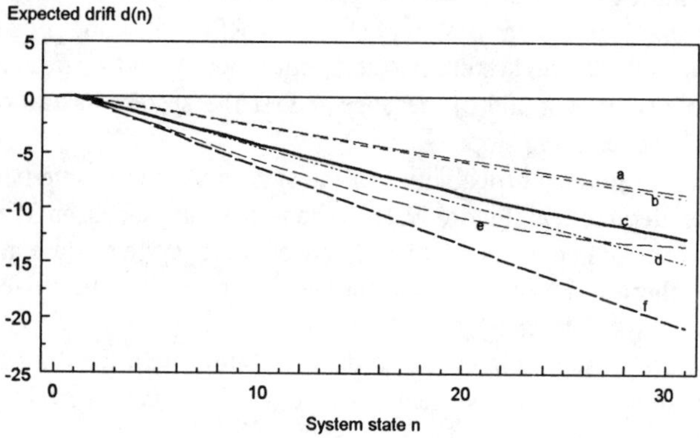

Figure 9.16 Expected drift $d(n)$ in system state n for MRC, with and without FEC codes for a Rice factor of R=11 dB, T_m=100 ns, N=127, N_d=128, M=2, and p_o=0.2:
for no coding, L=2, (a) p_r=0.3 (b) p_r=0.5 (c) p_r=0.7
for RS(15,9), L=4, (d) p_r=0.3 (e) p_r=0.5 (f) p_r=0.7.

Figure 9.17 shows the state occupancy probability of this system with fixed p_o=0.2.

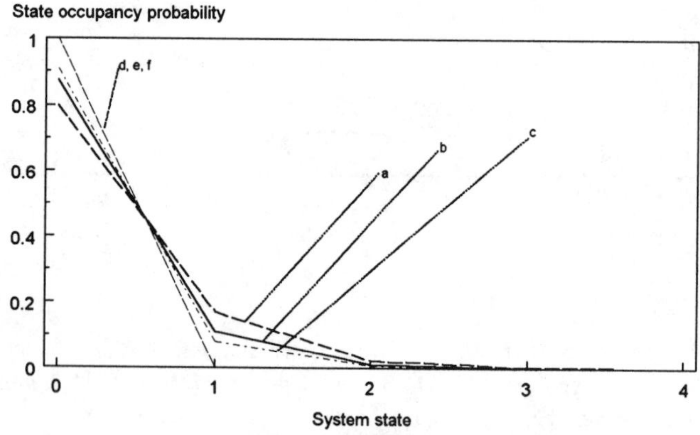

Figure 9.17 The state occupancy probability $m(n)$ in system state n for MRC, with and without FEC codes for a Rice factor of R=11 dB, T_m=100 ns, N=127, N_d=128, M=2, p_o=0.2:
for no coding, L=2, (a) p_r=0.3 (b) p_r=0.5 (c) p_r=0.7
for RS(15,9), L=4, (d) p_r=0.3 (e) p_r=0.5 (f) p_r=0.7.

9.4 CONCLUSIONS

Using a Markov chain model, the performance of a slotted CDMA in the indoor radio channel with DPSK modulation was studied. The Markov chain model offers various possibilities (e.g., throughput, delay, stability, average backlog) for performance analysis.

The computational results show that changes in the delay spread or data bit rate result in decreased performance only if the changes cause the number of resolvable paths to be higher. Therefore, the channel capacity decreases with an increase in (1) the packet length, (2) the system capacity, and (3) the number of resolvable paths, due to an increase in either delay spread or data bit rate. Calculations have also been carried out for code sequence lengths N=127 and N=255 by maintaining a fixed data rate and delay spread. It was found that the performance of the CDMA system was enhanced due to the increase in the code sequence length, resulting in a higher value for the number of resolvable paths. The application of FEC codes shows that not all FEC codes improve the performance. The best improvement is obtained by Reed Solomon (15,9) code [RS(15,9) code].

The system with RS(15,9) is more stable than the system without coding. The system state delay is made stable by using RS(15,9) codes. Rice factor R enhances the maximum throughput. When p_0 and p_r are different, it yields different performance characteristic.

CDMA systems with optimal performance may be designed with the model, by choosing optimal values for p_r, for values of p_0 based on user activity. It was also found that fast fading channels deteriorate the transmitted signals more than slow fading channels.

REFERENCES

[1] D. Raychaudhuri, "Performance analysis of random access packet-switched code division multiple access systems," *IEEE Trans. on Comm.*, Vol. COM-29, pp. 895-901, June 1981.

[2] R. Prasad and C.A.F.J. Wijffels, "Performance analysis of a slotted CDMA-system for indoor wireless communication using a Markov chain model," *Proceedings GLOBECOM'91*, Phoenix, Arizona, pp.1953-1957, December 1991.

[3] J.G. Proakis, "*Digital Communications,*" McGraw-Hill Book Company, New York, 2nd edition, 1989.

[4] R. Prasad, H.S. Misser, and A. Kegel, "Indoor radio communication in a Rician channel using direct-sequence spread-spectrum multiple access with selection diversity," *Proceedings of IEEE Symposium on Spread-Spectrum Techniques and Applications*, King's College, University of London, pp. 6-11, 24-26 September 1990.

[5] F.C. Schoute, *"Prestatie-Analyse van telecommunicatie-systemen,"*, Deventer, Clair Technische Boeken B.V., The Netherlands 1989.

[6] G. Pujolle, D. Seret, D. Dromard, and E. Horlait," *Integrated Digital Communications Networks,"* John Wiley & Sons, New York, 1988.

[7] M. Kavehrad and P.J. McLane, "Performance of low-complexity channel coding and diversity for spread-spectrum in indoor wireless communication," *AT&T Tech. J.*, Vol. 64, pp. 1927-1965, October 1985.

[8] G. George Clark Jr. and J. Bibb Cain, *"Error Correction Coding for Digital Communications,"* Plenum Press, New York, 1981.

Chapter 10
Hybrid CDMA/ISMA Protocol

10.1 INTRODUCTION

This chapter presents the description of a protocol, that is a combination of direct-sequence code division multiple access (DS-CDMA) and p-persistent inhibit sense multiple access (ISMA) protocols, known as the hybrid CDMA/ISMA protocol. Its performance is analyzed with a Markov model considering differential phase shift keying (DPSK) modulation.

The hybrid CDMA/ISMA protocol is highly suitable for indoor wireless computer communications. A typical indoor wireless computer communication system using hybrid CDMA/ISMA protocol has been simulated by computer simulation to verify the performance results obtained by the Markov model. These results are obtained in terms of the throughput and delay.

The ISMA protocol was discussed in Chapter 2. Chapters 5, 6, and 9 describe the CDMA protocol in detail. A combination of DS-CDMA and p-persistent ISMA protocol has been reported in [1-6]. The hybrid CDMA/ISMA protocol has two key features, namely, (1) it solves the hidden terminal problem by routing all traffic via a central base station, and (2) it allows many users with a relatively less short transmission code onto the same computer network by having users share transmission codes. The protocol can be used in a wide range of applications where users are semi-stationary, such as wireless patient monitoring in hospitals or flexible network access for mobile salespersons who check in several times a day with portable computers.

The protocol is first analyzed using a Markov model assuming perfect power control and neglecting the thermal noise and fading. Results are also presented for a noisy fading channel. Because it is well known that CDMA enhances the survival chance of packets and ISMA limits the contention in the channel, it would be typical to expect that the performance of a hybrid CDMA/ISMA network would be quite good, and at least better than the performance of separate protocols alone, which is illustrated in Figure 10.1.

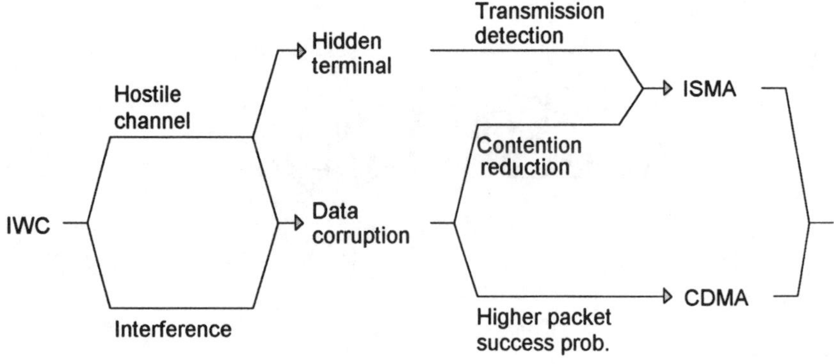

Figure 10.1 Combination of CDMA and ISMA protocols for indoor wireless communications.

For practical systems, implementing full CDMA would mean that the number of codes used equals the number of terminals in the network. In large networks this can become quite costly in terms of receiver hardware. For each code a separate receiver is needed. To reduce the number of receivers, we have investigated whether it is feasible to let a number of terminals share the same code. The number of terminals sharing the same code use the ISMA protocol. Thus, for a given indoor wireless network, N_t users are divided into n groups with different codes, where each group has $N_{t/n}$ users with the same code.

This chapter is organized as follows. Section 10.2 describes the protocol. The Markov model is presented in Section 10.3. The physical model of the system is discussed in Section 10.4. Section 10.5 treats the computer simulator of the protocol. In Section 10.6 the results obtained from the Markov model and computer simulation are compared, and Section 10.7 contains the concluding remarks.

10.2 PROTOCOL DESCRIPTION

As mentioned before, we are concentrating on indoor wireless office communications. The office system we have in mind consists of a single-story building in which users work together in groups. These users generate terminal traffic.

The terminals are stationary and communicate with each other by radio transmission using a random access protocol. On several wire communications networks such as Ethernet, the CSMA/CD protocol [7] is popular. If a terminal has to send data, it listens to determine when the line is quiet and then transmits its data. Due to propagation delays in the wire two or more terminals may transmit simultaneously, in which case information is lost. Fortunately terminals can detect such collisions and they abort the transmission and reschedule their transmissions to a later time. If the propagation delay is not too large (i.e., the channel state information is not too old), collisions are not frequent and this protocol performs well.

In radio communications, however, there is no guarantee that all terminals will always detect each other's transmissions. Obstacles might be blocking the radio waves resulting in severe performance degradation. This problem can be solved by routing all traffic via a central base station, which can be reached and heard by all terminals. As soon as the base station senses that a transmission is in progress, it issues a busy tone on a separate channel to forbid all new transmissions until the current transmission has finished. It is still possible for multiple terminals to start transmissions that collide, but the number of simultaneous transmissions is greatly reduced. This is the ISMA protocol.

To improve protocol performance, we could find a way to increase the survival chances of colliding packets. For instance, a packet may "capture" the receiver if it is much louder than its competitors. This can happen when terminals use the same transmission power, but at different distances from the receiver. This is the near-far effect. Although the near-far effect may improve the overall throughput in some cases, it is usually undesirable because it leads to unfairness among transmitters. Transmitters near the base station get better service because their packets have a better survival chance. The near-far effect can be eliminated by assuming perfect power control. This is an assumption we have also considered. Perfect power control means the transmission power of all transmitters is adjusted so that their received powers are all equal. The receiver stations could transmit a pilot tone to all terminals. Since the channel is reciprocal, the terminals can adjust their transmission power according to the pilot strength. What we want is an equal opportunity for all packets to be received correctly. This is where spread spectrum comes into play. If all messages are coded with special signature sequences, the receiver at the base station can tune in to the first inbound packet and receive it successfully, even if it is followed by many other transmissions just a fraction of a second later. The base station in our model has only one return channel. A consequence is that only one packet can be serviced at a time. Because this return channel is much simpler to implement (the base station is the only user, so there is no contention), we have ignored it in our study. Because ISMA limits the contention in the channel and because CDMA improves the survival chance of packets, it would be logical to expect that the performance of a hybrid CDMA/ISMA network will be quite good, and at least better than the performance of the separate protocols alone.

Looking at the practical systems, implementing full CDMA would mean that the number of codes used would equal the number of terminals in the network. For small networks this may be fine, but in large networks this can become quite costly in terms of receiver hardware. For each code a separate receiver is needed. To reduce the number of receivers, we have investigated whether it is feasible to let a number of terminals share the same code. It is not very likely that two or more transmitters using the same code may start simultaneously. Moreover, if the transmissions start more than one code bit or chip apart, then the receiver still can receive the first transmission successfully.

10.2.1 Performance Measures

We are interested in two performance quantities: throughput S and delay D. The throughput can be seen as a measure of efficiency. We define it as the fraction of the time that the return channel is used to transport correct packets. The delay is the time it takes for a packet, from the moment of its arrival at the terminal, to reach its destination. Often a high throughput also means high delays, so a balance between the two must be found. In our system the performance depends on a large number of parameters, which can be divided into three main groups: ISMA protocol related parameters, CDMA protocol related parameters, and physical system parameters. The parameters among the first group are offered traffic G and transmission probability p_{tr}. The CDMA related parameters are code type used, code length, total number of terminals, and the number of terminals sharing the same code. The physical system parameters include bit rate, packet size, channel characteristics such as delay spread and path attenuation, and order of diversity.

Now how are we going to find the performance parameters? We have used two approaches. First we have a theoretical approach, the Markov chain model, which is a statistical model that gives results in terms of averages. Second, we have made use of a simulation program to mimic the actual processes taking place in our system in real life. By keeping scores on certain events we can determine the system performance.

10.2.2 Terminal and Traffic Model

Because real-life systems are far too complex to model in detail, certain assumptions must be made to simplify the system and to ensure that insight is not lost in the complexity of the system. Below the assumptions are listed along with their motivation. We assume that all terminals share the same characteristics, except for their relative positions. At the terminals data packets arrive, generated by the users. These arrivals take place as if they were generated by a Poisson process. This assumption is required for using the Markov model. Each terminal has two states: free or blocked. Upon arrival of a packet at a free terminal, the terminal goes to the blocked state immediately and ignores all further incoming packets until it has serviced the current packet successfully. The terminal has no packet buffers. When the terminal is in a blocked state it will use the slotted p-persistent ISMA protocol depicted in Figure 10.2.

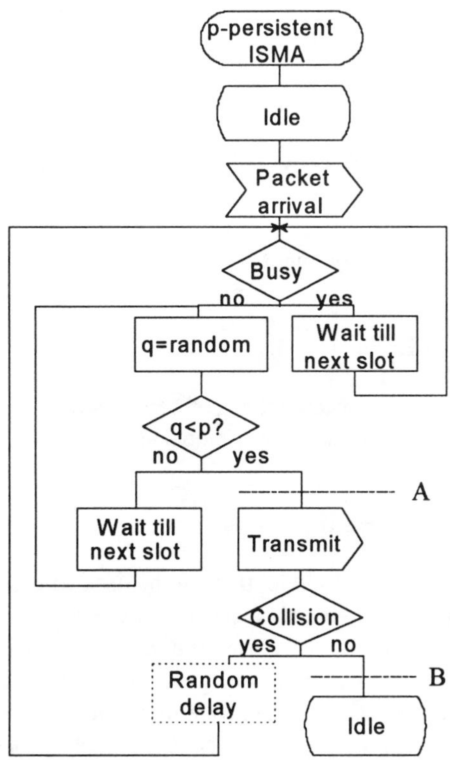

Figure 10.2 Terminal transmission protocol.

The protocol is slotted to limit contention and to simplify the modeling. In a slotted protocol, transmitted packets, if they collide, will overlap completely instead of partially overlapping other packets. In this way, the vulnerable period of packets is reduced from a twopacket duration to a onepacket duration.

A terminal in the blocked state will listen to the busy tone transmitted by the base station to determine when the channel becomes idle. As soon as the channel is idle the terminal performs a random binary experiment to determine whether it will really attempt to transmit the packet. In this way the number of active terminals is reduced, thus reducing the probability of a collision. If the probability of transmission is p_{tr}, then on average a blocked terminal will transmit every p_{tr}^{-1} time slots. Two points are different from what is usually assumed in the literature [7]. In the flowchart of the protocol shown in Figure 10.2 a box labeled "Random delay" is drawn. The point of this box is to ensure that if a transmission attempt was unsuccessful, the Poisson nature of the traffic offered to the channel is maintained. When the channel is congested and this box is not present, the terminal will try to transmit quite often, depending on p_{tr}. We have set the random delay in this box to zero because it would be cumbersome to include into our Markov model. Despite its exclusion this protocol performs quite well, even at high loads. The other thing that is different is our traffic reference. In our case we are interested in the physical traffic over the channel, because it causes interference with other packets and actually influences protocol performance. Therefore we define the physical channel traffic T_p as the traffic flowing across line A in the flowchart. The traffic flowing across line B will be regarded as the throughput S.

10.3 MARKOV MODEL

In this section we create a Markov model of the network. First we discuss Markov models in general. Then the specifics of our model are reviewed.

10.3.1 General Markov Model Description

The Markov model is a statistical model that can be used to describe a large number of processes, such as population dynamics, queuing systems, and economical processes. One of the main characteristics is that those processes have discrete states, for instance, the size of a population. The dynamics are governed by transition probabilities. They describe the likelihood that a system will change from one state to another within a given length of time. If this time is infinitely short, the model becomes a continuous time model and we speak of transition intensities instead of probabilities. If transitions can only occur to neighboring states, we have a birth-death model. This model can be depicted as a chain of states as shown in Figure 10.3.

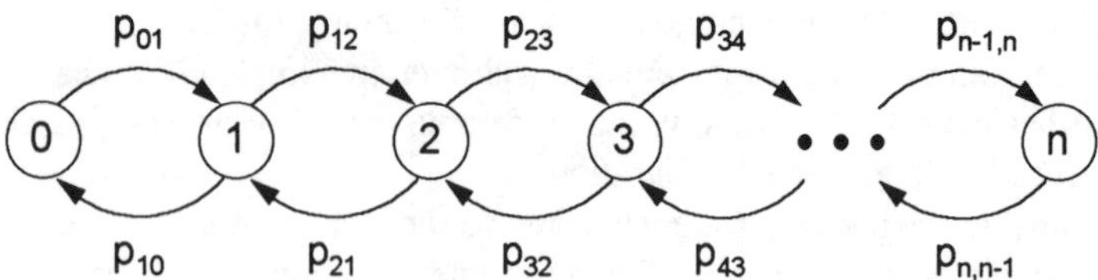

Figure 10.3 Markov chain model.

An important assumption for Markov models is that the system is memoryless, that is to say, future developments of the system are determined by the current state only. This implies that arrivals in a queuing system should be a Poisson process: the interval times are negatively exponentially distributed and the arrival time is independent of the time already spent waiting. The transition probabilities are grouped together in a matrix *P*. In this matrix the element row *I* and column *j* denotes the transition probability from state *I* to state *j*. The elements of each row must add up to 1. If the current state probability distribution of the system is the row vector $\vec{v}_o$,then it can be found by multiplying the previous state by the transition matrix *P*. If the model is finite, has no loops, and each state can eventually be reached from any other state, the Markov model will reach a steady state after an infinite number of steps. Here we are interested in the steady-state behavior of the system, which is not difficult to find. Steady state means that going to the next point in time leaves the state of the system unchanged. The following condition must be met: $\vec{v} = \vec{v}P$ or $\vec{v}_{steady\ state}(P - I) = \vec{0}$. The set of equations given by the last expression is not independent. To find the steady-state vector, we will have to make use of the fact that the sum of the elements of $\vec{v}_{steady\ state}$ equals 1. We construct a matrix $P' = P\text{-}I$, then we replace the last column of P' with ones, and construct a row vector $\vec{E} = [0, 0, ..., 0, 1]$. The steady-state vector is $\vec{v}_{steady\ state} = \vec{E}\,P^{-1}$ [8].

10.3.2 Markov Model for CDMA/ISMA Protocol

The state variable we use to create this model is the average number of blocked terminals μ, because it makes it relatively easy to derive channel traffic, throughput, and delay from this state variable. The physical channel traffic T_p, for example, equals the number of blocked terminals multiplied by the probability that they will transmit:

$$T_p = \mu\, P_{tr} \tag{10.1}$$

The other quantities are equally easy to determine. If the total number of terminals is N_T, the average number of blocked terminals is μ, and each terminal has an arrival probability *p*, then the throughput *S* is given by:

$$S = p(N_T - \mu) \tag{10.2a}$$

The reasoning behind (10.2) is that if, on average, μ terminals are blocked then apparently the other N_T-μ terminals are getting their arrivals serviced at the same rate at which they arrive.

The packet delay can be found by analyzing the possible paths a terminal can follow in the protocol. Looking back at Figure 10.2 we see that there are three places where a packet can be delayed. The packet first experiences a delay when arriving at an idle terminal, on average half a time slot. The second type of delay occurs during the transmission of the packet, whether it is successful or not. And then we have the delay waiting for the next time slot if a terminal decides not to transmit.

The delay model and its simplification are shown in Figure 10.4., where p, q, s, and t represent the probability that a certain branch is chosen, p is the transmission probability p_{str}, q is its complement, $s = S/T_p$ is the probability that a transmission is successful, t is the complement of s, and T is the duration of one time slot. The probability that a delay of 0.5+n time slots is encountered is $ps(q + pt)^{n-1}$. The average delay D is then:

$$D = \frac{1}{2} + \frac{T_p}{(p_{str}S)} \tag{10.2b}$$

where $p_{str} = ps$ is the probability of a successful transmission occurring during a time slot.

Now that we have established the performance parameters we can start creating the Markov model. In the network we have a number of terminals at which packets arrive and a base station that can only serve one packet per time slot at most. Examination of this network leads to five possible scenarios per time slot:

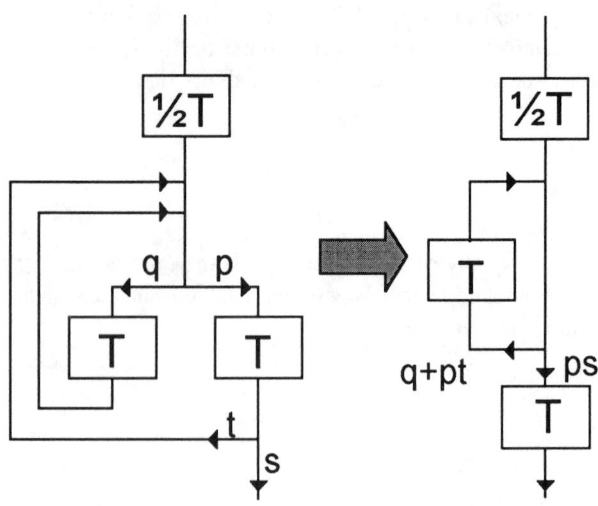

Figure 10.4 Packet delay model.

1. The network happens to be empty and zero or more arrivals occur;
2. The system state decreases by 1 because of a successful transmission and no new arrivals occur;
3. The system stays in the same state or moves up to any state but the highest due to multiple arrivals and a successful transmission;
4. The system jumps from the current state to the highest state because of multiple arrivals and no successful transmission;
5. State decreases by more than 1 (probability 0).

The division between scenarios 3 and 4 might seem a bit artificial, but it is not. Scenario 3 can be caused by two mutually exclusive events. If the system is in state i, it can jump to state $i+n$ by either n arrivals and no success or $n+1$ arrivals and 1 success. If a success takes place, then the highest state is unreachable. In scenario 4 no successes take place and the system jumps to the highest state. Scenario 5 is included to define the rest of the transition probabilities. The scenario may become clearer if we look at the expression for the transition probabilities:

$$p_{i,j} = \begin{cases} \binom{N_T}{j} p^j q^{(N_T-j)} & i = 0 \wedge 0 \le j \le N_T \\ p_{str} q^{(N_T-i)} & 1 \le i \le N_T j = i-1 \\ p_{str}\binom{N_T-i}{j-i+1} p^{(j-i+1)} q^{(N_T-j-1)} + (1-p_{str})\binom{N_T-i}{j-i} p^{(j-i)} q^{(N_T-j)} & 1 \le i \le N_T - 1 \wedge 1 \le j \le i \\ (1-p_{str}) p^{N_T-i} & 1 \le i \le N_T - 1 \wedge j = N_t \\ 0 & 2 \le i \le N_T \wedge 0 \le j \le i-2 \end{cases} \tag{10.3}$$

where N_T is the number of terminals in the network, i is the current state of the system, j is the next state of the system, p is the arrival probability per terminal, q equals $1\text{-}p$, and p_{str} is the probability that a successful transmission takes place during a time slot.

Since the number of terminals is discrete and finite, and since all terminals are equal, it is no surprise that the transition probabilities are governed by binomial distributions. Further insight into the matrix structure may be gained by looking at Figures 10.5 and 10.6, which show the model state diagram and the matrix regions. The problem now is that the whole model hinges completely on the probability that a successful transmission takes place during a time slot p_{str}. This problem turns out to be complex enough to warrant a section of its own.

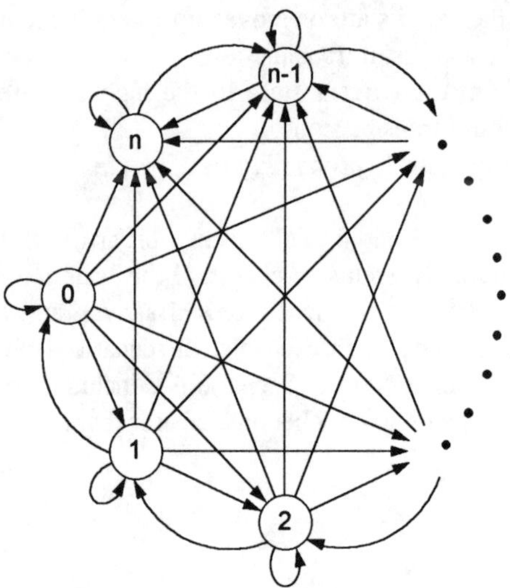

Figure 10.5 Markov model for ISMA/CDMA protocols. Transitions can occur to any higher state, the same state, and one state down.

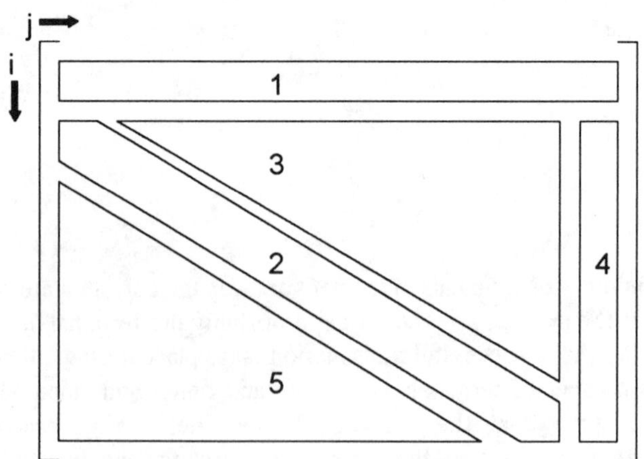

Figure 10.6 Matrix regions.

10.3.3 Probability of a Successful Transmission

The determination of the probability of a successful transmission is complicated by the following facts. In the first place it is not possible to talk about the value of p_{str}, because it depends on the system state and the probability that a blocked terminal will transmit p_{tr}. In the second place p_{str} will depend on the behavior of all blocked terminals. The blocked terminals may or may not transmit depending on p_{tr}. In the third place, the code sharing enforces a two-level hierarchy on the system. First a packet has to reach the distributed receiver successfully and then it must be accepted by the base station. Finally there are three possible events per blocked terminal per time slot: it is either a successful transmission or an unsuccessful one, both types causing interference, or the terminal is silent, causing no interference. The formula for p_{str} is developed by first looking at the transmission process of each terminal group. Once we have the group probabilities we can calculate p_{str} by regarding each group as a new terminal in a supergroup transmitting to the base station. Note that since free terminals can only be quiet, these terminals can be regarded as nonexistent in our calculations.

Terminal Probabilities

We start with the probability p_{ts} that a transmitting terminal is successful. This depends on whether or not other blocked terminals transmit. Because the other blocked terminals usually are at different distances from the receiver belonging to the reference terminal, we have to take all possible combinations into account. Fortunately there are only two possibilities (transmission or no transmission), so we have a binomial distribution:

$$p_{ts}(\vec{\mu},p_{tr}) = \sum_{i_\mu=0}^{1}\cdots\sum_{i_1=0}^{1} p_{tr}^{x}(1-p_{tr})^{\mu-x}\left[1-p_{be}\left((\vec{\mu},p_{tr})\right)\right]^{L_p} \tag{10.4}$$

where L_p is the packet length, $\vec{\mu}$ is a certain configuration of blocked terminals, and x is the number of active terminals given by:

$$x = \sum_{v=1}^{\mu} i_v \tag{10.5}$$

The last factor in brackets in (10.4) represents the packet success probability. To determine this we need the expression for the DPSK bit error probability. The probability that a blocked terminal is quiet is the complement of the probability that it transmits, so $p_{tq} = 1-p_{tr}$. From the fact that a collision is the only other possibility follows $p_{tc} = 1-p_{ts}-p_{tq}$.

Group Probabilities

The distributed receiver of each group tunes into the signal of the first packet it receives and forwards it to the base station, whether it is a good packet or not. This capture process is modeled as an equiprobable selection of arriving packets. This means that if there are i successful packets and j unsuccessful packets, the probability that a good packet is sent to the base station is simply $i/(i+j)$. Because there are three possible events (success, collision, silence), we will obtain a trinomial distribution. The probability that a particular group is successful, p_{gs}, is a summation of the probabilities of all possible combinations of successful and unsuccessful transmissions and silence per group, weighted by the success probability for the group for that combination. Mathematically this is expressed as:

$$p_{gs} = \sum_{i=0}^{N_t} \sum_{j=0}^{N_t - i} \frac{i}{(j+i)} \frac{N_t!}{i!j!(N_t - i - j)!} p_{ts}^i p_{tc}^j p_{tq}^{N_t - i - j} \tag{10.6}$$

where N_t is the number of terminals per group. The factor $i/(i+j)$ in this expression blocks any simplification of this equation.

The probability that a group is silent is simply the probability that all terminals are silent:

$$p_{gq} = (1 - p_{tr})^{N_t} \tag{10.7}$$

As before, the probability that a collided packet is sent to the base station is $p_{gc} = 1 - p_{gs} - p_{gq}$.

Unlike the terminal probabilities, the group probabilities will differ. Within the group the terminal probabilities are the same because these terminals transmit to the same receiver. Because of the assumed perfect power control all the signals from within the group experience the same amount of interference. This is not so between groups in general.

Successful Transmission Probability

For the final group success probability, we have generalized (10.4) to:

$$p_{str} = \sum_{i_n=0}^{L_n} \sum_{j_n=0}^{L_n - i_n} \cdots \sum_{i_1=0}^{L_1} \sum_{j_1=0}^{L_1 - i_1} \left[\frac{\sum_{k=1}^{n} i_k}{\left(\sum_{k=1}^{n} i_k + \sum_{k=1}^{n} j_k \right)} \prod_{v=1}^{n} \frac{L_v}{i_v! j_v! [L_v - i_v - j_v)!]} p_{gs,v}^{i_v} p_{gc,v}^{j_v} p_{gq,v}^{L_v - i_v - j_v} \right] \tag{10.8}$$

where n is the number of groups with blocked terminals, L_v is the number of groups with blocked terminals with the same characteristics, and $\sum i_k / (\sum i_k + \sum j_k) = 0$ where even $\sum i_k = 0$ and $\sum j_k = 0$. A slight simplification is possible by realizing that all i_v, j_v, and L_v are either 0 or 1, so the fraction with the factorials may be replaced by 1.

Blocked Terminal Distribution

So far we have not said anything about the distribution of the blocked terminals and, in fact, at this point things are starting to get pretty ugly numerically speaking. The mathematically correct thing to do is to assume that all possible distributions of blocked terminals are equiprobable. Then we could average (10.8) over these terminal distributions. Even if the number of terminals is small the calculations will become enormous. For a network of 32 active terminals of which 16 are blocked the number of possibilities is roughly 600 million (32 over 16). We solve this problem by adopting a slightly pessimistic case scenario. In this scenario we start by picking the terminal closest to the base station as the terminal that will be blocked when the system is in state one. As the state increases the terminals blocked next are those terminals closest to the first terminal, under the constraint that the difference between the number of blocked terminals of any two groups may not be larger than 1. In other words, the blocked terminals are spread over the groups as uniformly as possible, but these terminals are "attracted" to the first blocked terminal. Figure 10.7 gives an example of a network and the terminal activation order.

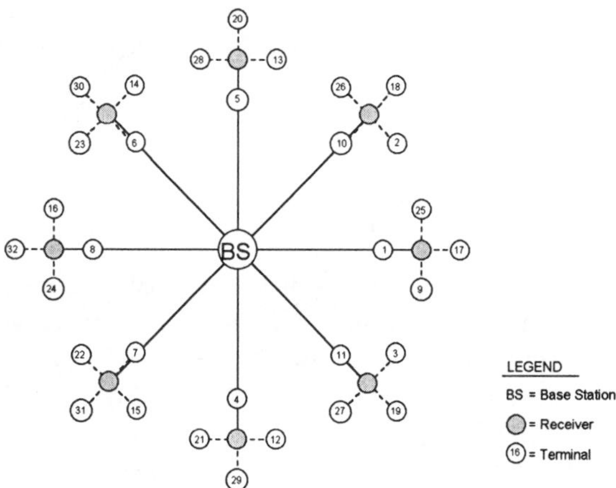

Figure 10.7 Example of network configuration. The numbers of the terminals indicate in which order they are assumed to become active.

10.4 SYSTEM MODEL

This section describes the transmission, channel, and receiver models. We also discuss reference systems of parameters.

10.4.1 Transmitter Model

Packet data are transmitted using DPSK. This is done on the assumption that the phase shifts occurring in a transmission channel are continuously changing. The changes are assumed to be too fast for successful coherent phase shift keying, but slow enough that the phase shift does not change significantly during two data bit intervals. DPSK is a relatively robust transmission method and is easy to implement. After DPSK modulation the signal is spread using the direct-sequence method with Gold codes.

10.4.2 Channel Model

The transmission channel we are considering is quite ideal. For instance, we assume that the channel does not suffer from fading. This assumption is made because the simulation of the network performance with fading would be too time consuming. We do, however, include multipath propagation in our model. The power from the receiver is assumed to be equally distributed over all paths. We assume no direct line of sight (LOS) path between transmitter and receiver, because a LOS path usually carries more power than specular paths. The transmitter would typically be placed behind the computer or at the far end of a desk, the signal reflecting against walls and computer metal casing.

The number of resolvable paths L is determined by the delay spread T_{max} and the chip duration T_c:

$$L = \left\lfloor \frac{T_{max}}{T_c} \right\rfloor + 1 \tag{10.9}$$

One of the effects of multipath propagation is that even if only one transmitter is active, its receiver will suffer from interference because several delayed versions of the same signal, with different phase shifts, will arrive at the receiver in short succession. The channel phase shifts are assumed to be uniformly distributed over $[0, 2\pi]$ and likewise the signal delays over $[0, T_b]$, where T_b is the bit duration. The channel is considered free of thermal noise for three reasons. As for the signal attenuation model, the far-field model was chosen because it is relatively easy to implement and easy to adapt to various environments. The path loss can be described as [3]:

$$S_r = \frac{g_T\ g_R}{\left(\dfrac{4\pi f R}{c}\right)^{\gamma}}\ S_t \qquad (10.10)$$

where S_r is the received power, S_t is the transmitted power, f is the transmission frequency, R is the distance between transmitter and receiver, g_T is the transmitter antenna gain, g_R is the receiver antenna gain, c is speed of light, and γ is the attenuation parameter.

Attenuation γ can vary between 1.8 and 6, depending on the environment. The former value corresponds with a transmitter and receiver in a hallway, where the hallway acts like a waveguide. The latter value corresponds to transmitters and receivers located in rooms perpendicular to a hallway with the doors closed. For ease of implementation γ equals 2 in our calculations, corresponding to free-space propagation.

If for example S_t = 1 W, f = 1.7 GHz, $g_t = g_r = 1$, and $c = 3.10^8$ m/s, then the received power at 15 m would be 0.88×10^{-6} W, a considerable path loss of 60.57 dB. The transmitter antenna gain in practice could very well be 1 because the user should not have to worry about the orientation of the antenna or lose a connection by moving equipment. A gain of 3 dB could be obtained by fixing the receiver antenna to the ceiling and making it sensitive to signals from below. This also reduces possible interference from floors above. Of course, the receiver antenna gain does not reduce multipath propagation as the interfering signals experience the same gain. In practice, however, it does give an advantage against the thermal noise.

The far-field model is valid as long as the following condition is met:

$$R > \frac{2\,r^2}{\lambda} \qquad (10.11)$$

where r is the size of the antenna and λ is the wavelength. In our system this would not be very restrictive because for an antenna of 20 cm transmitting at 1.7 GHz the distance above which the far-field model is valid is about 46 cm. Incorporating other propagation models should not be too difficult. The calculation programs have been written with this in mind.

An important assumption about the interference is that during transmission the interference power remains essentially fixed. This assumption is quite realistic because the transmission protocol and the negligible propagation delay force terminals to start their transmissions within much less than one bit duration from each other.

10.4.3 Receiver Model

The basic block diagram for the receiver is given in Figure 10.8. After the signal is received at the antenna, it is translated to an intermediate frequency. From there it is passed through a filter, which is matched to the code to be received. This is where the signal despreading takes place. After that the signal is DPSK demodulated. The receiver does not use diversity techniques other than the inherent diversity of the spread spectrum.

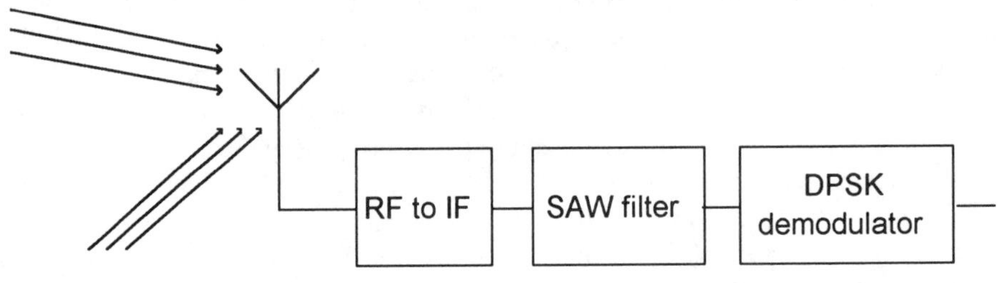

Figure 10.8 Receiver block diagram.

The demodulated data signal is now sent by wire to the base station. It is advantageous to put the receiver as close as possible to the terminals instead of at the base station. Although this does not make the system completely wireless, it does improve the terminal to receiver SIR and reduces the transmission power needed. This in turn reduces interference to the system itself and to other systems. This can become quite significant in large systems, since in most countries the maximum transmission power for IWC is limited to 1 W [9]. Once the base station notices that one of its receivers is active, all other incoming packets are ignored until the current packet has been handled. Just as with the receivers, the base station is a relatively unintelligent entity. It will service the packet, whether it is good or not. The packet is broadcast to all terminals through a separate transmission channel and the destination terminal can recognize it by the addressing bits in the packet header. The base station broadcast channel is not taken into consideration in our model because it is much simpler in nature. The only user of the broadcast channel is the base station itself and hence there is no contention for this channel. We also have not concerned ourselves with packet acknowledgment. Acknowledgments could be piggybacked onto return packets. We have only looked at raw data bit performance. The net network performance will depend on the error correction, addressing, and acknowledgment schemes. The base station controls the incoming traffic by transmitting a busy tone. As long as this tone is heard by the terminals no new transmissions are allowed to start. In our system this tone is used to indicate the time slots. At the beginning of each slot the busy tone is interrupted long enough to allow new transmissions to commence. In a practical system the busy tone would be replaced by a free tone to improve the system reliability. If the busy tone were not heard by a terminal due to fading or an object blocking the signal, it could accidentally start a transmission and interfere with other transmissions. If the free tone happened to be blocked, a transmission would not take place and less damage would be done.

10.4.4 Reference System Parameters

To evaluate the network models, a reference network was designed. To simplify the calculations the network has been kept as symmetrical as possible. This means that the receivers have been placed in a circle at 30 m from the base station and the terminals have been grouped around their receivers at a distance of 5 m, also in a circle. The network geometry for an eight-code network is shown in Figure 10.9.

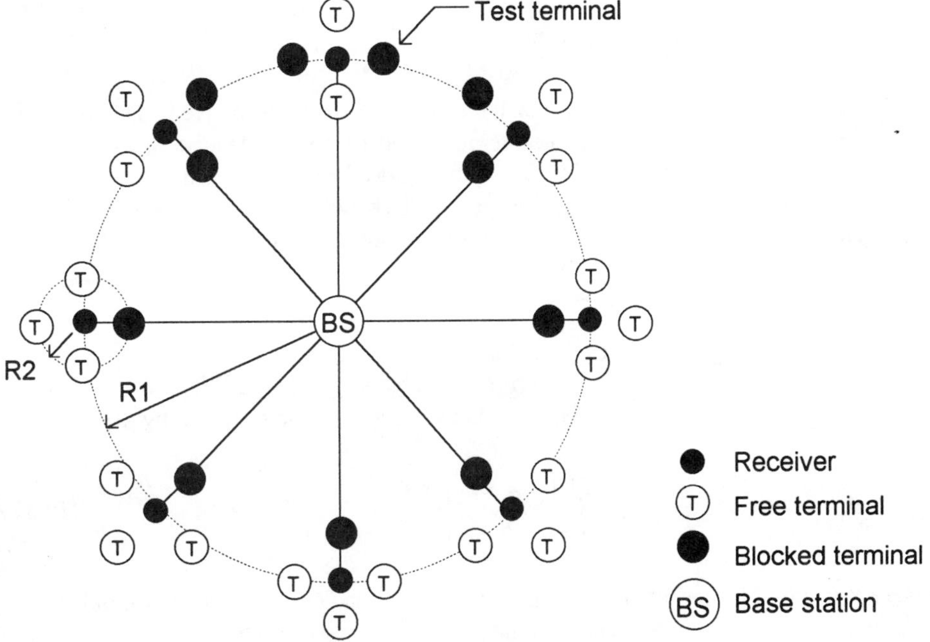

Figure 10.9 Reference network geometry.

To investigate the effect of code sharing, the number of terminals is chosen as a power of two. In this way we can start with one code for all terminals and then double the number of codes and halve the group size for each new configuration. The code lengths will be 31 and 127 chips. Anything more than 127 chips requires too much memory storage and leads to long calculation times. An arbitrary data rate is chosen to be 256.1024 bits/ or 0.26 Mb/s.

We assume a delay spread of 100 ns here. It may seem a bit large, but this way we can determine the effect of multipath propagation. According to (10.9) this gives us one resolvable path in the case of 31 chip codes and 4 paths for 127 chip codes.

The packet lengths may vary, but for now they are usually set to 64 bits. Even without error correction coding this packet length gives acceptable success probabilities. This demonstrates the robustness of spread-spectrum transmission.

The center frequency of the transmitter is 1.7 GHz. The bandwidth of the transmitted signal is 16.25 MHz for 31 chip codes and 66.6 MHz for 127 chip codes. The fractional bandwidth is 0.95% and 3.9% of the center frequency, so it is not far from the truth to assume the same path loss for all frequencies.

10.5 BIT ERROR PROBABILITY

In this section an expression for the bit error probability is derived. The bit error probability is readily available in the literature and of course depends heavily on the received signal-to-noise ratio (SNR). The better part of this section is devoted to finding the noise power, or rather, interference power since we will be excluding the thermal noise. Here it will become clear why proper spread-spectrum coding is essential.

10.5.1 BEP with DPSK

The basic expression for the bit error probability p_{be} for DPSK with multipath propagation in M time invariant frequency-nonselective channels is given by [3]:

$$p_{be} = \frac{1}{2^{(2M-1)}} \exp(-\gamma_b) \sum_{k=0}^{M-1} \left[\frac{\gamma_b^k}{k!} \sum_{n=0}^{M-1-k} \binom{2M-1}{n} \right] \qquad (10.12)$$

where γ_b is the SNR per bit. This formula is valid on condition that the interfering signals can be treated as independent Gaussian noise.

10.5.2 Signal-to-Interference Ratio with Multiuser Interference

As mentioned in the previous section the signal-to-interference ratio (SIR) γ_b is the key parameter to the bit error probability. In our analysis of γ_b we start with the signal $r_u(t)$, a distributed receiver belonging to user u sees at its input:

$$r_u(t) = \sum_{k=1}^{K} \sum_{l=1}^{L} \sqrt{\frac{2 P_k s_{ku}^2}{L}} \, a_k(t - \tau_{kul}) \, b_k(t - \tau_{kul}) \cos(\omega_c t + \theta_k + \phi_{kul}) \qquad (10.13)$$

where P_k is the power transmitted by user k, s_{ku} is the amplitude gain over path k-u for ($0 \le s_{ku} \le 1$), a_k is the kth user code, b_k is the kth user data bit, τ_{kul} is the delay of path kul for ($0 \le \tau_{k,l} \le T_b$), ϕ_{kul} is the phase shift of path kul for ($0 \le \phi_{k,l} \le 2\pi$), Θ_k is the initial carrier phase of user k for ($0 \le \Theta_k \le 2\pi$), and L is the number of paths.

To clarify the above symbols we have drawn a part of the network in Figure 10.10. It shows multipath propagation between terminals u and k and their corresponding receivers. The interference from u to k does exist but is not shown for the sake of clarity. Because it is assumed that the transmitted power is divided equally over all paths, we do not need to add an extra path number to the path gain s. Without loss of generality we assume that the initial carrier phase Θ_k is absorbed by the path shift ϕ_{kul}. Equation (10.13) can be written in a lowpass equivalent form:

$$r_u(t) = x_u(t)\cos(\omega_c t) - y_u(t)\sin(\omega_c t) \qquad (10.14)$$

with

$$x_u(t) = \sum_{k=1}^{K} \sum_{l=1}^{L} A_{ku} a_k(t - \tau_{kul}) b_k(t - \tau_{kul}) \cos(\phi_{kul}) \qquad (10.15)$$

$$y_u(t) = \sum_{k=1}^{K} \sum_{l=1}^{L} A_{ku} a_k(t - \tau_{kul}) b_k(t - \tau_{kul}) \sin(\phi_{kul}) \qquad (10.16)$$

$$A_{ku} = \sqrt{\frac{2 P_k s_{ku}^2}{L}} \qquad (10.17)$$

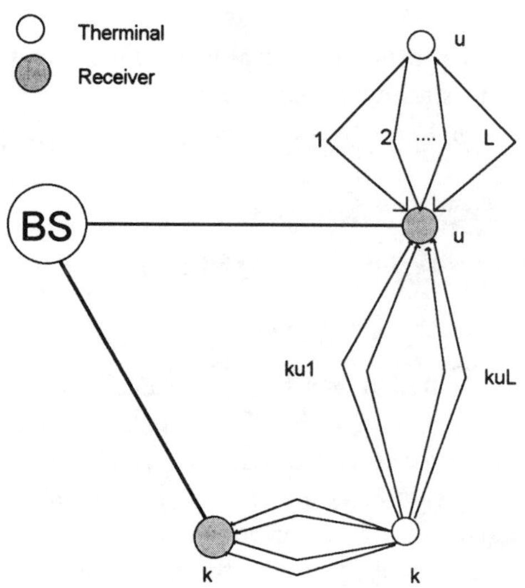

Figure 10.10 Multipath propagation between terminals and receivers.

So the in-phase component is $x_u(t)$ and the quadrature component is $y_u(t)$. In (10.16) and (10.18) we have omitted a factor of 0.5 stemming from the frequency conversion. Since this factor affects both signal and interference this does not change the SIR.

Let us assume that our reference is path h of user u. The in-phase and quadrature components after correlation at their sampling time T_b can be described as:

$$g_x(T_b) = \sum_{k=1}^{K}\sum_{l=1}^{L} A_{ku}\cos(\phi_{kul})\left[b_k^{-1} R_{uk}(\tau_{kul}) + b_k^{0}\hat{R}_{uk}(\tau_{kul})\right] \tag{10.18}$$

$$g_y(T_b) = \sum_{k=1}^{K}\sum_{l=1}^{L} A_{ku}\sin(\phi_{kul})\left[b_k^{-1} R_{uk}(\tau_{kul}) + b_k^{0}\hat{R}_{uk}(\tau_{kul})\right] \tag{10.19}$$

in which $R_{uk}(\tau_{kul})$ and $\hat{R}_{uk}(\tau_{kul})$ are partial cross-correlations:

$$R_{uk}(\tau) = \int_0^{\tau} a_u(t-\tau)\, a_k(t)\, dt \tag{10.20}$$

$$\hat{R}_{uk}(\tau) = \int_{\tau}^{T_b} a_u(t-\tau)\, a_k(t)\, dt$$

Assuming that $\tau_{uuh} = 0$ and $\phi_{uuh} = 0$, the current sample value of the complex envelope of this lowpass equivalent signal $g_x(t) + j\, g_y(t)$ can be written as:

$$Z_0 = A_{uu} T_b b_u^0 + \sum_{k=1}^{K}\left(b_k^{-1} X_k + b_k^0 \hat{X}_k\right) + j\sum_{k=1}^{K}\left(b_k^{-1} Y_k + b_k^0 \hat{Y}_k\right) \tag{10.21}$$

Similarly we have for the complex envelope at the previous sampling instant:

$$Z_{-1} = A_{uu} T_b b_u^{-1} + \sum_{k=1}^{K}\left(b_k^{-2} X_k + b_k^{-1} \hat{X}_k\right) + j\sum_{k=1}^{K}\left(b_k^{-2} Y_k + b_k^{-1} \hat{Y}_k\right) \tag{10.22}$$

where

$$\begin{aligned}
X_u &= \sum_{\substack{l=1\\ l\neq h}}^{L} A_{uu} R_{uu}(\tau_{uul})\cos(\phi_{uul}) & \hat{X}_u &= \sum_{\substack{l=1\\ l\neq h}}^{L} A_{uu} \hat{R}_{uu}(\tau_{uul})\cos(\phi_{uul})\\
Y_u &= \sum_{\substack{l=1\\ l\neq h}}^{L} A_{uu} R_{uu}(\tau_{uul})\sin(\phi_{uul}) & \hat{Y}_u &= \sum_{\substack{l=1\\ l\neq h}}^{L} A_{uu} \hat{R}_{uu}(\tau_{uul})\sin(\phi_{uul})\\
X_k &= \sum_{l=1}^{L} A_{ku} R_{ku}(\tau_{kul})\cos(\phi_{kul}) & \hat{X} &= \sum_{l=1}^{L} A_{ku} \hat{R}_{uk}(\tau_{kul})\cos(\phi_{kul})\\
Y_k &= \sum_{l=1}^{L} A_{ku} R_{ku}(\tau_{kul})\sin(\phi_{kul}) & \hat{Y}_k &= \sum_{l=1}^{L} A_{ku} \hat{R}_{uk}(\tau_{kul})\sin(\phi_{kul})
\end{aligned} \tag{10.23}$$

All the terms of Z_0 and Z_{-1} except the first one in (10.21) and (10.22) are regarded as interference. From now on this interference signal is denoted as Z_i. The decision variable for the DPSK modulator is $\xi = \mathrm{Re}[Z_0 \cdot Z_{-1}^*] = \mathrm{Re}[\,(A_{uu}T_b b_u^{0} + Z_{0,i}) \cdot (A_{uu}T_b b_u^{-1} + Z_{-1,i})^*\,]$ where * denotes complex conjugation.

The data bits are assumed to be equiprobable, giving the probability of bit error p_{be} by [3]:

$$P_{be} = Pr\{\,\xi < 0 \mid b_u^0 b_u^{-1} = 1\,\} = Pr\{\,\xi > 0 \mid b_u^0 b_u^{-1} = -1\,\} \tag{10.24}$$

For the calculation of the interference power we will assume that $b_u^{-1} \cdot b_u^0 = 1$, all other bit combinations such as $b_k^{-1}.b_i^0$ $((k=u \wedge i=u))$ are assumed to take on the values 1 and -1 with equal probability. This means that $E[b_i^v b_j^w] = E[b_i^v] E[b_j^w] = 0$ except for $i=j=u$, $v=-1$, and $w=0$. We use this property in calculating the expectation of the interference power over many samples.

The interference power per bit equals $E[Z_{o,i} \cdot Z_{0,i}^*]$. By evaluation of this expression for the interference power, we developed an expression for the conditional interference power $N_u|(\{\tau_{kul}\}\{\phi_{kul}\})$.

With the expression for the interference power, we can develop an expression for $\gamma_b|(\{\tau_{kul}\},\{\phi_{kul}\})$. In (10.21) the signal amplitude we want is $A_{uu}\ T_b b_u^0$. Therefore the signal power we want is $A_{uu}^2\ T_b^2$. This leads to:

$$\gamma_b \mid (\{\tau_{kul}\},\{\phi_{kul}\}) = \frac{A_{uu}^2\ T_b^2}{N_u \mid (\{\tau_{kul}\},\{\phi_{kul}\})} \tag{10.25}$$

The removal of the conditioning on τ_{kul} and ϕ_{kul} requires integration over these variables, which is not possible to do analytically. Instead, we used Monte Carlo integration to obtain values for our Markov model. An important assumption is that the bit error probability remains the same throughout the packet. This is justified because the busy tone emitted by the base station forces the terminals to start within one bit duration of each other.

For the calculation of the bit error probability in noisy fading channels, we have used the formula given in Chapters 5 and 6 for Rayleigh fading channels with first-order selection diversity.

10.6 PROTOCOL SIMULATION

In this section we describe how protocol performance is simulated. We begin with a general description of the simulation. Then we discuss the random number generator, an important part of the simulation program. Next the simulation program SIMIWC is described.

10.6.1 Simulation Principle

The goal of the simulation program is to get performance parameters by mimicking the processes that would take place in a real network. The performance of the network can be determined by recording certain events such as the arrival or successful servicing of a data packet. Here it is not necessary to average a large amount of noise samples to arrive at an average bit error probability. This is because the simulation considers separate transmission events. Every time a packet is sent, the data signal on its respective paths will suffer a certain delay and a certain phase shift, which will be picked at random and kept fixed during the transmission of the packet. Finally the results are averaged over all time slots. The results we are interested in are throughput and delay as a function of the mean physical channel traffic. The throughput S is the fraction of the transmission slots carrying a successful transmission. The mean delay D is the number of time slots a packet must wait until it is successful plus one-half. The half comes from the fact that T_p is the number of transmission attempts made by the terminals per time slot.

10.6.2 Random Number Generation

Random number generation forms the basis of most network simulation programs and therefore a good random generator is essential. There are several types of random number generators (e.g., the congruential generator, the Fibonacci generator, and the additive number generator) [10,11]. In practice the design of random number generators is far from trivial. It is not easy to generate a statistically sound random sequence from a deterministic process. The Turbo Pascal programming language has a built-in random number generator based on the congruential generation method. In this method we start with an initial value X_0 and generate a random number sequence according to the formula $X_{n+1} = (a\, X_n + c) \bmod m$. The quality of the generator is highly dependent on the choice of a, c, and m and good values are difficult to find. For instance Knuth [10] asserts that for certain choices of these constants all points of the set $\{(x_0, x_1, x_2), (x_3, x_4, x_5), \ldots\}$ lie on four planes in three-dimensional space. Definitely not an independent random sequence! Because the exact parameters of the Turbo Pascal random generator are not known and because the generator has only 16-bit precision, the FSU Ultra generator is used. It has 64-bit accuracy and an extremely long period of 10^{356} as opposed to roughly 65×10^3 for the Turbo Pascal generator. The basis of the FSU generator is described in [11].

10.6.3 Simulation Program

SIMIWC was written using Borland's Turbo Pascal 6.0 and the FSU Ultra random number generator version 1.05.

General Structure of SIMIWC

In Figure 10.11 the structure chart of the simulation program is given. The procedure INITCODES reads the Gold codes to be used from a file as a set of text strings. Then the codes are passed to the procedure INITCCAT which sets up a catalog of cross-correlation functions. This point requires some explanation.

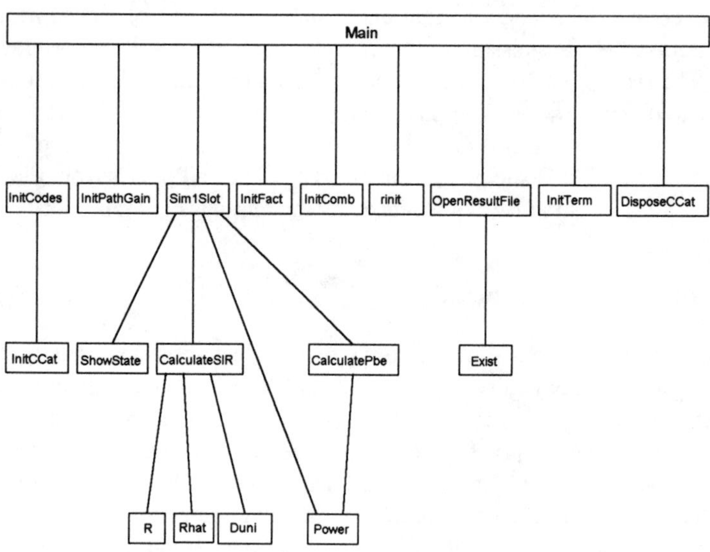

Figure 10.11 Structure chart of simulation program SIMIWC.

In (10.20) the time continuous cross-correlation functions $R_{uk}(t_{kul})$ and $R_{uk}(t_{kul})$ are defined. The cross-correlations are piecewise linear functions. All that is required to reconstruct the correlation functions is their values at every chip time of one period. According to [9], if the discrete aperiodic cross-correlation function $C_{k,i}$ is defined by:

$$C_{k,i}(l) = \begin{cases} \sum_{j=0}^{N-1-l} a_k(j)a_i(j+l) & 0 \le l \le N-1 \\ \sum_{j=0}^{N-1+l} a_k(j-l)a_i(j) & 1-N \le l < 0 \\ 0 & |l| \ge N \end{cases} \tag{10.26}$$

where N is the length of the code and a_k and a_i are the code sequences, then the cross-correlation functions can be found from:

$$R_{k,i}(\tau) = C_{k,i}(l-N)T_c + [C_{k,i}(l+1-N) - C_{k,i}(l-N)]\cdot(\tau - lT_c)$$
$$\hat{R}_{k,i}(\tau) = C_{k,i}(l)T_c + [C_{k,i}(l+1) - C_{k,i}(l)]\cdot(\tau - lT_c) \tag{10.27}$$

Because normalization affects both signal and interference in the same way, it makes no difference to the SIR. However, care must be taken if thermal noise is included in the calculations at a later stage. INITCCAT calculates all values of the aperiodic cross-correlation function for all code combinations occurring in the network and stores them in arrays as dynamic variables, and the return value of INITCCAT is a matrix of pointers to these arrays. The procedure INITPATHGAIN calculates all terminal positions and creates a matrix of path gains between transmitter i and the receiver belonging to transmitter j. Because we are mainly examining code sharing between terminals there will be fewer receivers than terminals, resulting in duplicate entries in the gain matrix. In our calculations we have assumed a far-field free-space propagation model. If another model is required, then the calculations must be modified here. CALCULATESIR computes the signal-to-interference ratio based on the configuration of the transmitting terminals. It calls the functions R and RHAT to calculate the partial cross-correlations. It also makes use of the double-precision uniform distribution (DUNI) function belonging to the random number generator unit. CALCULATEPBE uses the result of CALCULATESIR to determine the bit error probability according to (10.12). The POWER function calculates a^b and has little to do with signal powers.

10.7 ANALYTICAL AND SIMULATION RESULTS

In this section we present numerical results. A comparison is made between the Markov model results and the simulation results. These results are based on small networks, and we present some simulation results of larger networks.

For the results in this section, we use the following parameters: delay spread of 100 ns, a fixed terminal-to-receiver distance, and a fixed receiver-to-base station distance for networks using the hybrid protocol, random terminal positioning for CDMA-only networks, free-space signal attenuation ($\gamma = 2$), 31-chip Gold codes, a data rate of 0.26 Mb/s, an SNR of 20 dB in noisy channels, and a packet length of 64 bits unless stated otherwise.

We begin with a comparison of the Markov model and simulation results for eight terminal networks without retransmission delay, with noiseless nonfading channels (Figure 10.12).

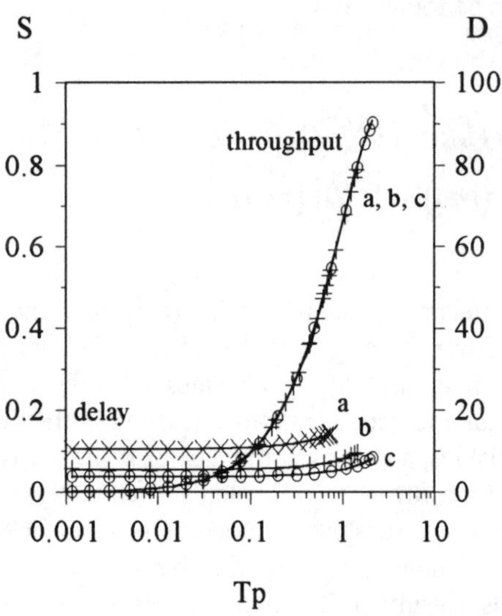

Figure 10.12 Comparison between Markov model (solid line) and simulation results (markers) of an eight-terminal network using one 31-chip code for a noiseless nonfading channel. (a-c) Performance for p_{tr} = 0.1 (X), 0.2 (+), 0.3 (O).

Terminal-to-receiver distance is 5 m, receiver-to-base station distance is 20 m. Both Markov model and simulation results agree quite well. Because the performance was virtually the same for all network geometries (code reuse schemes), we show only the results of using one code for different transmission probabilities. In this case, code sharing does not incur a performance penalty. The throughput curves are the same for all three values of the transmission probability p_{tr}, for both simulation and calculation. The only difference is that, for higher p_{tr}, the maximum throughput is higher. This result can be explained by the fact that the network is not heavily loaded for these values of p_{tr} and eight users. The delays experienced are due to the randomization of transmission attempts. Most of the transmissions succeed at the first attempt. Consequently, the performance is quite good.

The following results are for noisy Rayleigh fading channels. The transmission probability is 0.7 and the geometrically distributed retransmission delay is on average 4 and 8 time slots for 16 and 32 terminal networks, respectively. For the hybrid protocol, the terminal-to-receiver distance is 5 m and the receiver-to-base station distance is 30 m. The ripples in the simulation curves are caused by statistical variance of the simulation results.

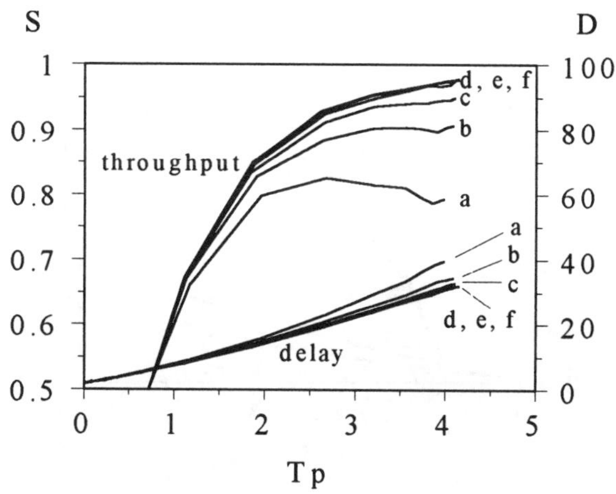

Figure 10.13 Effect of coding on 32-terminal network performance with a fixed transmission probability of 0.7 for a noisy fading channel. (a-f) 1, 2, 4, 8, 16, and 32 codes. Code length = 31.

Figure 10.13 presents the effects of code sharing for a 32-terminal network with distributed receivers. The influence of the number of transmission codes used is only visible at high traffic levels. Using 4 codes would suffice for this network. Computational results were also obtained to study the influence of multipath propagation on 32-terminal network performance by changing the code sequence length and data rate, for both the hybrid protocol with only 4 codes and conventional CDMA with 32 codes. Under equal circumstances, the hybrid protocol performs better than CDMA.

Finally, to demonstrate the robustness of the hybrid protocol, we compare network performance when the number of terminals is increased from 16 to 32 for CDMA and the hybrid protocol. For both hybrid protocol networks, 4 codes are used.

Figure 10.14 shows the simulation results for high traffic loads. The CDMA performance shows signs of degradation when the number of terminals is doubled, whereas the hybrid protocol continues to provide a high throughput with less delay than CDMA. This result suggests that a higher number of users can be supported by the hybrid protocol than by CDMA when using the same bandwidth.

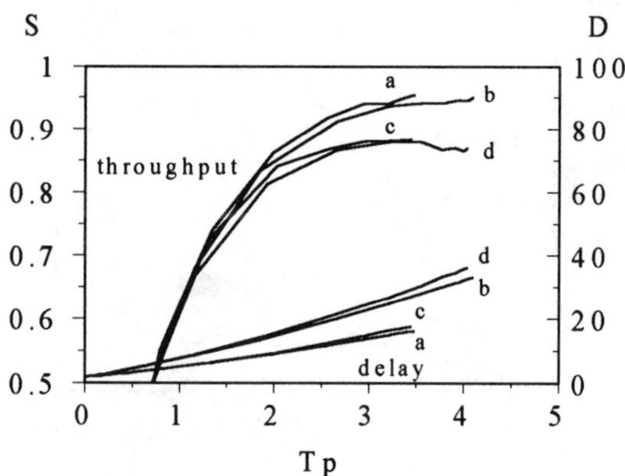

Figure 10.14 Effect on performance of increase in network size for CDMA and hybrid protocols for a noisy fading channel: (a) 16-terminal hybrid network, (b) 32-terminal hybrid network, (c) 16-terminal CDMA network, (d) 32-terminal CDMA network.

10.8 CONCLUSIONS

The throughput and delay have been evaluated for the proposed hybrid CDMA/ISMA protocol considering direct-sequence spread-spectrum CDMA and p-persistent ISMA with a DPSK modulation scheme for data communications in ideal and Rayleigh fading indoor environments.

A Markov model of the proposed hybrid CDMA/ISMA protocol is developed. The results of this model are compared with the simulation results for 8 terminals. Close agreement has been observed between analytical and simulation results. Therefore it is recommended to evaluate the performance of the hybrid CDMA/ISMA protocol using computer simulation for large numbers of terminals, because it gives results fast. Computational results further show that for the 32-terminal networks, 4 codes yield quite good performance. In a normal CDMA system, to achieve similar performance one needs 32 different codes.

In a Rayleigh fading channel, the computational results show that performance is enhanced by increasing the code sequence length, but decreases when the data rate is increased. It is worth mentioning that the performance is reduced only when the increase of the data rate results in more resolvable paths.

Code sharing proves to be a feasible technique to support many users using a limited number of short transmission codes. Further simulation results indicate that the hybrid protocol can support a higher number of users more successfully than ordinary CDMA.

REFERENCES

[1] R. Prasad, H.R.R. van Roosmalen and J.A.M. Nijhof, "A novel hybrid CDMA/ISMA protocol for indoor wireless computer communications," *Int. Conf. on Wireless Computer Communications; Emerging Business Opportunities,* Bombay, India, 15-17 December, 1994.

[2] H.R.R. van Roosmalen, J.A.M. Nijhof and R. Prasad, "Performance of a hybrid CDMA/ISMA protocol in the multiple return channels and buffering," in *Proceedings PIMRC'94*, The Hague, The Netherlands, pp. 1024-1024, 18-22 September, 1994.

[3] H.R.R. van Roosmalen, J.A.M. Nijhof and R. Prasad, "Performance analysis of a hybrid CDMA/ISMA protocol for indoor wireless computer communications," *IEEE J. Selected Areas Comm.*, Vol. 12, pp. 909-916, June 1994.

[4] R. Prasad, H.R.R. van Roosmalen and J.A.M. Nijhof, "CDMA, ISMA and for IWCC," *Proceedings IEE Colloquium on Spread-Spectrum Technique for Radio Communication Systems* London, UK, pp. 11/1-11/7, 15 April, 1994.

[5] H.R.R. van Roosmalen, J.A.M. Nijhofand R. Prasad, "A hybrid CDMA/ISMA protocol for computer communications in an indoor radio environment," *Proceedings PIMRC'93,* Yokohama, Japan, pp. 627-631, September 1993.

[6] F. Harmsze and R. Prasad, "Hybrid ISMA/CDMA and CDMA/ISMA protocols for IWCC," *Proc. Fourth IEEE International Conference on Universal Personal Communications,* Tokyo, Japan, pp. 733-737, November 1995.

[7] L. Kleinrock and F.A. Tobagi, "Packet switching in radio channels, part I: carrier sense multiple access modes and their throughput-delay characteristics," *IEEE Trans. Comm.*, Vol. COM-23, No. 12, pp. 1400-1416, December 1975.

[8] D. Raychaudhuri, "Performance analysis of random access packet-switched code division multiple access systems," *IEEE Trans. Comm.,* Vol. COM-29, No. 6, pp. 895-901, June 1981.

[9] J.G. Berline and E. Perratore, "Portable, affordable, secure: wireless LANs," *PC Magazine*, Vol. 11, pp. 291-314, February 1992.

[10] D.E. Knuth, "*The Art of Computer Programming, Vol. 2, Seminumerical Algorithms,*" Addison-Wesley Publishing Company, Reading, Mass., 1969.

[11] G. Marsaglia and A. Zaman, "A new class of random number generators," *Ann. Appl. Probability,* Vol. VI, pp 462-480, 1991.

Chapter 11
CDMA for Personal Communications

11.1 INTRODUCTION

In the previous chapters DS-CDMA has been considered for the indoor, outdoor, and landmobile satellite systems. To enhance its performance, diversity techniques and error correction codes are introduced. However, in many cases even after introducing combined diversity and coding schemes, system performance is not satisfactory. Therefore, several schemes are proposed to combat the problems inherent in the CDMA schemes to further enhance performance. They are interference cancellation [1-3], joint detection [4-6], hybrid CDMA [7-9], and superhybrid CDMA [10,11].

Further, the results presented in the previous chapters are optimistic because the effects of code and carrier synchronization on the system performance parameters are neglected in the analysis. There are several publications on carrier synchronization (e.g., [12-14]. But the information on the influence of multipath fading on code tracking is scarce. Some initial results on the code tracking are discussed in [15].

This chapter is organized as follows. Section 11.2 presents the influence of code synchronization errors on the performance of DS-CDMA systems. The effect of multiuser interference cancellation is discussed in Section 11.3. Section 11.4 concentrates on the overlay CDMA system with narrowband interferences, which gives a basic idea of the coexistence of CDMA systems in the presence of existing TDMA systems. Joint detection CDMA is introduced in Section 11.5. There are several types of hybrid CDMA schemes (e.g., hybrid DS/SFH, hybrid CDMA/ISMA, hybrid CDMA/TDMA, and multicarrier CDMA). Hybrid DS/SFH and hybrid CDMA/ISMA have already been discussed in Chapters 8 and 10, respectively. Section 11.6 presents a brief description of MC-CDMA and computational results.

11.2 SYNCHRONIZATION

Most papers on spread-spectrum synchronization errors due to multipath propagation and interference [16] focus on ranging and timing applications. For communication applications, however, the situation for code tracking is different. In the case of ranging or timing we want to track the delay of the line of sight (LOS) and/or dominant path signal; for code tracking in communication applications, the key parameter is the delay that maximizes the correlation output.

Figure 11.1 shows an example of a correlation function of an input signal consisting of two multipath components with a relative spacing smaller than one chip worth of time. The output correlation function is drawn for the cases of in-phase and out-of-phase signals.

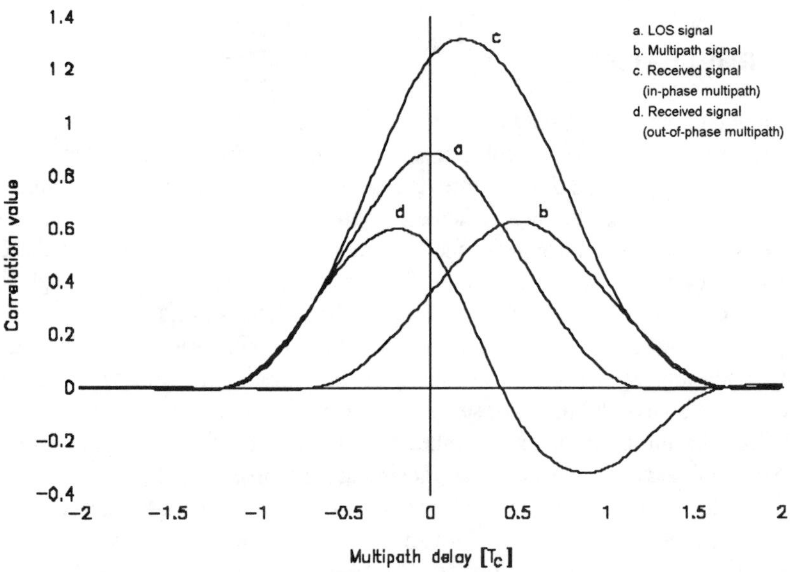

Figure 11.1 Correlation function with in-phase and out-of-phase multipath signals.

Due to fading, the peak value of the correlation function and the delay that maximizes the correlation output vary significantly (in the case of Figure 11.1 the received signal varies between curves c and d). To minimize the loss of signal power, the code tracking loop should track the instantaneous delay of the correlation function, which requires the loop bandwidth to be much larger than the fading bandwidth. As long as the fading bandwidth is smaller than the loop bandwidth (slow fading), the assumption that the influence of the synchronization errors due to fading on the system performance can be ignored is valid. However, there is an upper limit to the tracking loop bandwidth: an increase in the loop bandwidth also causes the loop noise to increase. For a carrier tracking loop, the calculation of the optimal loop bandwidth, which minimizes the total effect of both noise and fading, is described in [17].

For a tracking loop bandwidth smaller than the fading bandwidth, synchronization errors occur: a delay error is caused because the tracking loop is not able to track the instantaneous delay. As explained in the next section, by using a noncoherent delay lock loop (DLL) to estimate the code delay, a tracking bias is introduced in a fast fading environment. This bias can have a considerable impact on system performance.

Little is known about the influence of these code tracking errors on the performance of CDMA communication systems. In this section, insight is gained into the performance degradation of a direct-sequence spread-spectrum system with BPSK modulation due to the tracking bias of a noncoherent delay lock loop. This is done for two cases: the case of an unfiltered input signal consisting of a (LOS) component and just one multipath signal and the case of an unfiltered input signal consisting of a LOS component and multiple multipath signals. For both cases the correlation output is determined from the calculated tracking bias. This is done in a deterministic way for the single multipath case, and for the multiple multipath case, a Gaussian distribution is assumed. With the correlation output the bit error probability is obtained. The system performance parameters calculated with tracking bias have been compared to the performance parameters determined when this error is ignored. In these analyses, the relative levels of performance are far more important than the absolute levels. This research does not apply only to satellite-based systems. It applies to CDMA systems in a fast fading environment in general.

11.2.1 Code Tracking Errors due to Multipath Propagation

The received signal can be written as

$$r(t) = \sum_{i=0}^{M} a_i(t) p[t - \tau_i(t)] b[t - \tau_i(t)] \cos[\omega t + \theta_i(t)] \tag{11.1}$$

where M is the number of multipath signals that are taken into account, $p(t)$ is the spread-spectrum code, $b(t)$ is the data signal, ω is the angular frequency of the LOS signal, and $a_i(t)$, $\tau_i(t)$, and $\theta_i(t)$ are the time-dependent amplitude, delay, and phase of the, i^{th} signal, respectively. From now on, for simplicity of notation, the time dependence of the parameters a_i, τ_i, and θ_i is left out. Noise and multitransmitter interference are left out in (11.1), since the primary interest is the influence of synchronization errors caused by multipath signals. The data signal $b(t)$ has not been used in the remainder of this section since its influence on tracking is removed by envelope detection in a noncoherent DLL.

Notice that, in general, all signals can have different frequencies $[\omega+\delta\theta_i(t)/\delta t]/2$. The bandwidth spread of these frequencies is called the fading bandwidth, B_F, which is a crucial parameter in the analysis of multipath tracking errors. The fading bandwidth depends on the change of the transmitter-reflector-receiver geometry. If, for instance, the receiver velocity is 10 m/s, the fading bandwidth becomes 100 Hz for a carrier frequency of 1.5 GHz.

To track the desired signal delay τ_d, for communication applications the delay that maximizes the correlation output, the input signal (11.1) is downconverted and correlated with an "early" and a "late" code. These are replicas of the received spread-spectrum code with a delay of plus or minus $dT_c/2$ seconds compared to a "prompt" code, respectively. T_c is the chip time and the parameter d is often referred to as the early-late spacing. For communication applications, d=1 is a common value.

In the case of a noncoherent DLL, the resulting early and late correlation functions are first squared and then subtracted to produce the "S-curve" $S_{nc}(\hat{\tau})$, thereby removing the influence of the data and the carrier phase:

$$S_{nc}\left(\hat{\tau}\right) = \left\|\sum_{i=0}^{M} a_i R\left(\hat{\tau} - \tau_i + \frac{d}{2}T_c\right)\exp(j\theta_i)\right\|^2 - \left\|\sum_{i=0}^{M} a_i R\left(\hat{\tau} - \tau_i - \frac{d}{2}T_c\right)\exp(j\theta_i)\right\|^2 \qquad (11.2)$$

where $R(\tau)$ is the correlation function of the spread-spectrum code and $\hat{\theta}$ and $\hat{\tau}$ are the estimates of the carrier phase θ and the code delay, respectively. For an unfiltered ideal code, $R(\tau)$ is equal to $1-|\tau/T_c|$ for $|\tau|\le T_c$ and equal to zero elsewhere. The DLL tracks that value of $\hat{\tau}$ for which $S_{nc}(\hat{\tau})$ is zero while its slope $\delta\, S_{nc}(\hat{\tau})$ is negative.

The carrier tracking loop tracks the phase of the received signal after correlation with the prompt code. The carrier phase estimate can be expressed as

$$\hat{\theta} = \arg\left[\sum_{i=0}^{M} a_i \exp(j\theta_i) R\left(\hat{\tau} - \tau_i\right)\right] \qquad (11.3)$$

To understand the effects of multipath propagation on code tracking, it is important to distinguish between two different cases; the fading bandwidth B_F is large compared to the tracking loop bandwidth B_L or B_F is small compared to B_L.. For a noncoherent DLL also the predetection bandwidth B_p, which has to be greater than or equal to the data bandwidth in order to let the signal pass, has to be considered. In the slow fading case B_F is small compared to B_L and also to B_p, since B_p is always larger than B_L. If B_F is large compared to B_L, but small compared to B_p, all multipath signals pass the predetection correlation. This is fast fading. Finally, there can be a second type of fast fading for which B_F is large in comparison to both B_L and B_p.

A Slow Fading

If B_F is small compared to B_L, then the averaging in the delay lock loop has no influence on the tracking of the delay of the summed LOS and multipath signals; the DLL simply locks on the value of τ_d. For a small B_F/B_L ratio the receiver will "see" the instantaneous delay and no tracking errors due to multipath fading occur.

B Fast Fading

If B_F is large compared to B_L, the noncoherent delay lock loop tracks the zero crossings of the time averaged *S*-curve. In the case of fast fading when all multipath signals pass the predetection correlation, the time average is, because of the squaring operations in (11.2), equal to

$$S_{nc}\left(\hat{\tau}\right) = \sum_{i=0}^{M}\left[a_i R\left(\hat{\tau} - \tau_i + \frac{d}{2}T_c\right)\right]^2 - \left[a_i R\left(\hat{\tau} - \tau - \frac{d}{2}T_c\right)\right]^2 \qquad (11.4)$$

All cross-products are filtered out because of their relatively high frequencies. The resulting *S*-curve is simply the summation of M+1 different noncoherent *S*-curves, which causes a certain tracking bias compared to the LOS signal delay, τ_0, which is always positive.

Due to fading, the instantaneous delay varies around τ_0. For a positive instantaneous delay (e.g., curve c in Figure 11.1), the difference between τ_d and the biased estimated delay will be small. In the case of a negative value of the instantaneous delay (e.g., curve d in Figure 11.1), however, the tracking bias leads to a considerable tracking error and therefore to a substantial loss of signal power.

In the second case of fast fading, when B_F is large compared to both B_L and B_p, the noncoherent DLL will lock onto the dominant path and no tracking bias is introduced.

11.2.2 Performance Evaluation

This section presents a performance analysis of a direct-sequence spread-spectrum CDMA system with BPSK modulation, in terms of bit error probability (BEP). Assuming that the data bits -1 and 1 are equiprobable, the bit error probability P_e can be expressed as the probability that the correlation output z is negative while the transmitted data bit was positive:

$$P_e = P(z < 0 | b = 1) \tag{11.5}$$

In general, in the case of Gaussian noise P_e can be written as

$$P_e = \int_{\infty}^{\infty} p_e(xA)p_u(x)dx \tag{11.6}$$

where $p_u(x)$ represents the probability density function of the desired part of the correlation output z (see Chapter 7) and $p_e(xA)$ the bit error probability, given a signal amplitude A and a factor x caused by the path gain and a signal loss due to tracking errors. The value of $p_e(xA)$ is given by

$$p_e(xA) = \frac{1}{2}\operatorname{erfc}\left(\frac{xA}{\sigma\sqrt{2}}\right) \tag{11.7}$$

where erfc(x) is the complementary error function and σ^2 is the total variance of the Gaussian noise, the sum of the noise variance N_0/T_b and the interference power σ_i^2.

Single Multipath Signal Case

First, the case of an unfiltered input signal consisting of a line-of-sight component and just one multipath signal is considered. Also we assume only one user, so there is no multitransmitter interference. As we saw in Section 11.2.1, in the slow fading case there are no tracking errors. In the fast fading case, where the estimated carrier phase equals the LOS phase, the correlation output can be written as

$$z = a_0 R(\hat{\tau}) + a_1 \cos(\theta_1) R(\hat{\tau} - \tau_1). \tag{11.8}$$

Because the signal-to-multipath ratio is assumed to be greater than one, the peak of the correlation output always has the same delay as the peak of the correlation function of the LOS signal. So in this case, independent of the multipath delay and phase, the code delay that maximizes the correlation output is equal to the LOS delay ($\tau_d = 0$). For a noncoherent DLL, however, tracking biases are introduced. By determining the zero crossing of the S-curve with negative slope, for $d = 1$ the resulting tracking bias τ_e can be obtained from (11.4):

$$\tau_e = \begin{cases} \dfrac{a_1^2}{a_0^2 + a_1^2}\tau_1, & 0 \le \tau_1 \le T_1 \\ T_c\dfrac{\tau_1}{T_c} - 1.5 + \dfrac{a_0^2}{a_1^2} - \sqrt{\dfrac{a_0^2}{a_1^2}\left(2\tau_1 - 3 + \dfrac{a_0^2}{a_1^2}\right)} & T_1 \le \tau_1 \le 1.5T_c \end{cases} \tag{11.9}$$

with $T_1 = 0.5T_c\left(a_1^2 / a_0^2 + 1\right)$. With the signal-to-multipath ratio (SMR) as a parameter, the biases are depicted in Figure 11.2. If the SMR is large enough, the tracking error is small. In the case of strong multipath signals, however, the tracking error can become considerable; for instance $\tau_e = 0.25\ T_c$ for $SMR = 3$ dB and $\tau_1 = 0.75T_c$.

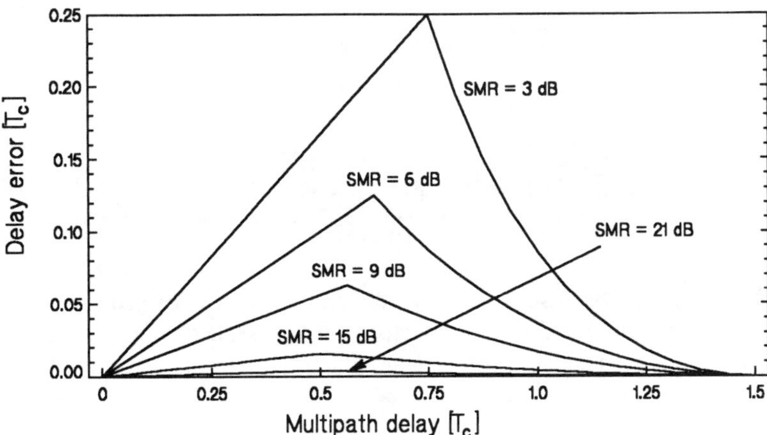

Figure 11.2 Bias of a noncoherent DLL for early-late spacing: $d = 1$.

From (11.8) we can see that for a certain multipath delay τ_1, the correlation output will vary due to the varying multipath phase θ_1. In Figure 11.3 the instantaneous correlation output as a function of θ_1 is shown for a multipath delay of $0.7T_c$ and SMR of 5 dB. The correlation output is plotted for $\hat{\tau} = \tau_e$, the delay as estimated by the noncoherent DLL, and for $\hat{\tau} = 0$, if the tracking bias is not taken into account and the correlation output is maximum. In the case shown, the loss of signal power due to the tracking error amounts to almost 3 dB in the out-of-phase situation. In the worst case situation (maximum delay error) for $SMR = 3$ dB, namely, $\tau_1 = 0.75T_c$, we can calculate that the tracking bias causes a loss of signal power of more than 6 dB for out-of-phase signals.

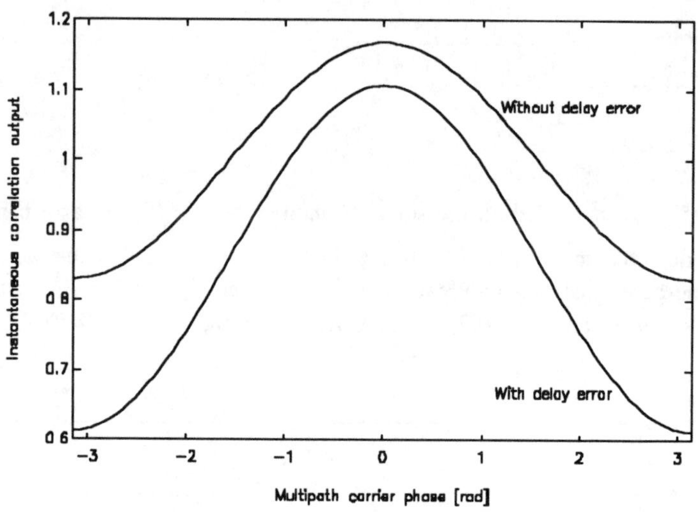

Figure 11.3 Instantaneous correlation output as a function of the multipath carrier phase θ_1, for *SMR* = 5 dB, multipath delay of $\tau_1 = 0.7\ T_c$, and early-late spacing of $d = 1$.

The loss of signal power results in a degradation of the bit error probability. The BEP, given signal amplitude A and gain/loss factor x, is represented by (11.7). The factor x in this expression for p_e is equal to the normalized ($a_0 = 1$) correlation output (11.8), which is a function of the multipath phase θ_1:

$$x(\theta_1|\tau_1) = R(\hat{\tau}) + a_1 \cos(\theta_1) R(\hat{\tau} - \tau_1). \tag{11.10}$$

Because we consider only one user, the variance of the Gaussian noise, σ^2, is just the noise variance N_0/T_b. With the bit energy-to-noise ratio defined as $\frac{E_b}{N_0} \underline{\Delta} \frac{A^2 T_b}{2N_0}$, for a uniform distributed multipath phase θ_1, the BEP given a multipath delay τ_1, becomes:

$$P_{e,\tau_1} = \frac{1}{2\pi} \int_{-\pi}^{\pi} \frac{1}{2} \operatorname{erfc}\left[x(\theta_1|\tau_1) \sqrt{\frac{E_b}{N_0}} \right] d\theta_1 \tag{11.11}$$

We computed the bit error probabilities for $\tau_1 = 0.3T_c$ and for $\tau_1 = 0.7T_c$ for a typical value of the SMR: *SMR* = 5 dB. P_e was calculated for the cases in which the signal delay is estimated with and without code tracking bias. It was observed that the code tracking error results in a degradation of the bit error probability.

If the bias is taken into account, to maintain a bit error probability of 10^{-3}, the signal-to-noise ratio (SNR) has to be increased by about 1.5 dB for $\tau_1 = 0.3T_c$ and by more than 2 dB for $\tau_1 = 0.7T_c$ as compared to the bias-free situation. So, given a bit error probability, there is a loss in SNR due to the tracking bias. For $P_e = 10^{-3}$, the increase in SNR as compared to the situation where the code delay error is ignored, is depicted in figure 11.4. In the case of the maximum delay error for *SMR* = 3dB ($\tau_1 = 0.75T_c$), the signal-to-noise ratio has to be increased by about 5.5 dB.

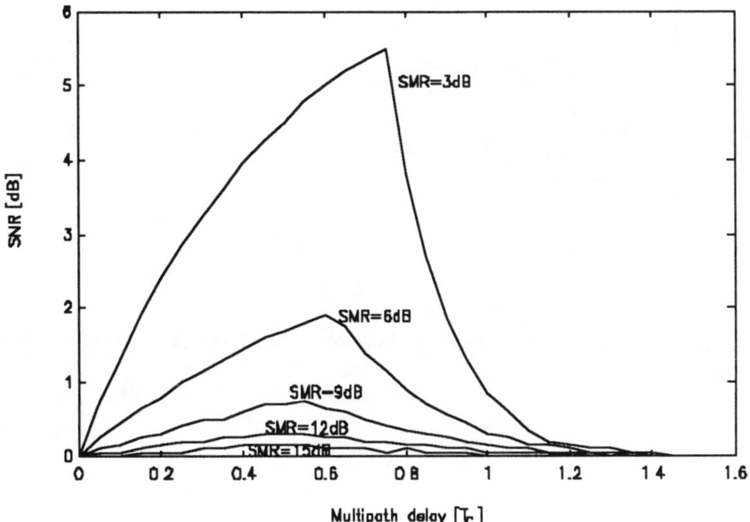

Figure 11.4 SNR increase to maintain a bit error probability of 10^{-3} as compared to the situation where the code delay error is ignored for $d = 1$.

To visualize the degradation of the bit error probability due to the tracking bias, for a signal-to-noise ratio of 10 dB, the BEP calculated with bias divided by the BEP without bias is plotted versus the multipath delay τ_1 (Figure 11.5), with *SMR* as a parameter. This ratio of the BEP probability calculated with bias and the BEP without bias is called the degradation ratio. For a SMR-multipath ratio of 3 dB and a multipath delay of $0{,}75T_c$, the bit error probability calculated with delay bias is 300 times larger than the probability if the tracking error is not taken into account.

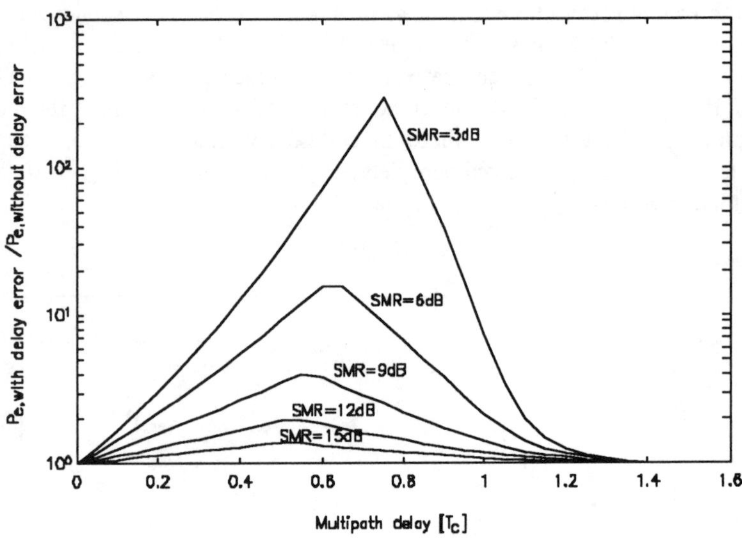

Figure 11.5 P_e obtained with delay error divided by P_e obtained without delay error for $E_b/N_0 = 10$ dB and $d = 1$.

Multiple Multipath Signal Case

Now that we have gained insight into the influence of code tracking errors on the bit error probability for the case of an input signal consisting of a LOS component and just one multipath signal, we extend our case to multiple multipath signals and more performance parameters. Again we consider unfiltered signals.

To model the multipath signals we will use the power-delay profile. For rural and suburban environments, the average power-delay profile of multipath signals is approximately exponential (Chapter 7):

$$P(\tau) = \frac{1}{T_s} b_0 \exp(-\frac{1}{T_s}\tau) \tag{11.12}$$

where T_S is the delay spread. For a rural environment, a typical value of T_S is 0.65 μs. As for the case of one multipath signal, we consider multipath signals with a delay of up to $1.5T_c$. If we take M_{chip} signals between $\tau = 0$ and $\tau = T_c$, so the total number of multipath signals M will be $1.5 \cdot M_{chip}$, the power-delay profile can be used to calculate the average multipath power for each signal. For signal number m, the multipath power b_{m0} is approximated as the power between the delay values $m \cdot T_c/M_{chip}$ and $(m+1) \cdot T_c/M_{chip}$:

$$b_{m0} = b_0 \left[1 - \exp\left(-\frac{T_c}{M_{chip} T_s} \right) \right] \exp\left[-m \frac{T_c}{M_{chip} T_s} \right] \tag{11.13}$$

By knowing the average multipath power for each signal, the S-curve (11.4) can be calculated. Determining the zero crossing of the S-curve with negative slope gives the tracking bias, τ_{bias}, of the noncoherent DLL in a fast fading environment. In Figure 11.6 the biases are depicted with SMR as a parameter for $M_{chip} = 10$ and $d = 1$. For large values of T_s/T_c the tracking bias becomes negligible: if spread-spectrum modulation is used with a chip time T_c that is less than the delay spread T_s of the channel, then the multipath power is partially reduced by the correlation operation in the receiver.

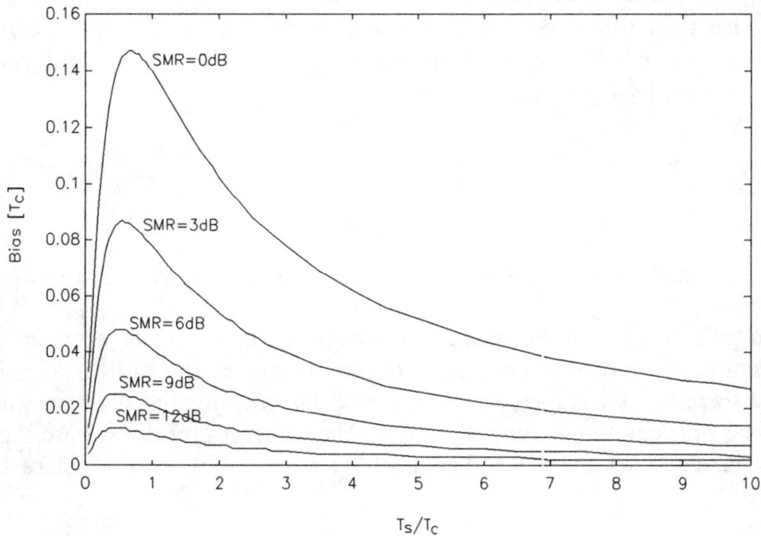

Figure 11.6 Bias of a noncoherent DLL for $M = 15$ and $d = 1$.

Notice that the bias is not equal to the instantaneous delay error, $\hat{\tau} - \tau_d$. Due to the varying multipath phases θ_i, the code delay that maximizes the correlation output is not equal to the LOS delay ($\tau_d = 0$). Because it is impractical to determine the instantaneous delay error and because the code tracking error is caused mainly by the bias, from now on we consider only the tracking bias.

To calculate the bit error probability (11.6), $p_u(x)$, the probability density function (pdf) of the desired part of the correlation output z, has to be known. Since the received signal can be considered as the sum of a LOS signal and a number of multipath signals with random varying phases, the correlation output can be described by a Gaussian pdf. This assumes that the fading bandwidth is large in comparison with the tracking loop bandwidth of the code and carrier tracking loops, which causes the receiver to track the average vector component (the LOS signal), whereas the multipath components, multiplied by the cosine of a random phase, will contribute a Gaussian component to the correlation output. The mean of the pdf is equal to the amplitude of the LOS signal times the output of the correlation function of the LOS signal: $a_0 \cdot R_{LOS}(\hat{\tau})$. The variance of the Gaussian pdf is determined by the residue of the multipath power after the correlation operation in the receiver and therefore also depends on the estimated code delay $\hat{\tau}$.

To obtain the bit error probability given A and x, $p_e(xA)$ (11.7), the total variance of the Gaussian noise, which consists of the noise variance N_0/T_b and the interference power σ_i^2, has to be determined. In Chapter 7, a closed-form expression was derived for the variance σ_i^2 of K interfering products. When log-normal shadowing is not taken into account, the normalized equation is

$$\sigma_i^2 = \frac{2K}{3N}\left(b_0 + \frac{1}{2}\right) \tag{11.14}$$

Here, N is the length of the spreading code and b_0 is the average scattered power due to multipath.

To gain insight into the effects of the tracking bias on the bit error probability, first, we consider the case of one user. We have compared the BEP for the situation where the tracking bias is ignored $(\hat{\tau} = \tau_0)$ with the BEP for the situation where the tracking bias is taken into account $(\hat{\tau} = \tau_{bias})$. The bit error probabilities are shown for two T_s/T_c values, 0.65 and 6.5, in Figure 11.7 for SMR=3 dB, and in Figure 11.8 for SMR=6 dB.

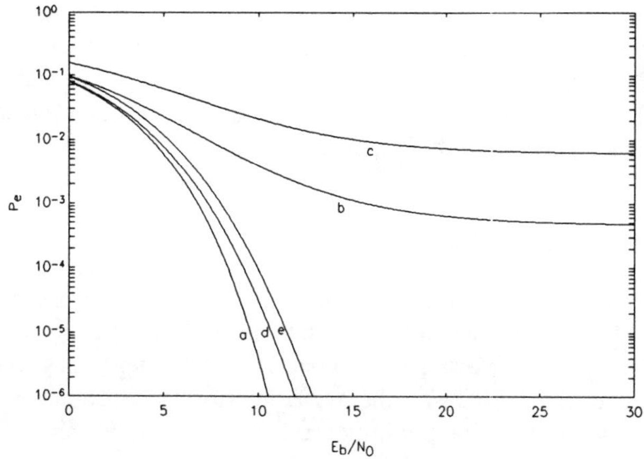

Figure 11.7 Bit error probability for *SMR* = 3 dB: (a) multipath free; (b) without bias, $T_s/T_c = 0.65$; (c) with bias, $T_s/T_c = 0.65$; (d) without bias, $T_s/T_c = 6.5$, (e) with bias, $T_s/T_c = 6.5$.

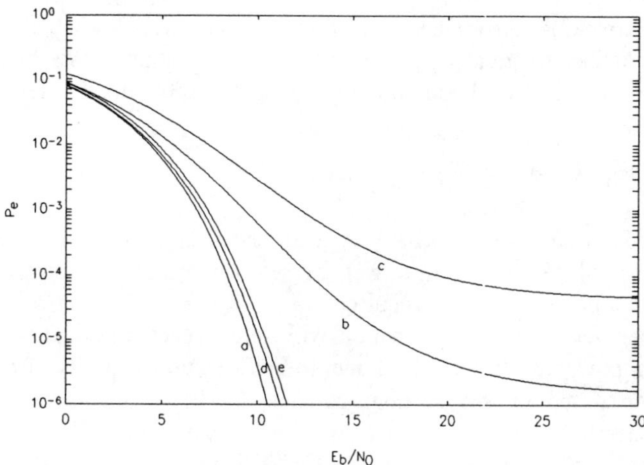

Figure 11.8 Bit error probability for *SMR* = 6 dB: (a) multipath free; (b) without bias, $T_s/T_c = 0.65$; (c) with bias, $T_s/T_c = 0.65$;(d) without bias, $T_s/T_c = 6.5$, (e) with bias $T_s/T_c = 6.5$.

For $T_s/T_c = 6.5$, the BEP degradation is small, for $T_s/T_c = 0.65$, however, the degradation is considerable. As we can see in Figure 11.8 (*SMR* = 6 dB), for $E_b/N_0 = 10$ dB, the BEP obtained with bias is 4.6 times worse than the BEP obtained without bias; for $E_b/N_0 = 20$ dB it is almost 21 times larger. If the bias is taken into account, to maintain a bit error probability of 10^{-3} and 10^{-4}, the SNR has to be increased by about 3 dB and by almost 7 dB, respectively, as compared to the bias-free situation.

11.2.3 Conclusions

If the tracking bias is not taken into account, the results of the system performance analysis of a CDMA system are too optimistic in some situations. For receivers with an early-late spacing of $d = 1$ using a noncoherent DLL, depending on the received multipath power and the ratio T_s/T_c, the system performance calculated with tracking bias can be much worse than the performance determined when this error is not taken into account.

To reduce the loss of signal power and the degradation of the performance due to the tracking bias, the early-late spacing could be decreased. In [16] it is shown that a smaller early-late spacing can result in a smaller tracking bias, which results in a smaller bit error probability. In practice, however, this only helps if the spread-spectrum bandwidth is equal to or larger than $1/d\, T_c$.

Because a coherent DLL has no tracking bias for large B_F/B_L values, for situations where fast fading can be expected this type of DLL is preferable in order to minimize code tracking errors and therefore the bit error probability degradation. Further, it is possible to greatly reduce code synchronization errors by using special tracking loops designed to deal with multipath or spread-spectrum interference.

11.3 INTERFERENCE CANCELLATION

A drawback of DS-CDMA is its sensitivity to the near-far effect. A deviation of only 1 dB in the mean of the received power causes the system performance to crash. Therefore, perfect power control is required.

Multiuser detection in combination with interference cancellation (IC) reduces the demands on power control and is shown to increase user capacity. Two types of IC methods are discussed in this section: parallel IC (PIC) and successive IC (SIC) schemes. In general, parallel IC schemes are of a complex nature. An alternative to the parallel scheme is the successive IC (SIC) scheme. Overall system performance improvement of the SIC scheme compared to the conventional scheme has been reported in [18]. This approach has been shown to be more robust in cancellation [19] than the parallel scheme, but it requires complex hardware.

11.3.1 Parallel Interference Cancellation (PIC)

A theoretical model for the performance of a DS-CDMA system with BPSK modulation and multiuser interference cancellation in terms of average bit error probability has been presented in [20]. In [20] the receiver synchronizes to a single path. In this section we present an improved version of the theoretical model presented in [20]. The performance of the system after multiuser interference cancellation proposed in [20] depends heavily on the initial performance of the system (before cancellation). So to improve the performance of the system after multiuser interference cancellation, the initial performance of the system has to be improved. Therefore, first estimation of the data bits is achieved by using a multipath combiner receiver [21]. The multipath combiner receiver combines the different paths' signals in contrast to the receiver in [20], which synchronizes to a single path.

System Model

In this subsection the model of the communication system is described. The system topology is that of a star connected communication system with K transmission stations, which can simultaneously, asynchronously transmit over their individual multipath fading channel to a central base station. This communication system is shown in Figure 11.9.

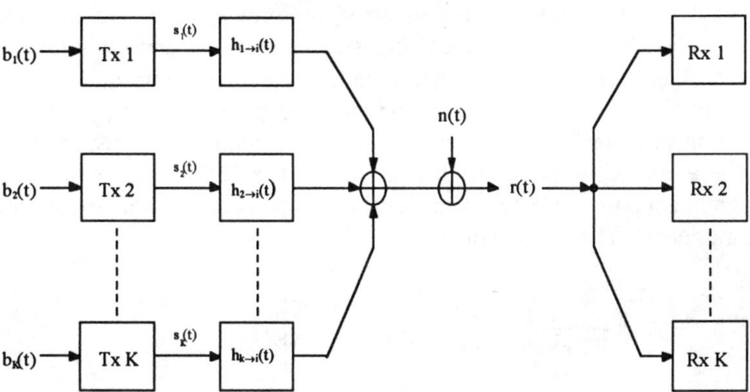

Figure 11.9 Communication system.

In Figure 11.9 $b_k(t)$ is the data waveform, Tx K is the transmitter of user k, $s_k(t)$ is the transmitted signal by user k, $h_{ki}(t)$ is the impulse response of the bandpass channel for the link between the kth transmitter and the ith receiver, $n(t)$ is a white zero mean Gaussian process with a two-sided noise power spectral density of $N_0/2$ at the receiver input, and $r(t)$ represents the total signal received by the base station.

The system model consist of a transmitter model, a channel model, and a receiver model. The transmitter model is the same as the one discussed in Chapter 5. As in Chapter 5 the channel is modeled as a discrete multipath slow fading channel, which means that there are L discrete multipath links between each user and the receive antenna at the base station. It also implies that the different random variables do not change considerably over the duration of one bit. Note that we have assumed that the signal bandwidth W is much larger than the coherence bandwidth of the channel $[W>>(Df)_c]$. This is the criterion for the existence of multiple resolvable paths. The lowpass equivalent impulse response of the bandpass channel for the link between the kth user and the ith receiver can be denoted as:

$$h_{k\to i}(t) = \sum_{\lambda=1}^{L(k,i)} g(k,i;\lambda)\delta[t-\tau(k,i;\lambda)]\exp\{j\theta[(k,i;\lambda)]\} \tag{11.15}$$

where $L(k,i)$ represents the number of propagation paths that exist from the kth transmitter to the ith receiver and $d(t)$ represents the delta function. Furthermore $g(k,i;l)$, $t(k, l\ i;)$ and $q(k,i;l)$ represent the gain, delay, and carrier phase associated with the lth replica at the ith receiver of a signal from the kth transmitter, respectively. The path gain $g(k,i;l)$ is subject to fading. The random variables $g(k,i;l)$, $t(k,i;l)$, $q(k,i;l)$, and $L(k,i)$, together with the data symbols, are assumed to form a set of mutually independent random variables. Each path phase $q(k,i;l)$ is assumed to be uniformly distributed over the interval $[0,2p]$. The path delay $t(k,i;l)$ is also assumed to be uniformly distributed over the interval $[0,T_b]$. Further we assume that each path gain $g(k,i;l)$ is an independent Rician random variable. This is in contradiction to the model presented in [20] where the path gain is assumed to be Rayleigh distributed. The Rician distribution is applicable when a significant part of the received signal envelope is due to a constant path (such as a LOS component). The Rician pdf is:

$$p_{g(k,i;\lambda)}(r) = \frac{r}{\sigma^2}\exp\left[-\frac{r^2+s^2}{2\sigma^2}\right] I_0\left[\frac{sr}{\sigma^2}\right] \quad \text{with} \begin{Bmatrix} r>0 \\ s>0 \end{Bmatrix} \tag{11.16}$$

where $I_0(\cdot)$ is the modified Bessel function of the first kind and zero order. An important parameter is the Rician parameter, R:

$$R = \frac{s^2}{2\sigma^2} \tag{11.17}$$

The Rician parameter represents the ratio of the power associated with the direct LOS component $s^2/2$ and the scattered components s^2 (ratio of the LOS power to the scattered power). To confine the power transmitted over all paths (passive channel), we set s to 1.

The last point that should be made clear in the channel model is how to determine the number of resolvable paths, L. The number of resolvable paths can be obtained from the delay spread T_{max} and the chip duration T_c:

$$L = \left\lfloor \frac{T_{max}}{T_c} \right\rfloor + 1 = \lfloor T_{max} R_c N \rfloor + 1 \tag{11.18}$$

From (11.18) we can see that L depends on the minimum resolution of the code. This channel model for the case of three paths (L=3) is depicted in Figure 11.10.

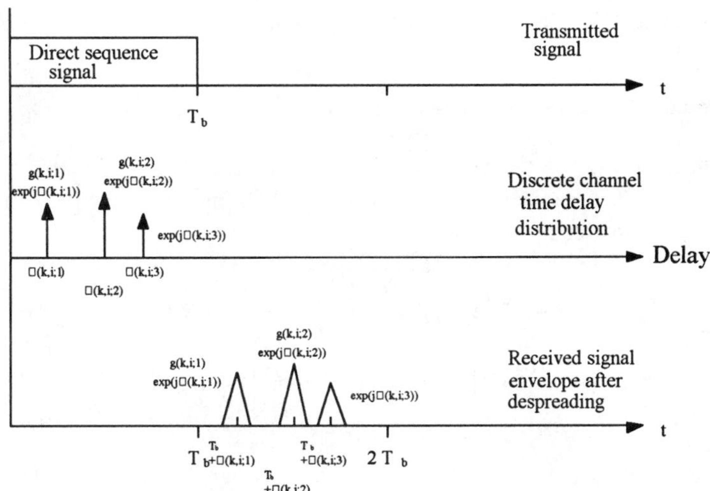

Figure 11.10 Discrete multipath channel model.

In the following section we give a description of the receiver. The receiver model we adopt consists of a matched filter (each user has its own particular spread-spectrum code), a transversal filter, a BPSK demodulator, a lowpass filter (LPF), and multiuser interference canceller. In contradiction to the model presented in Chapter 5, where diversity techniques were used, a transversal filter is used to combine the several paths.

The structure of this receiver is shown in Figure 11.11.

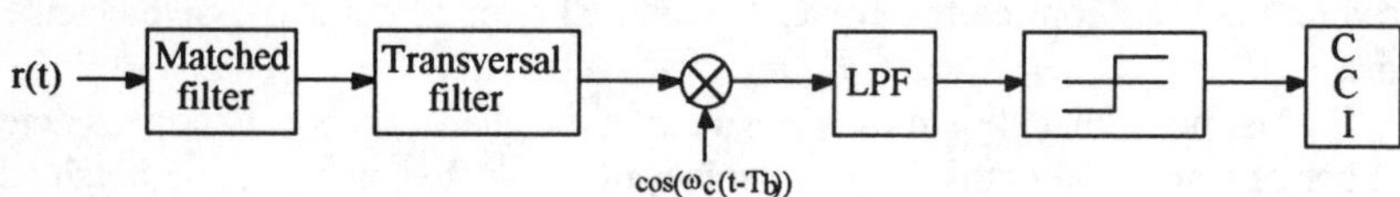

Figure 11.11 Receiver block structure.

When a bit is coded with the ith users' code and transmitted, the filter matched to that user code will yield output peaks corresponding to each channel path. Bits that are coded with other users' codes do not produce large peaks. The transversal filter coherently combines the output peaks. For detection of the signal, the output of the transversal filter is multiplied by 2 cos $[\omega_c(t-T_b)]$. The resulting product is passed through a LPF, sampled, and compared to a threshold to make a decision on the bits.

Performance of DS-CDMA system with Multiuser Interference Cancellation

In this section we discuss the multiuser interference cancellation. We also discuss how multiuser interference cancellation is achieved and its influence on performance. The received signal is written as

$$r(t) = \sum_{k=1}^{K}\sum_{\lambda=1}^{L(k,i)} A_k\, g(k,i;\lambda) b_k \{t-[\tau(k,i;\lambda)+T_k)]\} a_k \{t-[(k,i;\lambda)+T_k]\} \exp\left(j\{\theta(k,i;\lambda)+\omega_c[t-\tau(k,i;\lambda)-T_k]+a_k\}\right)+n(t) \tag{11.19}$$

If we define

$$\phi_{k,\lambda} = \alpha_k + \theta(k,i;\lambda) - \omega_c\left[\tau(k,i;\lambda)+T_k\right] \tag{11.20}$$

the received signal can be written as:

$$r(t) = \sum_{k=1}^{K}\sum_{\lambda=1}^{L(k,i)} A_k\; g(k,i;\lambda) b_k \{t-[\tau(k,i;\lambda)+T_k]\} a_k \{t-[\tau(k,i;\lambda)+T_k]\} \exp\left[j(\omega_c t+\phi_{k,\lambda})\right]+n(t) \tag{11.21}$$

For ease in notation, we omit the time argument, *t,* and define

$$b(k,i;\lambda) = b_k \left\{t - \left[\tau(k,i;\lambda) + T_k\right]\right\} \tag{11.22}$$

and

$$x(k,i;\lambda) = A_k\, g(k,i;\lambda) a_k \left\{t - \left[\tau(k,i;\lambda) + T_k\right]\right\} \exp\left[j(\omega_c t + \phi_{k,\lambda})\right] \tag{11.23}$$

In Figure 11.12 we see how initial data estimation was achieved.

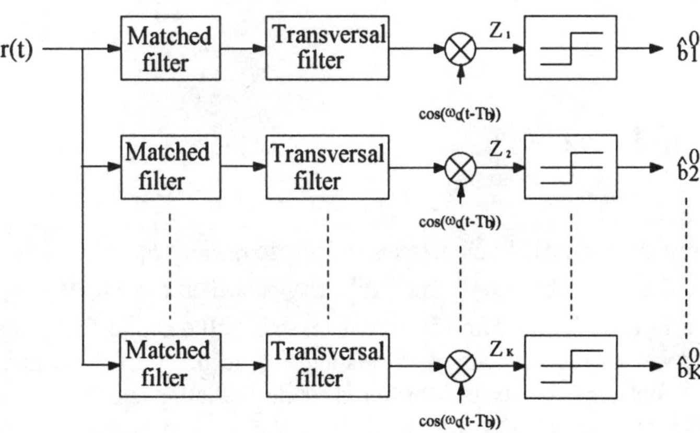

Figure 11.12 Initial data estimation.

In this analysis we have assumed that correct channel estimates are available for the use at the transversal filters (TF). Under this ideal condition, if the bit decision for the ith transmitter is correct, a perfect replica of its received signal can be created. Consequently the only cause of errors lies in the occurrence of incorrect data estimates at the first part of the receiver (before multiuser interference cancellation). If we distinguish between the desired signal and the multiuser signal, the transmitted signal becomes:

$$r(t) = \sum_{\lambda=1}^{L(i,i)} b(i,i;\lambda)x(i,i;\lambda) + \sum_{\substack{k=1 \\ k\neq i}}^{K} \sum_{\lambda=1}^{L(k,i)} b(k,i;\lambda)x(k,i;\lambda) + n(t) \tag{11.24}$$

The first term is the desired signal from the ith transmitter, the second term is the multiuser interference, and the third term is the noise.

In the cancellation part, each data estimate $\hat{b}(k,i;\lambda)$ is first multiplied by its respective synchronized code and carrier signal to create a replica of the originally transmitted signal. Then the signal is passed through a TF, which emulates channel $h_{ki}(t)$. The output of the filter is given by:

$$F^{(k)} = \sum_{\lambda=1}^{L(k,i)} \hat{b}(k,i;\lambda)x(k,i;\lambda) \tag{11.25}$$

where

$$\hat{b}(k,i;\lambda) = \hat{b}_k \{t - [\tau(k,i;\lambda) + T_k]\} \tag{11.26}$$

is the estimated data signal. Each element of the sequence, $\hat{b}(k,i;\lambda)$, is an estimate of $b(k,i;\lambda)$. By adding all the other users' $F^{(k)}$, the canceller can create a replica of the cochannel interference (CCI). How closely it resembles the actual CCI depends on two things: the number of correct data estimates and the accuracy of the channel estimates. In this section, only the former is considered while the latter is assumed to be perfect. The CCI replica of each user is subtracted from the received signal, $r(t)$, to produce a "clean" received signal $r_i(t)$, for the ith user. The process described above is shown in Figure 11.13. Thus we obtain:

$$\begin{aligned} r_i(t) &= r(t) - \sum_{\substack{k=1 \\ k \neq i}}^{K} F^{(k)} \\ &= \sum_{\lambda=1}^{L(i,i)} b(i,i;\lambda)x(i,i;\lambda) + \sum_{\substack{k=1 \\ k \neq i}}^{K} \sum_{\lambda=1}^{L(k,i)} \left[b(k,i;\lambda) - \hat{b}(k,i;\lambda) \right] x(k,i;\lambda) + n(t) \end{aligned} \tag{11.27}$$

If we define

$$\tilde{b}(k,i;\lambda) = b(k,i;\lambda) - \hat{b}(k,i;\lambda) \tag{11.28}$$

(11.25) becomes

$$r_i(t) = \sum_{\lambda=1}^{L(i,i)} b(i,i;\lambda)x(i,i;\lambda) + \sum_{\substack{k=1 \\ k \neq k}}^{K} \sum_{\lambda=1}^{L(k,i)} \tilde{b}(k,i;\lambda)x(k,i;\lambda) + n(t) \tag{11.29}$$

Ideally, if all the other data estimates were correct $\left[\tilde{b}(k,i;\lambda) = 0\right]$, this cleaned signal should only contain its desired signal and noise regardless of whether its own data estimates were correct or incorrect.

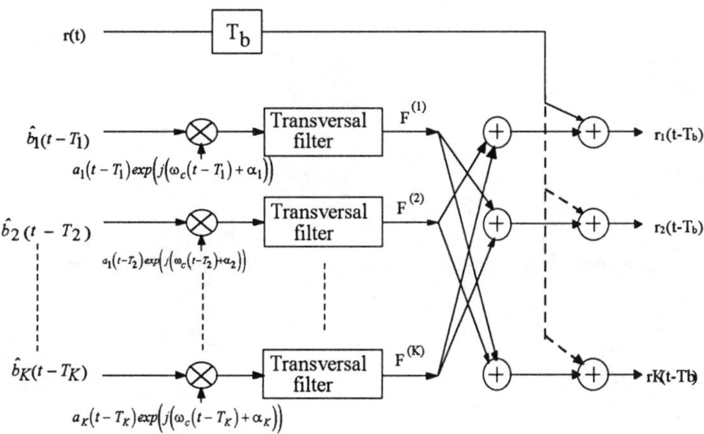

Figure 11.13 Multiuser interference canceller.

This process of cancellation can be repeated. The resulting difference is that only the noise variance resulting from the K–1 interfering transmitters is changed and is given by

$$N_k = 4E_b s^4 \left(R^{-1} + 1\right)^2 T_b \sum_{\substack{k=1 \\ k \neq i}}^{K} E\left[\tilde{b}_k^2\right] I_{MU} \tag{11.30}$$

The probability of a correct data estimate when $\hat{b}_k = b_k$ is:

$$P\left[\tilde{b}_k = 0\right] = 1 - P_e \tag{11.31}$$

The probability of an incorrect data estimate when $\hat{b}_k = -b_k$ is:

$$P\left[\tilde{b}_k = 2b_k\right] = P_e \tag{11.32}$$

Since the data symbols b_k are equally distributed, we get:

$$P\left[\tilde{b}_k = -2\right] = P\left[\tilde{b}_k = 2\right] = \frac{P_e}{2} \tag{11.33}$$

This results in:

$$E\left[\left(\tilde{b}_k\right)^2\right] = 4\,P_e \tag{11.34}$$

The noise variance resulting from the K-1 interfering transmitters can now be written as:

$$N_K = 16\,P_e\,E_b\,s^4\left(R^{-1}+1\right)^2 T_b \sum_{\substack{k=1\\k\neq i}}^{K} I_{MU} \tag{11.35}$$

In Figure 11.14 the average BEP of the system after multiuser interference cancellation is presented as a function of the number of transmitting users for two Rician factors (R = 6.8 dB and 11 dB).

From the plot it can be seen that the performance of the system after multiuser interference depends on the initial performance of the system. For example, if we have a system, which has a maximum allowable BEP of $1E^{-5}$ then 66 users can be admitted to the system instead of 4 users before cancellation.

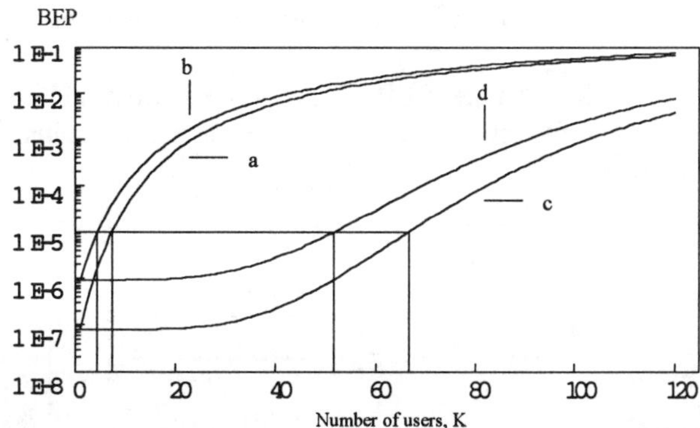

Figure 11.14 Performance of the system with multiuser interference cancellation as a function of the number of users for a bit rate of 32 kbits/s (L=2), E_b/N_0=10 dB, and N=127: (a) R = 6.8 dB, (b) R = 11 dB, (c) R = 6.8 dB (with cancellation), (d) R = 11 dB (with cancellation).

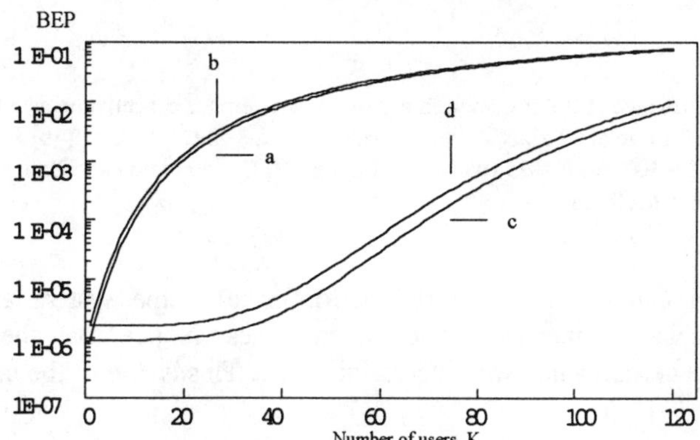

Figure 11.15 Performance of the system with multiuser interference cancellation as a function of the number of users for a Rician factor of 11 dB, E_b/N_0=10 dB and N=127: (a) 32 kbits/s (L=2), (b) 144 kbits/s (L=5), (c) 32 kbits/s (with cancellation), (d) 144 kbits/s (with cancellation).

In Figure 11.15 the average BEP of the system after multiuser interference cancellation is presented as a function of the number of transmitting users for two bit rates (R_c = 32 and 144 kbits/s).From the plot it can be seen that the performance of the system after multiuser interference cancellation for lower bit rates (resolvable paths are small) is much better.

In Figure 11.16 the average BEP of the system after multiuser interference cancellation is presented as a function of E_b/N_0 for four different numbers of transmitting users ($K = 1, 25, 50, 100$).

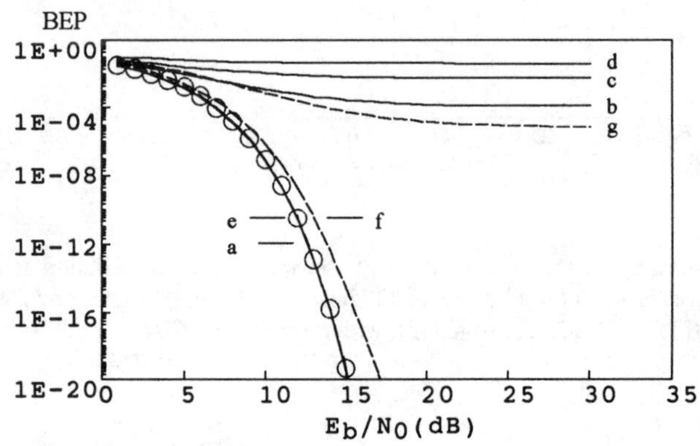

Figure 11.16 Performance of the system with multiuser interference cancellation as a function of E_b/N_0 for a bit rate of 32 kbits/s (L=2), R=6.8 dB, and N=127: (a) $K = 1$, (b) $K = 25$, (c) $K = 50$, (d) $K = 100$, (e) $K = 25$ (with cancellation), (f) $K = 50$ (with cancellation), (g) $K = 100$ (with cancellation).

From the plot it can be seen that the performance of the system after multiuser interference cancellation increases if E_b/N_0 increases. Notice that the performance convergences to a certain value with increasing E_b/N_0. This is due to the influence of the thermal noise.

Conclusions

In this section we have presented a performance analysis of a DS-CDMA system with BPSK modulation. The performance has been measured in terms of the average BEP. The following conclusions are drawn:

• Multipath combiner: A multipath combiner receiver provides better results than a receiver that only uses one single path. The multipath combiner receiver combines the correlator outputs of all the paths, instead of a single path. This involves an increase in the complexity of the receiver (components).

• Influence of the bit rate: We have seen that there is a relation between the bit rate and the number of resolvable paths. The number of resolvable paths increases with increasing bit rate. A lower bit rate may correspond to a lower number of resolvable paths. A lower number of resolvable paths correspond to a lower bit error probability. This is due to the fact that more paths cause more intersymbol and multiuser interference.

• Influence of the delay spread: If the delay spread T_m increases, the number of resolvable paths may increase. This means that the intersymbol and multiuser interferences also increase, which decreases the performance of the system.

• Influence of cancellation: The performance of the system after multiuser interference cancellation depends on the initial performance of the system. For example, if we have a system that has a maximum allowable BEP of $1E^{-5}$, then 66 users can be admitted to the system instead of 4 users before cancellation.

In the model used, we assumed that the receiver could ascertain the channel parameters perfectly. For further research we recommend these parameters be modeled with an error, which is a zero mean Gaussian variable, and determine how the performance of this system declines.

11.3.2 Successive Interference Cancellation (SIC)

This subsection concentrates on a groupwise SIC (GSIC) scheme, which is a generalization of the SIC scheme. The GSIC scheme feeds back and cancels the signals in groups. The reason for grouping the signals is that it reduces the hardware complexity considerably. In the GSIC scheme the hardware complexity increases to a power of two with the number of groups, and in the SIC scheme to a power of two with the number of users. Therefore, the GSIC scheme has lower implementation costs than the SIC scheme. The objective of this subsection is to examine the system performance of the GSIC scheme.

Ideally, the signals in a group are of equal strength, so that the processing gain is sufficient to perform the separation. However, in practice, this situation does not occur. It is, therefore, important to pay much attention to the group size.

The GSIC scheme does not require knowledge of the individual user power. The output of conventional correlation receivers is used to discriminate the signals by strength. The strongest outputs with similar signal strength are respread with the appropriate chip sequences, after which the group they belong to is cancelled. This process is repeated until the weakest group is demodulated.

11.3.2.1 GSIC Scheme Analysis

A GSIC scheme with coherent BPSK modulation is analysed in an asynchronous Rayleigh fading Channel. Figure 11.17 shows the basic scheme of multi-user detection with GSIC. As an example, the group size is taken as two. The received signal at baseband is fed into a bank of complex correlators. The chip sequences of all users are assumed to be known, however, no knowledge of the energies of the individual users is required. The two correlators with the strongest outputs are selected. They are re-spread with the appropriate chip sequences, after which they are cancelled simultaneously. This process is repeated until the weakest group is demodulated.

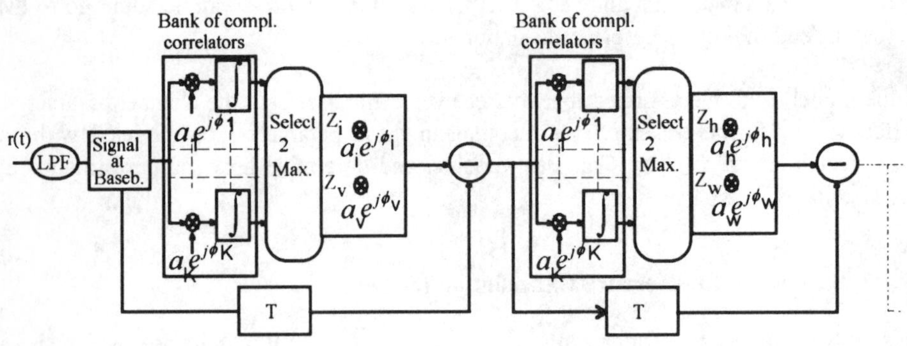

Figure 11.17 Receiver structure of a multi-user detector with a GSIC. Group size is equal to two.

From figure 11.17 it can be seen that the number of complex correlators, Ncc, increases to the power 2 with the number of users, K, and the amount of delay, D, increases linearly with the number of users:

$$\begin{aligned} Ncc &= K^2 \\ D &= K \cdot T \end{aligned} \tag{11.36}$$

where T = bit period.

A reduction factor, R_f, is introduced, which is equal to the number of users per group. The resulting reduction in the amount of complex correlators and delay, as a function of R_f, is given by

$$Ncc = \left\lceil \frac{K}{R_f} \right\rceil K$$
$$D = \left\lceil \frac{K}{R_f} \right\rceil T \qquad (11.37)$$

The modulation method used is coherent BPSK. The received signal, $r(t)$, is modeled as follows:

$$r(t) = \sum_{k=1}^{K} A_k \cdot a_k(t - \tau_k) \cdot b_k(t - \tau_k) \cdot \cos(\omega_c + \phi_k) + n(t) \qquad (11.38)$$

where K = total number of active users
A_k = amplitude of kth user
$b_k(t)$ = random bit sequence of kth user at bit rate R_b
$a_k(t)$ = spreading sequence of kth user at chip rate R_c
ω_c = carrier frequency
$n(t)$ = additive white Gaussian noise
N = processing gain.

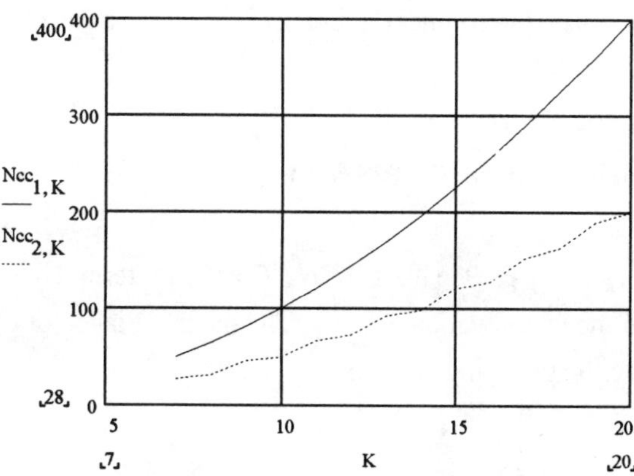

Figure 11.18 The number of complex correlators, *Ncc*, required for the GSIC (.....) and the SIC (—) scheme versus the number of users, *K*.

Figure 11.18 shows *Ncc* versus *K* for R_f=1 and R_f=2. It is clear that the GSIC scheme reduces the hardware complexity considerably compared to the SIC scheme.

The delay time τ_k and the phase ϕ_k of the kth user are uniform random variables in [0,*T*(=bit period)] and [0,2π], respectively, in the case of asynchronous detection. After cancellation of *j* signals, the decision variable for the *j*+*1*th user is given by

$$Z_{j+1} = \frac{1}{2} A_{j+1} b_{j+1} + C_{j+1} \tag{11.39}$$

When the group size is fixed and equal to R_f, C_{j+1} is given by

$$\begin{aligned} C_{j+1} = & \sum_{k=\left\lfloor \frac{j}{Rf} \right\rfloor \cdot Rf+1}^{K} A_k I_{k,j+1}(\tau_{k,j+1}, \phi_{k,j+1}) - A_{j+1} I_{j+1,j+1}(\tau_{j+1,j+1}, \phi_{j+1,j+1}) + (n^I_{j+1} + n^Q_{j+1}) \\ & - \sum_{i=1}^{\left\lfloor \frac{j}{Rf} \right\rfloor \cdot Rf} C_i I_{i,j+1}(\tau_{i,j+1}, \phi_{i,j+1}) \end{aligned} \tag{11.40}$$

The noise term, (11.40), consists of three parts. The first term is due to cross-correlation interference of uncancelled users; the second term, i.e., $n^I + n^Q$, is AWGN; and the last term* is cumulative noise due to imperfect cancellation (cross-correlation≠0). The cross-correlation term is given by

$$I_{k,i}(\tau_{k,i}, \phi_{k,i}) = \frac{1}{T} \int_0^T a_k(t - \tau_{k,i}) \cdot a_i(t) dt \cdot \cos(\phi_k - \phi_i) \tag{11.41}$$

The variance of $I_{k,i}(\tau_{k,i}, \phi_{k,i})$ equals 1/3*N*. The noise term C_{j+1} is assumed to be Gaussian with zero mean and variance η_{j+1}. The mean variance of C_{j+1} under Rayleigh fading is affected by the grouping as follows:

$$\eta_{j+1} = \frac{1}{3N} \sum_{k=\left\lfloor \frac{j}{Rf} \right\rfloor \cdot Rf+1}^{K} E[A_k^2] - \frac{E[A_{j+1}^2]}{3N} + \frac{N_o}{T} + \frac{1}{3N} \sum_{i=1}^{\left\lfloor \frac{j}{Rf} \right\rfloor \cdot Rf} \eta_i \tag{11.42}$$

---

* In [18] $I_{i,i+1}$ instead of $I_{i,j+1}$.

The analysis of the bit error probability performance under Rayleigh fading is done using order statistics [22]. Each user is assumed to have the same local mean power. The mean square value of the amplitude is used for simplicity. The probability density function of the ordered A_k (A_1 is the strongest) is given by

$$f_{A_k}(x) = \frac{K!}{(K-k)!(k-1)!} 2x(1-e^{-x^2})^{K-k} e^{-kx^2} \tag{11.43}$$

The BEP of the $j+1^{\text{th}}$ user is then given by

$$P_e^{j+1} = \int_0^\infty Q\left(\frac{x}{\sqrt{\eta_{j+1}}}\right) f_{A_{j+1}}(x)dx \tag{11.44}$$

The average BEP is then obtained as

$$P_e = \frac{1}{K}\sum_{j=1}^{K} P_e^j \tag{11.45}$$

where P_e is an ensemble average that is equal to the time average in this case, because the probability density functions of all users are equal.

Comparison Between GSIC and SIC

The SIC scheme (i.e., R_f=1), will be optimal with respect to the BEP performance of the DS-CDMA system. However, it is not optimal with respect to hardware efficiency. Therefore, the GSIC scheme is introduced. From (11.37) we see what the decrease in hardware complexity will be in terms of the reduction factor (=group size). We have examined the effect this grouping has on the different performance parameters. The performance parameters that are important to the system designer are:

1. BEP;
2. SNR;
3. User capacity.

Figures 11.19 and 11.20 show the factor by which the BEP increases for different group sizes, with respect to R_f=1, versus the E_b/N_0. In the case of 10 users, when R_f is 10, no cancellation is done. This corresponds to the conventional DS-CDMA receiver. It is clear that the GSIC scheme improves BEP performance dramatically. When E_b/N_0 is high, the GSIC scheme is particularly efficient.

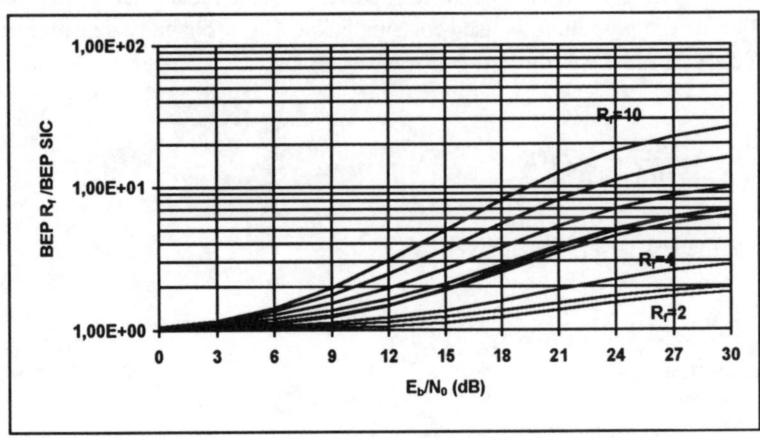

Figure 11.19 Factor with which BEP for R_f=1...10, BEP$_{Rf,}$ increases compared to BEP$_{SIC}$ versus E_b/N_0 under Rayleigh fading. Number of users =10 and processing gain = 31.

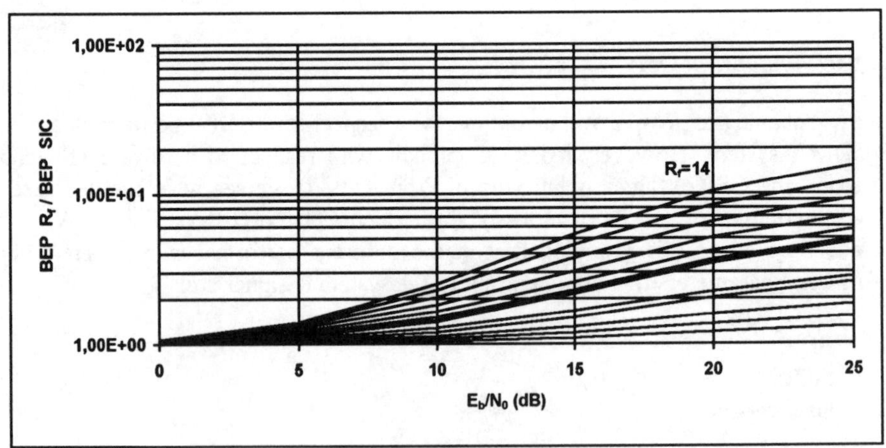

Figure 11.20 Factor with which BEP for R_f=1..,14, BEP$_{Rf,}$, increases compared to BEP$_{SIC}$ versus E_b/N_0 under Rayleigh fading. Number of users =14 and processing gain = 31.

Furthermore, the decrease in BEP for an R_f smaller than 5 will be smaller than a factor of 3 compared to R_f=1. In the case of 14 users R_f should be smaller than 7. This shows that with a marginal decrease in BEP performance a large decrease in hardware complexity is obtained.

From Figures 11.21 and 11.22 we see that the additional required SNR compared to the R_f=1 case is within reason for a BEP that is not to small. When the requirements for the BEP become more strict, the additional SNR increases, because, as can be seen from Figures 11.19 and 11.20, the difference in BEP performance increases when the CCI becomes dominant over the Gaussian noise. So, increasing the SNR has less influence on the BEP.

Next, we examine the effect of grouping on the user capacity. Figure 11.23 shows that only a reduction factor of 2 can keep track with the user capacity of R_f equal to 1. So, when this performance parameter is important R_f should not exceed 2. Also, the capriciousness in the relation between R_f and the number of users shows that the choice of R_f is an important step in the design process.

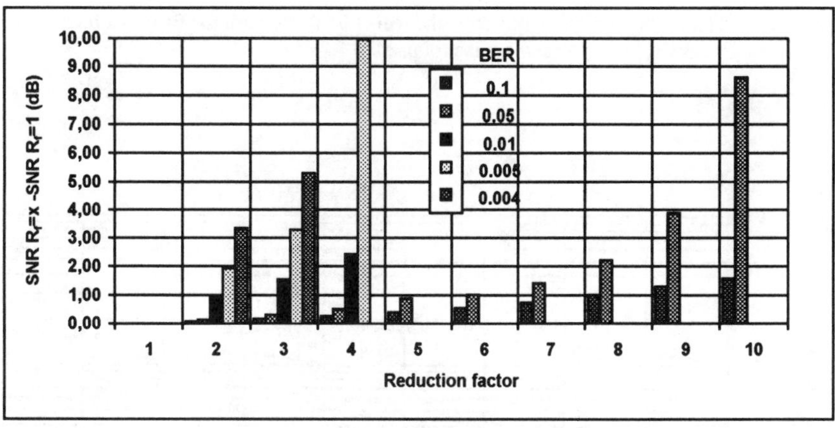

Figure 11.21 Additional required SNR versus the reduction factor with the BEP as a fixed parameter. Processing gain=31 and number of users = 10.

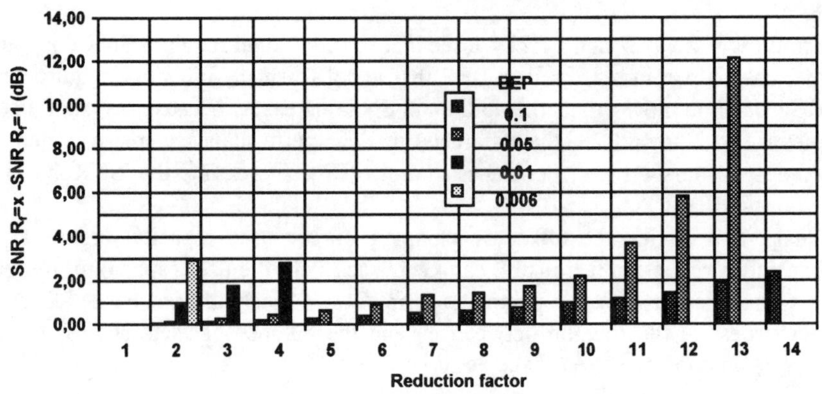

Figure 11.22 Additional required SNR versus the reduction factor with the BEP as a fixed parameter. Processing gain=31 and number of users=14.

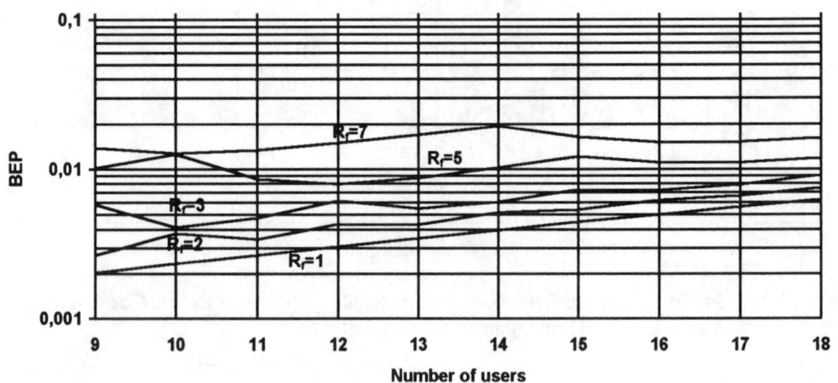

Figure 11.23 BEP versus number of users with R_f as fixed parameter. Processing gain=31 and SNR=25 dB.

Performance Improvement Strategy

Finally, we examine a way to improve the BEP performance. From (11.44) we see that the BEP is obtained by averaging the BEP over all outputs. The cumulative distribution function of the BEP of one user shows that the average BEP (E[BEP]=0.01149) is mainly determined by the periods when the signal is weak (Figure 11.24). Its BEP is smaller than E[BEP] for 80% of the time. A good strategy to combat this would be to have a retransmission protocol layered on the GSIC scheme. Identification of the weakest signals can be done by using the chip sequences. However, the retransmission protocol introduces a time delay in the system. This effect should be investigated in future research.

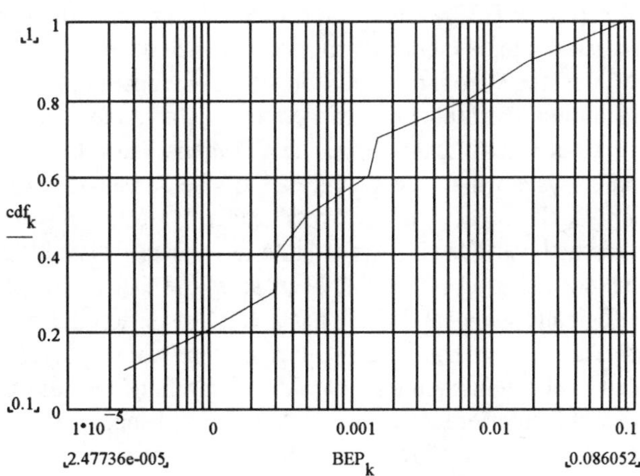

Figure 11.24 Cumulative distribution function of the BEP of all users. With N=31, number of users = 10, R_f=2, and E_b/N_0=15 dB. Average BEP = 0.01149.

Conclusions

This section introduced a multiuser receiver in combination with a groupwise SIC scheme for a DS-CDMA system. This scheme results in a considerable reduction of the hardware complexity of the receiver compared to the SIC scheme and other parallel IC schemes.

When the group size is not too large, the system performance parameters (BEP, E_b/N_0 and user capacity) are just marginally affected. However, a significant simplification of the hardware complexity is obtained when a group size of two is used.

Furthermore, a retransmission protocol has been suggested to increase system performance, because the average BEP is mainly determined by the weaker signals. Identification of the weaker signals is done by their chip sequences, which are assumed to be known.

11.4 COEXISTENCE OF CDMA AND TDMA

The question of coexistence of CDMA and TDMA systems is discussed in this section by presenting a performance analysis for a DS-CDMA system sharing a common spectrum with narrowband interferences. The performance is investigated in terms of the BEP of a DS-CDMA system overlaid by multiple BPSK interferers, and the BEP is evaluated as a function of BPSK and CDMA users. To enhance the performance of the receiver, a suppression filter is employed to reduce interference.

DS-CDMA systems are interesting for future personal radio communications for many reasons, including low probability of interception, multiple access, and interference rejection. One more attractive feature of using CDMA is that a CDMA network can be overlaid by another system that occupies a part of the spread-spectrum bandwidth. In fact, this will increase the capacity of the overall spectrum. Much research has been done on CDMA systems with various narrowband interference and related topics [23-44].

This bandwidth sharing is possible because of the fact that a CDMA waveform, spread over a sufficiently wide bandwidth, has a very low spectral density, in contrast to the narrowband user, which has a high spectral density in a small frequency band. However, such an application should be considered carefully, because CDMA users and narrowband users still interfere with each other. To enhance the performance of the CDMA receiver, one can consider employing a suppression filter matched to the interferences.

The objective of this section is to evaluate the performance of a BPSK direct-sequence spread-spectrum system which overlays multiple narrowband BPSK channels. The models are described and analyzed from the CDMA receiver's perspective. It is assumed that the system is operating over a frequency nonselective, slowly fading, Rayleigh channel with coherent demodulation, although the latter assumption might be a difficult job in a fading environment. Furthermore, we neglect the interference from adjacent cells. Because of the overlay situation, we will use a suppression filter to reduce the interference of the narrowband waveforms. The average bit error rate and the influence of various parameters on the performance are evaluated.

11.4.1 Model

The channel is a frequency nonselective Rayleigh fading channel with the random parameters b_i, t_i, and m_i as the gain, delay, and the phase of the ith signal at the receiver, respectively. The gain is an independent Rayleigh random variable with parameter $r=r_i=$

$E[b_i^2]/2$ for all i. The delay is also independent for each signal, and has a uniform distribution in $[0,T_b]$. Further, we assume that the phase m_i is an independent random variable, uniformly distributed in [0,2p]. First, the interference $J(t)$ is taken as one narrowband BPSK signal with an offset represented as:

$$J(t) = \sqrt{2J}\,j(t)\cos\left[2\pi\left(f_0 + \delta\varpi\right)t + \eta\right] \tag{11.46}$$

with dw as the offset between the carrier frequency f_0 of the CDMA system and the BPSK interferer, J as the interference power, and h as the phase. The information signal of the jamming is given by $j(t)$, which has the values -1 and +1. The parameter p is defined as the ratio between the interference bandwidth and the spread system bandwidth given by:

$$p \triangleq \frac{B_{interfering}}{B_{spreading}} \tag{11.47}$$

Finally, the received signal is given by

$$\begin{aligned} y(t) = {} & \sqrt{2P}\sum_{i=1}^{K}\beta_i b_i\left(t-\tau_i\right)PN\left(t-\tau_i\right)\cos\left(2\pi f_0 t+\phi_i\right)+\sqrt{2J}\,j(t) \\ & \cdot\cos\left[2\pi\left(f_o+\delta\omega\right)t+\eta\right]+n(t) \end{aligned} \tag{11.48}$$

where

$$\begin{aligned} \phi_i &= \theta_i - \mu_i - 2\pi f_0 \tau_i \\ P &= \frac{A_S^2}{2} \\ J &= \frac{A_{interference}^2}{2} \end{aligned} \tag{11.49}$$

and where P and K denote the transmitted power and the number of CDMA users. The pseudo-noise (PN) sequence is denoted by $PN(t)$. The amplitude of the transmitted CDMA signal and the interfering signal are denoted by A_S and $A_{interference}$, and $b_i(t)$ stands for the ith information signal with unity values and bit rate $1/T_b$, and q_i is the phase angle of the ith signal. The noise $n(t)$ is assumed to be additive white Gaussian noise (AWGN) with one-sided spectral density N_0. Notice that the interference is assumed unfaded. Figure 11.25 shows the DS-CDMA receiver considered in this section.

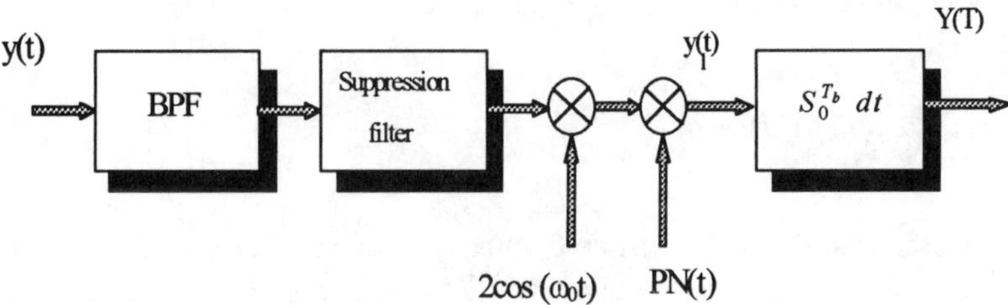

Figure 11.25 DS-CDMA receiver.

To simplify the performance analysis, we assume that the CDMA signals and the narrowband interference can pass through the bandpass filter (BPF) without distortion. The suppression filter is an adaptive double-sided transversal filter and is located between the BPF and the DS despreader. Its impulse response is:

$$h(t) = \sum_{m=-M1}^{M2} \alpha_m d(t - mT_c), \text{ with } \alpha_0 = 1 \tag{11.50}$$

where a_m, $M1$, and $M2$ stand for the mth tap, and the number of taps on the left and on the right sides of the center tap a_0. The parameter T_c denotes the duration time of one chip of the PN sequence. If $f_0 T_c$ is assumed to be an integer, the output of the suppression filter becomes

$$y_f(t) = \sum_{m=-M_1}^{M_2} \alpha_m y(t - mT_c) \tag{11.51}$$

The output of the demodulator gives rise to the decision variable $Y(l)$, in which l denotes the lth data bit of the reference user. This decision variable is represented by

$$Y(\lambda) = \int_{\lambda T_b}^{(\lambda+1)T_b} y_f(t) PN(t) 2\cos(2\pi f_0 t) dt \tag{11.52}$$

with $2pf_0$ as the recovered carrier frequency and T_b as the duration of one bit of the information signal. Without loss of generality, we can assume f_i and t_i equal zero. The decision variable $Y(l)$ is the sum of the desired signal, internal interferences, noise, narrowband interference, and multiple interference:

$$\sigma^2_{noise} = N_0 T_b \sum_{m=-M_1}^{M_2} \alpha^2_m \tag{11.53}$$

11.4.2 Bit Error Probability

According to the central limit theorem and the large number K of active users, we can approximate the sum of the noise, the narrowband interference term, and the multiple access interference terms by a Gaussian random variable. The variance of the signal can now be taken as the sum of the contributions due to the thermal noise, the narrowband interference, the multiple access interference, and the self-interference term

$$\begin{aligned}\sigma^2_{total} &= N_0 T_b \sum_{k=-M_1}^{M_2} \alpha^2_k + \frac{T_b^2 J}{L} \sum_{kl=-M_1}^{M_2} \sum_{k2=-M_1}^{M_2} \alpha_{kl}\alpha_{k2}\sigma_{interference}(k1-k2) \\ &+ \frac{(K-1)ST_b^2}{3L}\left(2\sum_{k=-M_1}^{M_2}\alpha^2_k + \sum_{k=-M_1}^{M_2}\alpha_k\alpha_{k+1}\right) + \frac{2T_b E_b}{L}\sum_{k=-M_1}^{-1}\left(2-\frac{M_1+k}{L}\right)\alpha^2_k\end{aligned} \tag{11.54}$$

Here L is the processing gain (number of chips per bit).

The variance of the multiple narrowband interference may be written as [24,27,34]

$$\sigma^2_{interference} = \frac{T_b^2 J}{L} \sum_{k=-\frac{G-1}{2}}^{\frac{G-1}{2}} \sum_{k1=-M_1}^{M_2} \sum_{k2=-M_1}^{M_2} \alpha_{k1}\alpha_{k2}\sigma^2_{interference,k}(k1-k2) \tag{11.55}$$

where G is the number of BPSK interferers symmetrically located on multiples of $2pDT_c$ around and on the carrier frequency of the CDMA system. Note that we assume that the interferers are totally uncorrelated with each other, so the cross-terms between the different interferers can be neglected. In an overlay TDMA system, the TDMA signals might show a certain correlation with each other.

The variance $\sigma_{interference,k}$ results in

$$\sigma^2_{interference,k}(z) = \int_{-1}^{1} R_{T_j}\left[(x-z)T_c\right]\cos\left[2\pi kq(x-z)\right]R_{T_c}(xT_c)\left(1-\frac{|x|}{L}\right)dx \tag{11.56}$$

where $q=GWT_c$ denotes the offset ratio.

Now, the SNR yields

$$\mathrm{SNR} = \left(\begin{array}{l} \dfrac{N_0}{<E_b>} \displaystyle\sum_{k=-M_1}^{M_2} a_k^2 + \dfrac{J}{LS} \sum_{k=-\frac{G-1}{2}}^{\frac{G-1}{2}} \sum_{k1=-M_1}^{M_2} \sum_{k2=-M_1}^{M_2} a_{k1} a_{k2} \sigma^2_{interference,k} (k1-k2) \\ + \dfrac{(K-1)}{3L} \left[\displaystyle\sum_{k=-M_1}^{M_2} 2a_k^2 + \sum_{k=-M_1}^{M_2} a_k a_k + 1 \right] + \dfrac{2}{L} \sum_{k=-M_1}^{-1} \left(2 - \dfrac{M_1+k}{L} \right) a_k^2 \end{array} \right)^{-1} \tag{11.57}$$

Finally, the probability of bit error is obtained by substituting the SNR in the BEP function [32,33]:

$$P_{\hat{e}} = \frac{1}{2}\left(1 - \sqrt{\frac{\mathrm{SNR}}{1+\mathrm{SNR}}} \right) \tag{11.58}$$

11.4.3 Filter Performance

Various techniques can be used to supplement the capability of a direct-sequence (DS) system to reject narrowband interference. With respect to the use of notch filters for this purpose, two techniques have received the most interest [30,31]: transform domain processing and adaptive filtering.

In this section we discuss adaptive filtering. Figure 11.26 shows the adaptive transverse filter structure. An adaptive filter for interference suppression uses a block of a received signal to estimate adaptively the interference. This estimate is subtracted from sample values of the desired signal plus interference, thereby cancelling the interference. The adaptive transversel filter multiplies each tap output by a weight except for the center tap. Since both the DS signal and the noise are wideband processes, their values cannot be predicted from their past values. On the other hand, if the interference is narrowband, the interference components of the tap outputs are correlated with each other and its future values can be predicted. The adaptive algorithm minimizes the filter output power by adjusting the weights, so that the interference is cancelled in the filter output, but the direct-sequence signal will be largely unaffected. There are three different criteria used to design the tap weights for the filter:

1. Whitening the entire received signal;
2. Whitening the noise and interference only;
3. Designing an infinite deep notch placed at the frequency location of the interfering tones.

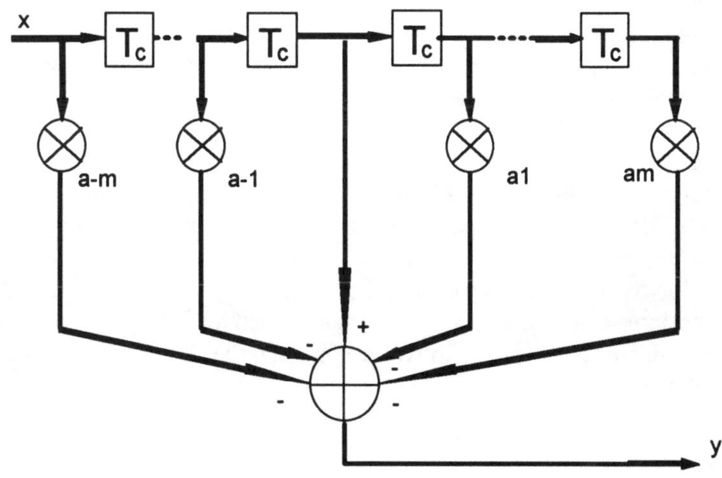

Figure 11.26 Double-sided filter.

In this subsection we only use criterion 1, which has been shown to provide better performance for a higher level of interference, due to the suboptimal balance achieved by the filter in terms of minimizing the degradation to the DS signal while maximizing the interference rejection. Note that for a large interferer, the performance of the system under any of these three interferers converges to the same result [24].

Consider the situation in the presence of multiple BPSK interferers. The autocovariance to any number of BPSK interferers is given by

$$R_y(lT_c) = \sum_{k=-\frac{G-1}{2}}^{\frac{G-1}{2}} R_{interference,k}(l) + P\delta(l) + \sigma_n^2\delta(l) \tag{11.59}$$

with G as the number of BPSK interferers and

$$R_{interference,k}(l) = J\,T_b^2(1-|l|\,p)\cos(2\pi klq) \tag{11.60}$$

11.4.4 Computational Results

Unless otherwise mentioned, the following parameters were held fixed in the results presented in this subsection: L=255, T_c=1, p=1/255, D=2/255 and J/S=20 dB.

The interference power is assumed to be constant for every BPSK interferer, and the interference-to-signal power ratio is set for every separate interferer, instead of all interfering power. The SNR is set to 1000 dB, which can be assumed to be infinite. Thus, it represents the best situation and optimistic results are obtained. The number of CDMA users K is taken as 10. Figures 11.27 and 11.28 depict the performances of the suppression filter with 3 and 9 taps, respectively, for different numbers of interferers. Because we use double-sided filters, the total number of taps is $2M$+1. Thus M=1 and 4 on each side stands for the total number of 3 and 9 taps, respectively. Note that the spacing between each interferer is constant, which means that for a larger number of interferers, the filter needs to be broader. So, we can assume the filter coefficients are calculated in the right way.

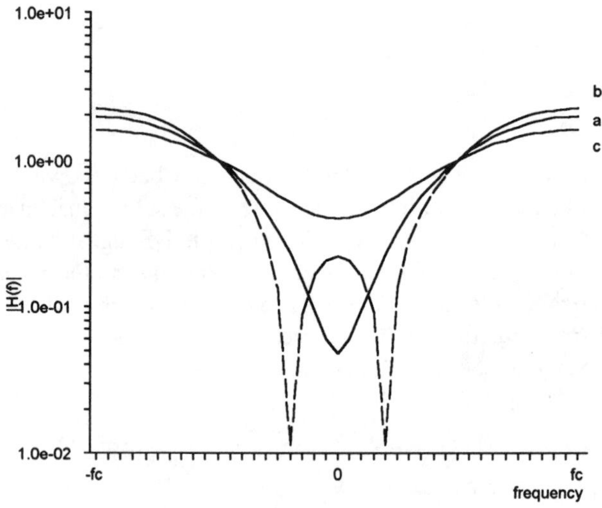

Figure 11.27 Filter with 3 taps for (a) 1 BPSK interfer, (b) 51 BPSK interferers, and (c) 101 BPSK interferers.

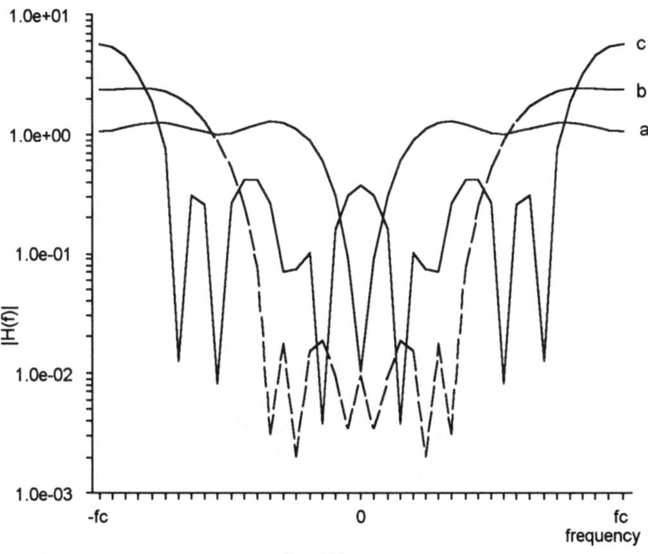

Figure 11.28 Filter with 9 taps for (a) 1 BPSK interferer, (b) 51 BPSK interferers, and (c) 101 BPSK interferers.

The computational results showed that for 1 interferer, the 9-tap filter has the steepest response, which is also valid for the situation with 101 interferers with a broader range. The 3-tap filter has an almost flat response for 101 interferers.

Using the analytical results from the previous section, we evaluated some numerical results for multiple BPSK interferers. In the first series of plots, self-interference is not taken into consideration. In Figure 11.29, the receiver performance is compared with multiple interfering BPSK signals for different interference-to-signal ratios. First, the coefficients of the suppression filter are set to zero, except for the center tap with the unity value.

It is clear that the performance with multiple BPSK interferers is less effective for higher interference-to-signal ratios. It is also obvious that the number of BPSK interferers increases the bit error probability.

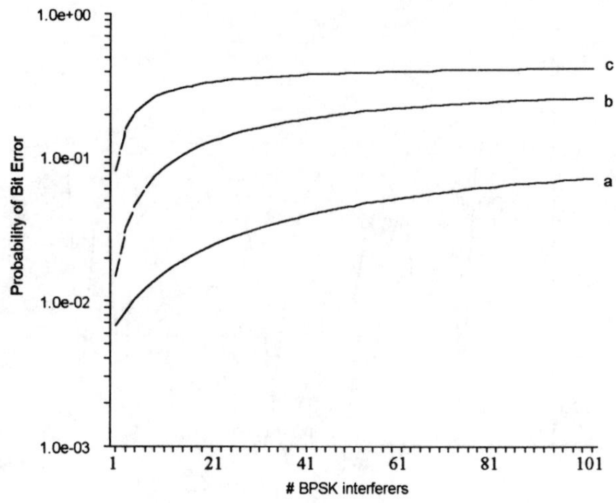

Figure 11.29 BEP versus number of BPSK interferers for 10 CDMA interferers: (a) *J/S*=0 dB, (b) 10 dB, and (c) 20 dB.

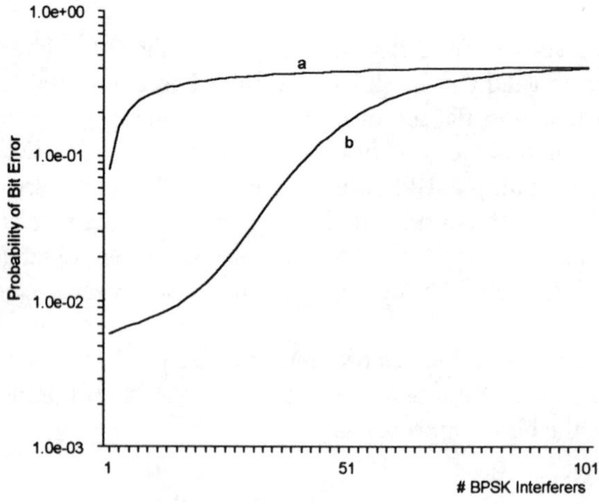

Figure 11.30 BEP versus number of BPSK interferers for *J/S*=20 dB (a) without and (b) with filter.

In Figure 11.30, the interference-to-signal ratio is fixed at 20 dB. The other parameters are held fixed as in Figure 11.29. The performance is plotted for a system using a double-sided filter with three taps. Note the better performance of the system using a suppression filter, and note that this effect decreases for a larger number of BPSK interferers. For more than 100 BPSK interferers, the suppression filter does not decrease the BEP significantly.

In Figure 11.31, system performance is plotted for a different number of CDMA users without a suppression filter. The interference-to-noise ratio is set at 0 dB. For an increasing number of CDMA users, the performance gets worse. Figure 11.31 should be compared with the results shown in Figure 11.32, which shows the bit error rate versus the number of other CDMA users. In Figure 11.32, each plot belongs to a different number of BPSK interferers. These graphs show that the performance of the system is a little bit more sensitive to the number of BPSK interferers than to the number of CDMA interferers.

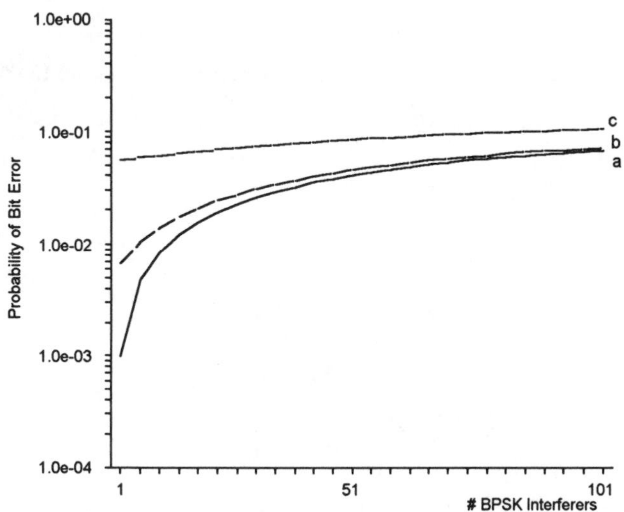

Figure 11.31 BEP versus number of BPSK interferers for (a) 1, (b) 10, and (c) 100 CDMA interferers.

In Figure 11.33, the system performance corresponding to a different number of taps on each suppression filter is compared to the situation without a filter. Again, the parameters are held fixed as in the previous figures. Note the better performance with an increasing number of taps. It is also shown that the improvement in performance for more taps decreases when using fewer taps in the filter.

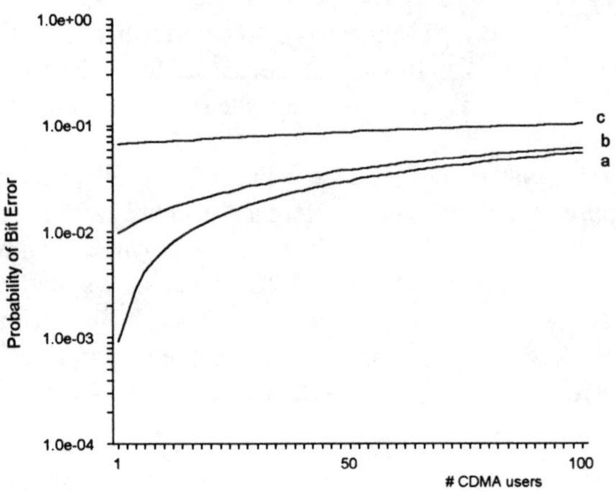

Figure 11.32 BEP versus number of CDMA interferers for (a) 1, (b) 10, and (c) 100 BPSK interferers.

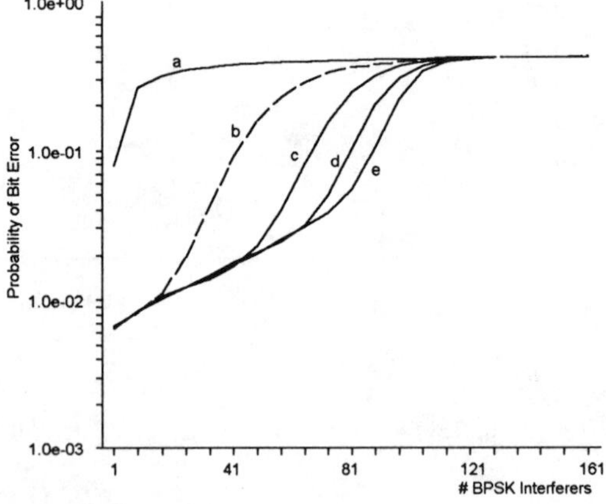

Figure 11.33 BEP versus number of BPSK interferers for (a) M=0, (b) 1, (c) 2, (d) 3, and (e) 5 taps per side.

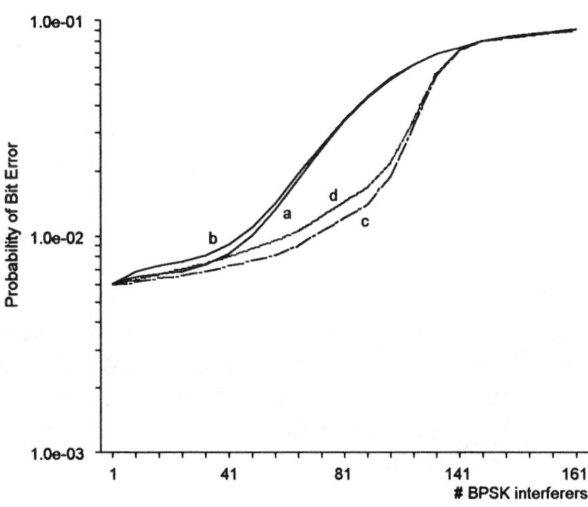

Figure 11.34 Performance for M=1 (a) without and (b) with self distortion, and for M=4 (c) without and (d) with self-distortion.

In Figure 11.34, self-interference due to distortion of the reference user is taken into consideration. In Figure 11.34, the results with and without self-distortion are compared. Note the bigger influence of self-distortion on the performance using a suppression filter with 9 taps. For a large number of BPSK interferers, the self-distortion can be neglected compared to the variance of the other signals.

11.4.5 Conclusions

In this section, we presented the results on the BEP performance of a DS-CDMA transverse system, employing an adaptive double-sided transversal filter, with an overlay of multiple BPSK interferers. It was shown that suppression filters increased the performance of a CDMA system overlaid with multiple BPSK interferers, although the filters do not show a significant improvement for a large number of BPSK interferers. It can also be concluded that the performance of a DS-CDMA system is shown to be a little bit more sensitive to the number of BPSK interferers than to the number of CDMA users. Finally, the results show that for a large number of BPSK channels, the internal interference of the reference signal is small compared to other distortion contributions.

11.5 JOINT DETECTION CDMA

Recently, the application of joint detection (JD) in CDMA systems has been studied in great detail; see, e.g., [45]-[54]. Important contributions were made by P.W. Baier, J.J. Blanz, P. Jung, A. Klein, M.M. Naßhan, A. Steil, and B. Steiner from the Research Group for RF Communications, University of Kaiserslautern, Germany, and also by G.K. Kaleh from the Département Communications, Ecole Nationale Supérieure des Télécommunications, Paris, France. The essence of these contributions is presented in this section.

In CDMA systems, several independent users are simultaneously active in the same frequency band, only discernible by different user-specific spreading codes. This situation is illustrated in Figure 11.35 for the uplink of a mobile radio system using CDMA with a number K users or mobile stations. Each data symbol of the data symbol sequence $\underline{\mathbf{d}}^{(k)}$ transmitted by user k is multiplied by the user-specific spreading code k, $k = 1...K$ [45,48,49]. The resulting signal of user k is transmitted over the time-varying multipath mobile radio channel. In the uplink, the mobile radio channels k, $k = 1...K$ users are in general different [48,49]. At the receiver of the base station, the superposition of the contributions of all K users appears. At the receiver, transmission over the time-varying multipath mobile radio channels entails both intersymbol interference (ISI) between the data symbols of one and the same user and multiple access interference (MAI) between data symbols of different users [48,49]. MAI is also termed interuser interference (IUI). The superposition signal is furthermore disturbed by a sequence $\underline{n}$ representing intercell interference and thermal noise [48,49]. The received sequence $\underline{e}$ containing samples at the chip rate has to be processed at the receiver by a data detection algorithm to determine estimates $\hat{\underline{d}}^{(k)}$ of the data symbol sequence $\underline{d}^{(k)}$, $k=1...K$, of all K users [45,48,49].

The conventional suboptimum single-user receiver is implemented as a bank of K matched filters or K Rake receivers, which consider each user as if it were the only one present. This is inefficient because MAI is treated as noise. In mobile radio systems using single-user receivers, measures such as very accurate power control combating fast fading, voice activity monitoring, cell sectorization, and softer handover are mandatory to obtain an acceptable performance. To reduce the requirement on the performance of these measures or even avoid some of these measures, the application of multiuser receivers, exploiting *a priori* knowledge about both ISI and MAI, is advisable. However, also when applying multiuser receivers, the above-mentioned measures can be beneficially deployed. The optimum multiuser receiver [55,56] is too complex to be implemented in third-generation mobile radio systems. Therefore, suboptimum multiuser receiver techniques were proposed and investigated; see, e.g., [45,48,49,52], [57]-[60]. These techniques can be subdivided into JD techniques [45,48,49,52,58] and IC techniques [15,16].

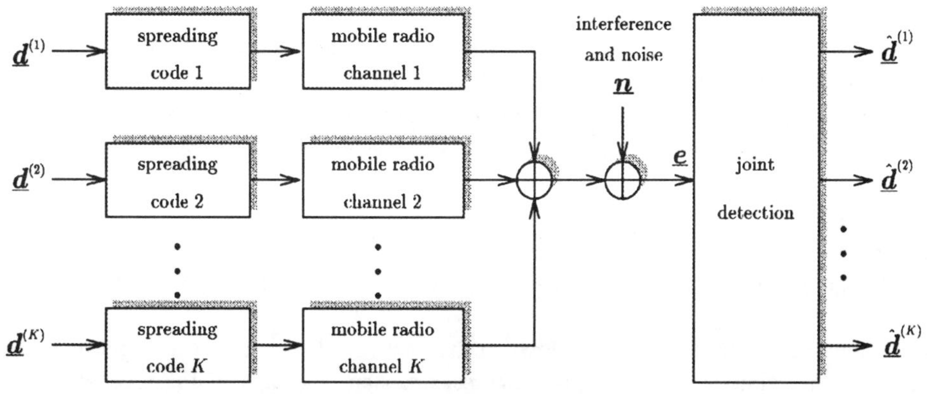

Figure 11.35 Discrete-time model for the uplink of a mobile radio system using CDMA with joint detection.

Concerning JD techniques, the four equalizers

- zero forcing block linear equalizer (ZF-BLE), [44,48,49,52];
- minimum mean square error block linear equalizer (MMSE-BLE), [48,49,52];
- zero forcing block decision feedback equalizer (ZF-BDFE), [48,49,52];
- minimum mean square error block decision feedback equalizer (MMSE-BDFE) [48,49,52]

were developed and extensively investigated. These four equalizers, which perform JD, are designed for multipath channels and burst transmission. The four equalizers are derived in what follows starting from the system model in Figure 11.35. The transmission in this system model can be described mathematically by the matrix-vector expression [45,48,49,52]:

$$\underline{e} = \underline{A}\,\underline{d} + \underline{n} \qquad (11.61)$$

with the received vector $\underline{e}$, which is a function of the combined data vector $\underline{d}$, the system matrix $\underline{A}$, and the interference and noise vector $\underline{n}$. The combined data vector $\underline{d}$ contains all transmitted data symbol sequences $\underline{d}^{(k)}$, k = 1...K users [45,48,49,52]. The system matrix $\underline{A}$ is determined by the spreading codes of all K users [45,48,49,52]. The mathematical description (11.61) is valid for the uplink as well as for the downlink, and for the case of one receiver antenna as illustrated in Figure 11.35 and for the case of more than one receiver antenna [45,48,49,52]. The derivation of (11.61) and the exact definitions of $\underline{e}$, $\underline{A}$, $\underline{d}$, and $\underline{n}$ can be found in [45,46,52] for the case of one receiver antenna and in [47,48,49] for the case of coherent receiver antenna diversity (CRAD).

Equation (11.61) represents a system of linear equations. At the receiver, this system has to be solved in order to determine an estimate $\hat{\underline{d}}$ of $\underline{d}$. To solve this system of linear equations, the following assumptions are prerequisites:

- The received vector $\underline{e}$;
- The spreading codes of all K users;
- The impulse responses of the mobile radio channels

and, thus, the system matrix $\underline{A}$ is known at the receiver. When using training sequences, the K mobile radio channel impulse responses can be estimated at the receiver by an algorithm for joint channel estimation which was described in, for example, [50,51]. The system of linear equations, (11.61), can be solved according to the criteria zero forcing (ZF) or minimum mean square error (MMSE), which lead to two different approaches for jointly determining an estimate $\hat{\underline{d}}$ of $\underline{d}$ [48,49,52]. Assuming AWGN of variance σ^2 and uncorrelated data, the ZF-BLE [45,48,49,52] yields the estimate

$$\hat{\underline{d}}_{ZF} = \left(\underline{A}^{*T}\underline{A}\right)^{-1}\underline{A}^{*T}\underline{e} \tag{11.62}$$

and the MMSE-BLE [45,48,49,52] gives the estimate

$$\hat{\underline{d}}_{MMSE} = \left(\underline{A}^{*T}\underline{A}+\sigma^2\mathrm{I}\right)^{-1}\underline{A}^{*T}\underline{e}. \tag{11.63}$$

The symbol $(\cdot)^{*T}$ denotes complex conjugation and transposition and **I** is the identity matrix. Equations (11.62) and (11.63) have the same structure. In the case of the MMSE-BLE, the variance σ^2 of the noise is taken into account, which requires that σ^2 be estimated at the receiver. The ZF-BLE for JD leading to (11.62) is a generalization of the single-user ZF equalizer. Also in the case of more than one receiver antenna and CRAD, (11.62) and (11.63) remain valid, as explained in [47-49]. Instead of the inversion of the matrices ($\underline{A}^{*T}\underline{A}$) and ($\underline{A}^{*T}\underline{A}+\sigma^2\mathbf{I}$) in (11.62) and (11.63), respectively, Cholesky decomposition or QR decomposition of these matrices can be applied, which considerably reduces computational complexity. Furthermore, when applying Cholesky decomposition, decision feedback (DF) can be introduced to improve performance without increasing the computational complexity [48,49,52]. Introducing DF in (11.62) and (11.63) leads to the ZF-BDFE or the MMSE-BDFE, respectively [48,49,52]. To avoid the impairing effect of error propagation when using DF, a method termed channel sorting has been proposed [48,49,52]. Details on the applications of Cholesky decomposition, DF including channel sorting, and the extension of the equalizers to the case of nonwhite noise are given in [52] and in [47-49] for the cases of one or more than one receiver antenna, respectively.

Concerning the computational complexity of the ZF-BLE, MMSE-BLE, ZF-BDFE, and MMSE-BDFE for JD, the most important results follow:

- All four equalizers require essentially the same computational complexity.
- The computational complexity required is independent of the size of the data symbol alphabet.

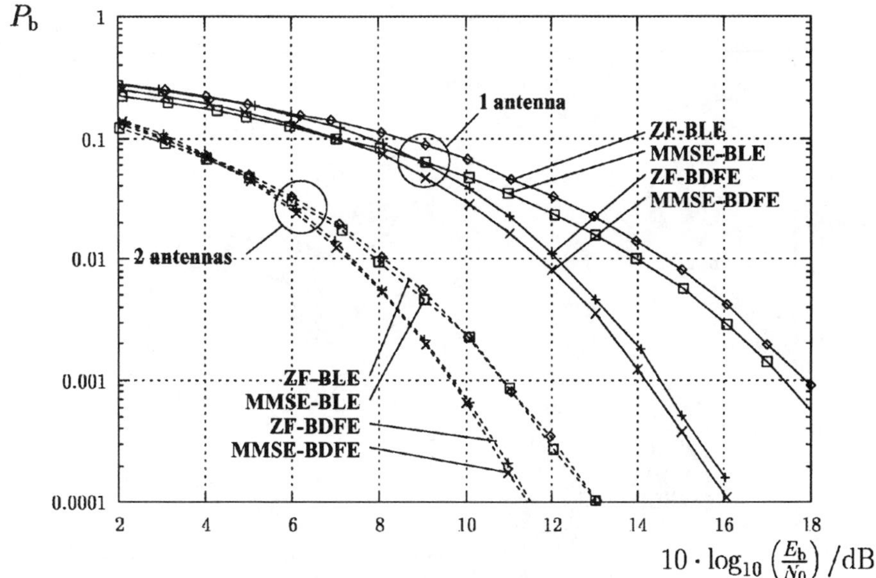

Figure 11.36 BEP performance of the ZF-BLE, MMSE-BLE, ZF-BDFE, and MMSE-BDFE equalizers; without FEC.

The computational complexity required for JD is larger than that required by conventional suboptimum single-use detectors. However, considering the potential of modern microelectronics, the computational complexity required for JD is already feasible today. Particularly in systems with a TDMA component, leading to a reduction of the number K of simultaneously active users, the computational complexity required for JD seems to be affordable; see [48].

The ZF-BLE, MMSE-BLE, ZF-BDFE, and MMSE-BDFE have been applied to the JD-CDMA system proposal described in [3,48,54,61]. The performance in terms of simulated bit error rate (BER) P_b versus SNR E_b/E_0 per bit and per receiver antenna of the ZF-BLE, MMSE-BLE, ZF-BDFE, and MMSE-BDFE for one and for two receiver antennas in the uplink is given in the following [48,61]. In Figure 11.36, cases without forward error correction (FEC) coding are considered. In Figure 11.37, cases with convolutional FEC coding with rate ½ and constraint length 5 are considered [53,54]. The system parameters used in the simulations are for example, given in, [48,53,54]. The Rayleigh fading multipath mobile radio channels have been modeled according to the COST 207 channel model bad urban (BU) area with a speed for the mobile stations of 30 km/h. A number of $K = 8$ mobile stations have been simultaneously active.

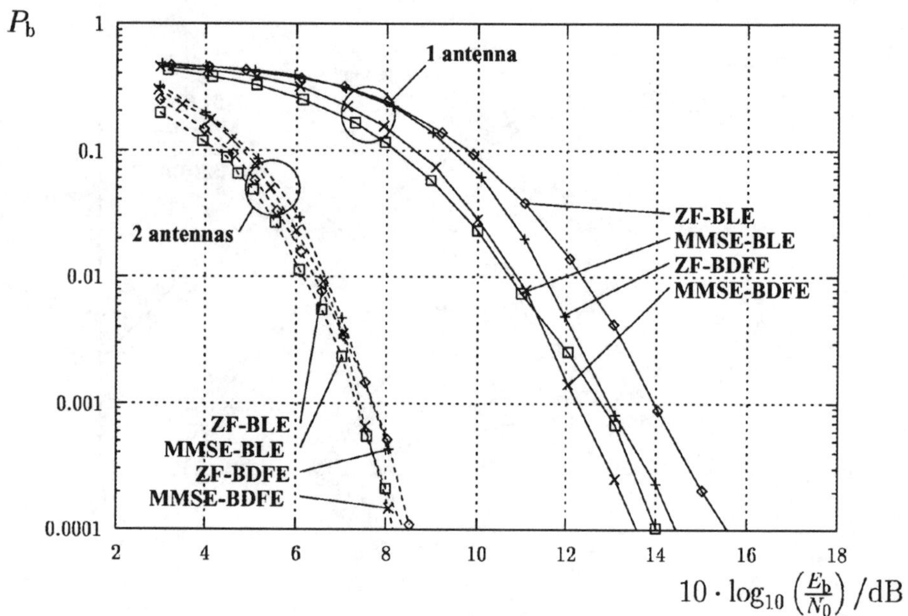

Figure 11.37 BEP performance of the ZF-BLE, MMSE-BLE, ZF-BDFE, and MMSE-BDFE equalizers; with FEC.

To obtain the results in Figures 11.36 and 11.37, channels have been estimated according to [50,51], so that the effect of imperfect channel estimation is included. More details about the simulations and more simulation results can be found in [48,61] for the cases without FEC and in [53,54] for the cases with FEC. The most important results are as follows [48,49,52,53,54]:

- The two equalizers ZF-BDFE and MMSE-BDFE provide better performance than the ZF-BLE and MMSE-BLE equalizers.
- The two MMSE equalizers, MMSE-BLE and MMSE-BDFE, perform slightly better than the two ZF equalizers, ZF-BLE and ZF-BDFE, respectively.
- CRAD, with two receiver antennas, facilitates the reduction of E_b/N_0 per receiver antenna by about 6 to 11 dB depending on the type of terrain as compared with the case of a single receiver antenna. Of this improvement, 3 dB are due to an energy gain; the additional improvement is due to diversity.
- When applying CRAD with, e.g., two receiver antennas, the performance difference of the four equalizers is smaller than in the case of one receiver antenna.
- When applying FEC, the performance difference of the four equalizers is smaller than when no FEC is applied.

11.6 FUTURE CDMA SCHEMES

To find the answer to the question "What will be the multiple access scheme for future wireless personal communication systems?," a number of new, hybrid, and superhybrid multiple access schemes in combination with CDMA have been proposed. They are multirate (MR-)CDMA, broadband (B-)CDMA, multicarrier (MC-)CDMA, multitone (MT-)CDMA, MC-FHCDMA, hybrid DS/SFH, hybrid CDMA/ISMA, hybrid CDMA/TDMA, hybrid CDMA/FDMA, superhybrid CD-CRMA, superhybrid joint detection (JD-)CDMA/TDMA, superhybrid CDMA/CRMA, and so on [36-46].

To provide details on each of the schemes, a separate book would have to be written. Here we briefly summarize multicarrier CDMA (MC-CDMA) [71,72].

11.6.1 Multicarrier CDMA (MC-CDMA)

This section presents the advantages and disadvantages of DS-CDMA and MC-CDMA systems in synchronous downlink mobile radio communication channels. Furthermore, the bit error probability performance is analyzed in frequency selective slow Rayleigh fading channels.

We theoretically derive the BEP lower bound for MC-CDMA systems, and propose a simple multiuser detection method. In the BEP analysis, we use the same multipath delay profiles for both DS-CDMA and MC-CDMA systems, and discuss the performance theoretically and by means of computer simulations. Finally, we theoretically prove that the time domain DS-CDMA Rake receiver is equivalent to the frequency domain MC-CDMA Rake receiver for the case of one user.

The DS-CDMA technique has been considered to be a candidate to support multimedia services in mobile communications, because it has its own capabilities to cope with the asynchronous nature of multimedia data traffic, to provide higher capacity over conventional access techniques such as TDMA and FDMA, and to combat the hostile channel frequency selectivity.

Recently, another CDMA technique based on a combination of CDMA and orthogonal frequency division multiplexing (OFDM) signaling has been reported in [1,2]. This technique is called the multicarrier CDMA (MC-CDMA) technique, and much attention has been paid to it, because it is potentially robust to the channel frequency selectivity with good frequency utilization efficiency.

In the BEP analysis of a DS-CDMA system, an independent fading characteristic at each received path is assumed; on the other hand, an independent fading characteristic at each subcarriers is considered in the BEP analysis of MC-CDMA system [1,2]. When the multipath channel is a wide sense stationary uncorrelated scattering (WSSUS) one, the assumption for the DS-CDMA system is correct, however, the assumption for the MC-CDMA-system is not correct. In the BEP analysis of the MC-CDMA system, we should take into account the frequency correlation function determined by the multipath delay profile of the channel, and in the BEP comparison, we should make a fair assumption for both DS-CDMA and MC-CDMA systems using the same channel frequency selectivity, that is, the same multipath delay profile. To the best of the authors' knowledge, no paper has been reported on the fair BEP comparison.

11.6.2 DS-CDMA and MC-CDMA Systems

DS-CDMA Systems

A DS-CDMA transmitter spreads the original signal using a given spreading code in the time domain (Figure 11.38). The capability of suppressing multiuser interference is determined by the cross-correlation characteristic of the spreading codes. Also, a frequency selective fading channel is characterized by the superimposition of several signals with different delays in the time domain [34]. Therefore, the capability of distinguishing one component from other components in the composite received signal (time resolution) is determined by the auto-correlation characteristic of the spreading codes.

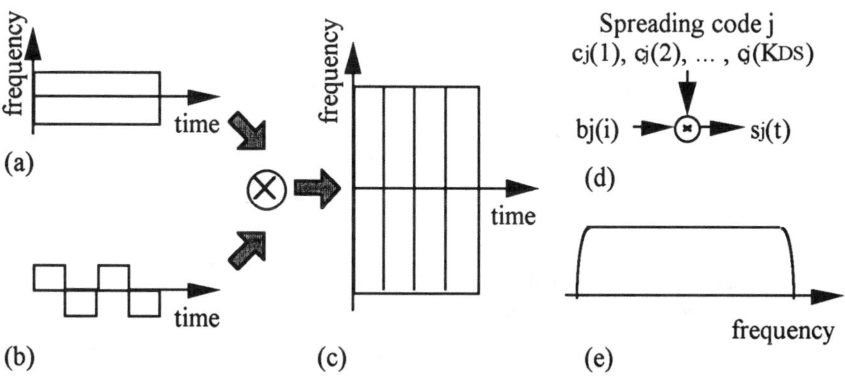

Figure 11.38 Concept of the DS-CDMA system: (a) Time-frequency representation of original signal; (b) Spreading code in time domain; (c) Time-frequency representation of DS-CDMA signal; (d) DS-CDMA transmitter of user j; (e) Power spectrum of transmitted signal.

Figure 11.38(d) shows the DS-CDMA transmitter of the jth user for the binary PSK modulation/coherent demodulation (CBPSK) scheme. The complex equivalent lowpass transmitted signal is written

$$s_j(t) = \sum_{i=-\infty}^{\infty} \sum_{k=1}^{K_{DS}} b_j(i) c_j(k) p(t - kT_c - iT_s), \tag{11.64}$$

where $b_j(i)$ and $c_j(k)$ are the ith information and the kth bit of the spreading code with length K_{DS} and chip duration T_c, respectively, T_s is the symbol duration, and $p(t)$ is the pulse waveform defined as:

$$p(t) = \begin{cases} 1 & 0 \le t \le T_c \\ 0 & \end{cases} \tag{11.65}$$

The BEP is determined by the path diversity strategy in the receiver, and the diversity order depends on how many fingers the Rake receiver employs. Usually, a diversity order of 1 (non-Rake), 2, or 3 is used depending on hardware limitations. Furthermore, when Nyquist filters are introduced in the transmitter and receiver for baseband pulse shaping, the Rake receiver may wrongly combine paths. This is because noise causing distortion in the autocorrelation characteristic often results in an incorrect correlation. Finally, it is difficult for the DS-CDMA Rake receiver to use all the received signal energy scattered in the time domain.

MC-CDMA Systems

The MC-CDMA transmitter spreads the original signal using a given spreading code in the frequency domain (Figure 11.39). It is crucial for multicarrier transmission to have frequency nonselective fading over each subcarrier. Therefore, if the original symbol rate is high enough to become subject to frequency selective fading, the signal first needs to be serial-to-parallel converted before spreading over the frequency domain. Also, in a synchronous downlink mobile radio communication channel, we can use the Hadamard Walsh codes as an optimum orthogonal code set, because we do not have to pay attention to the autocorrelation characteristic of the spreading codes.

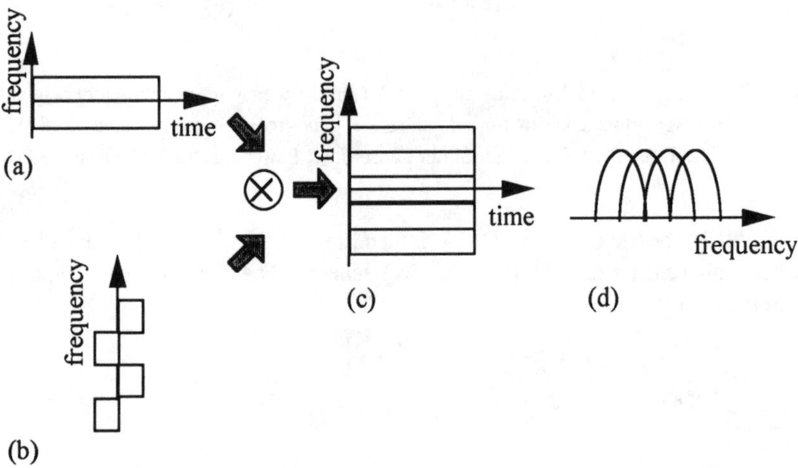

Figure 11.39 Concept of the MC-CDMA system: (a) Time-frequency representation of original signal; (b) Spreading code in frequency domain; (c) Time-frequency representation of MC-CDMA signal; (d) Power spectrum of transmitted signal.

Figure 11.40(a) shows the MC-CDMA transmitter of the jth user for the CBPSK scheme, where the input information sequence is converted into P parallel data sequences $[a_{j,1}(i), a_{j,2}(i), \ldots, a_{jP}(i)]$. The complex equivalent lowpass transmitted signal is written

$$s_j(t) = \sum_{i=-\infty}^{+\infty} \sum_{p=1}^{P} \sum_{m=0}^{K_{MC}-1} a_{j,p}(i) d_m^j p(t - iT_s) e^{j2\pi \Delta f (m + \frac{p-1}{P})t} \tag{11.66}$$

$$\Delta f = \frac{1}{T_s} \tag{11.67}$$

where $\{d_1^j, d_2^j, \ldots, d_{KMC}^j\}$ is the Hadamard Walsh code for the jth user (the length is K_{MC} and Δf is the subcarrier separation for $a_{j,p}(i)$. Also, the total number of subcarriers is $P \times K_{MC}$.

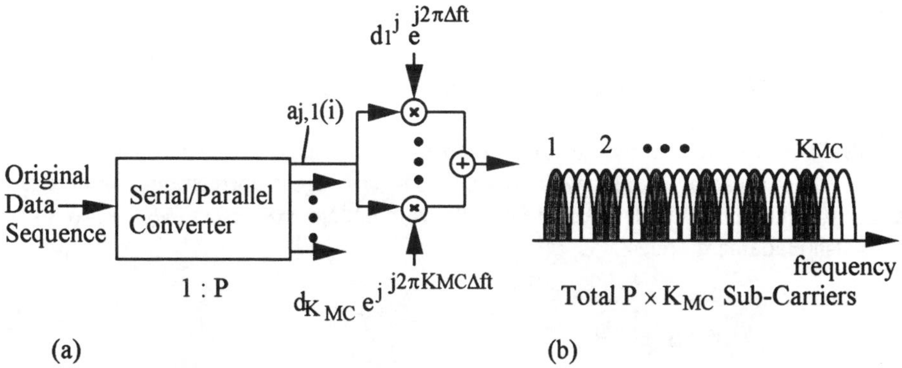

Figure 11.40 MC-CDMA transmitter of user j: (a) MC-CDMA transmitter of user j; (b) Power spectrum of transmitted signal.

Advantages of the MC-CDMA System over the DS-CDMA System

We define the following parameters:

- Transmission rate: R ($=1/T_s$) [bits/sec];
- Processing gain: K_{DS} (DS-CDMA), K_{MC} (MC-CDMA);
- Maximum number of users: M_{DS} (DS-CDMA), M_{MC} (MC-CDMA);
- Number of subcarriers: N ($=P \times K_{MC}$).

The required frequency bandwidth (main lobe) for the DS-CDMA system becomes

$$B_{DS} = 2 \cdot R \cdot K_{DS} \tag{11.68}$$

On the other hands, for the MC-CDMA system, it is

$$B_{MC} = R \cdot K_{MC} \cdot (N+1)/N \approx R \cdot K_{MC} \tag{11.69}$$

From (11.68) and (11.69),

$$K_{MC} = 2 \cdot K_{DS} \quad \text{if} \quad B_{MC} = B_{DS} \tag{11.70}$$

Equation (11.70) shows that the processing gain of the MC-CDMA system is twice as large as that of the DS-CDMA system for a given frequency bandwidth. Furthermore, the DS-CDMA system cannot accommodate K_{DS} users ($M_{DS} < K_{DS}$) because we need to choose the spreading codes with good auto- and cross-correlation characteristics carefully, while the MC-CDMA system can accommodate K_{MC} users ($M_{MC} = K_{MC}$) using the Walsh Hadamard codes:

$$M_{MC} > 2 \cdot M_{DS} \quad \text{if} \quad B_{MC} = B_{DS} \tag{11.71}$$

Disadvantages of the MC-CDMA System over the DS-CDMA System

At the MC-CDMA receiver, we have to make every effort in the fast Fourier transform (FFT) window position synchronization, the frequency offset compensation, and the coherent detection at each subcarrier. Also, the MC-CDMA transmitter requires a large input backoff in the amplifier, because it is very sensitive to nonlinear amplification.

We also point out that the BEP of multicarrier modulation itself is inferior to that of single-carrier modulation mainly due to the power loss associated with guard interval insertion [62-64].

11.6.3 BEP Analysis

Frequency Selective Slow Rayleigh Fading Channel Model

We assume a WSSUS channel model [34] with L received paths in the complex equivalent lowpass time-variant impulse response:

$$h(\tau;t) = \sum_{l=1}^{L} g_l(t)\delta(\tau - \tau_l). \tag{11.72}$$

where t and τ are the time and the delay, respectively, $\delta(t)$ is the Dirac delta function, $g_l(t)$ is the complex envelope of the signal received on the lth path, which is a complex

Gaussian random process with zero mean and variance σ_l^2, and τ_l is the propagation delay for the lth path.

Figure 11.41 shows the corresponding multipath delay profile given by

$$\begin{aligned}\phi_c &= \frac{1}{2}E[h^*(\tau;t)\cdot h(\tau;t)] \\ &= \sum_{l=1}^{L}\sigma_l^2\delta(\tau-\tau_l),\end{aligned} \tag{11.73}$$

where $E[\cdot]$ is the expectation.

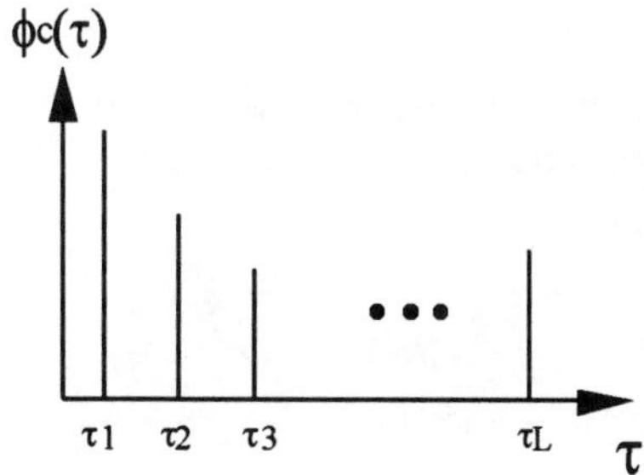

Figure 11.41 Multipath delay profile.

System Model

We consider a synchronous downlink communication channel, where the signal is transmitted in a burst format with a preamble and a postamble. In this section, we assume that the receiver can correctly estimate all the channel state information.

BEP of DS-CDMA Systems

The transmitted signal for a total of J users is written by

$$s(t) = \sum_{j=1}^{J} s_j(t) \tag{11.74}$$

The received signal through the frequency selective slow Rayleigh fading channel given by (11.72) is written

$$r(t) = \int_{-\infty}^{+\infty} s(t-\tau) * h(\tau;t)d\tau + n(t) \\ = \sum_{l=1}^{L} r_l(t) + n(t) \tag{11.75}$$

$$r_l(t) = \sum_{l=1}^{L} s(t-\tau_l) g_l(t) \tag{11.76}$$

where *n(t)* is the complex additive Gaussian noise. Defining $\mathbf{r_t}$ as the received signal vector, the time domain covariance matrix $\mathbf{M_t}$ is given by

$$\mathbf{r_t} = [r_1, r_2, \ldots, r_L]^T \tag{11.77}$$

$$\mathbf{M_t} = \frac{1}{2} E[\mathbf{r_t} \cdot \mathbf{r_t^T}], \\ = \begin{bmatrix} \sigma_1^2 & 0 & \cdots & 0 \\ 0 & \sigma_2^2 & & \vdots \\ \vdots & & \ddots & 0 \\ 0 & \cdots & 0 & \sigma_L^2 \end{bmatrix} \tag{11.78}$$

where T is the transpose. In the above equation, we assume a perfect autocorrelation characteristic for the spreading codes.

The BEP of a time domain *l*-finger DS-CDMA Rake receiver is uniquely determined by the eigenvalues of $\mathbf{M_t}$ (in this case, the eigenvalues are clearly σ_1^2, σ_2^2, ... , σ_L^2) [65]. For example, when σ_l^2 ($l=1, \ldots, L$) are different from each other, the BEP is expressed as

$$BEP_{DS} = \sum_{l=1}^{I} w_l \cdot \frac{1}{2}\left\{1 - \sqrt{\frac{\sigma_l^2 / N'}{1+\sigma_l^2 / N'}}\right\} \tag{11.79}$$

$$w_l = \frac{1}{\prod_{\substack{n=1\\n\neq l}}^{I}\left(1-\frac{\sigma_n^2}{\sigma_l^2}\right)} \tag{11.80}$$

$$E_b = \sum_{l=1}^{L} \sigma_l^2 \tag{11.81}$$

$$N' = N_0 + \frac{2(J-1)}{3K_{DS}} E_b \tag{11.82}$$

where E_b and N_0 are the signal energy per bit and the noise spectral density, respectively. Also, when σ_l^2 $(l=1, \dots, L)$ are all the same $(=\sigma^2)$ [37],

$$\begin{aligned} BEP_{DS} &= \left(\frac{1-\mu_{DS}}{2}\right)^I \sum_{l=1}^{I-1} \binom{I-1+l}{l} \left(1+\frac{\mu_{DS}}{2}\right)^l \\ \mu_{DS} &= \sqrt{\frac{\sigma_l^2 / N'}{1+\sigma_l^2 / N'}} \end{aligned} \tag{11.83}$$

Equations (11.79) and (11.83) are based on the Gaussian approximation for multiuser interference.

BEP of MC-CDMA Systems

The transmitted signal for a total of J users is written

$$s(t) = \sum_{j=1}^{J} s_j(t) \tag{11.84}$$

The received signal through the frequency selective slow Rayleigh fading channel is written

$$r(t)=\sum_{i=-\infty}^{+\infty}\sum_{p=1}^{P}\sum_{m=0}^{K_{MC}-1}\sum_{j=1}^{J} z_{m,p}a_{j,p}(i)d_m^j p(t-iT_s)e^{j2\pi\Delta f(m+\frac{p-1}{P})t}+n(t) \tag{11.85}$$

where $z_{m,p}$ is the complex received signal at the $(mP+p-1)^{th}$ subcarrier.

Figure 11.42 shows the MC-CDMA receiver of the j'^{th} user, where after the serial-to-parallel conversion using the FFT, the mth subcarrier component for the received data $a_{j,p}(i)$ is multiplied by the gain G_m and despreading code $d_m^{j'}$ to combine the energy of the received signal scattered in the frequency domain. The decision variable is given by (we can omit the subscriptions p and i without loss of generality)

$$D^{j'}(t)=\sum_{m=0}^{K_{MC}-1}\sum_{j=1}^{J} G_m d_m^{j'}\{z_m a_j d_m^j+n_m(t)\} \tag{11.86}$$

where $n_m(t)$ is the complex additive Gaussian noise at the mth subcarrier.

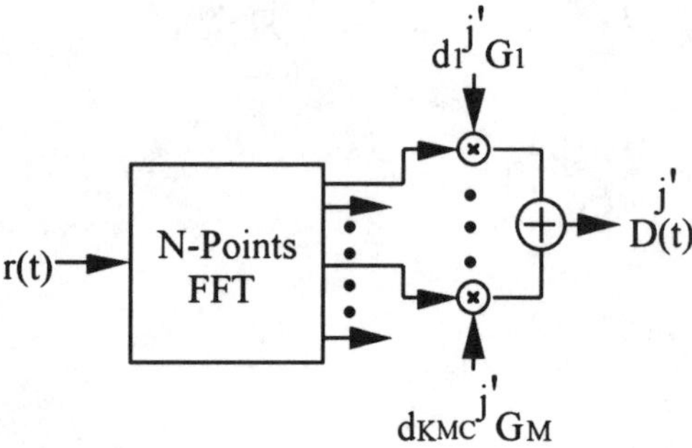

Figure 11.42 MC-CDMA receiver of the j'^{th} user.

The BEP is determined by a detection method such as the orthogonality restoring method [7] (a single-user detection scheme), the maximum ratio combining method (without a simultaneous other user) [34], and a multiuser detection method.

Orthogonality restoring single-user detection method

Choosing the gain G_m as

$$G_m = z_m^* / |z_m|^2 \tag{11.87}$$

the receiver can eliminate the multiuser interference perfectly. However, low-level subcarriers tend to be multiplied by the high gains, so the BEP degrades due to noise amplification [66,67] (a weak signal suppression method using a detection threshold is presented in [67]).

In this case, we need to resort to computer simulation to analyze the BEP, because the closed BEP expression has not been obtained.

Frequency domain rake receiver with no simultaneous other user

When there is no simultaneous other user, the frequency domain MC-CDMA Rake receiver based on the maximum ratio combining method in the frequency domain ($G_m = z_m^*$) can achieve the best BEP performance (the BEP lower bound) [34].

Defining $\mathbf{r}_f$ as the received signal vector, the frequency domain covariance matrix $\mathbf{M}_f$ is given by

$$\begin{aligned} \mathbf{r}_f &= [z_1, z_2, \cdots, z_{K_{MC}}]^T \\ \mathbf{M}_f &= \frac{1}{2} E\left[\mathbf{r}_f \cdot \mathbf{r}_f^T\right] = \{m_f^{a,b}\} \end{aligned} \tag{11.88}$$

$$m_f^{a,b} = \Phi_c((a-b)\Delta f) \tag{11.89}$$

where $m_f^{a,b}$ is the a-b element of $\mathbf{M}_f$, and $\Phi_C(\Delta f)$ is the spaced frequency correlation function defined as the Fourier transform of the multipath delay profile given by (11.73):

$$\Phi_c(\Delta f) = \int_{-\infty}^{+\infty} \phi_c(\tau) e^{-j2\pi\Delta f\tau} d\tau \tag{11.90}$$

Defining $\lambda_1, \lambda_2, \ldots, \lambda_{KMC}$ as the eigenvalues of $\mathbf{M}_f$, the BEP is given by a form similar to (10.79) or (10.83) [5]. For example, when λ_m ($m=1, \ldots, K_{MC}$) are different from each other,

$$BEP_{MC} = \sum_{m=1}^{K_{MC}} v_m \cdot \frac{1}{2}\left\{1 - \sqrt{\frac{\lambda_m / N_0}{1 + \lambda_m / N_0}}\right\} \tag{11.91}$$

$$v_m = \frac{1}{\prod_{\substack{n=1 \\ n \neq m}}^{K_{MC}} \left(1 - \frac{\lambda_n}{\lambda_m}\right)} \tag{11.92}$$

Also, when λ_m ($m=1, \ldots, K_{MC}$)are all the same (=λ),

$$BEP_{MC} = \left(\frac{1-\mu_{DS}}{2}\right)^{K_{MC}} \sum_{m=0}^{K_{MC}-1} \binom{M-1+m}{m} \left(1+\frac{\mu_{MC}}{2}\right)^m$$
$$\mu_{MC} = \sqrt{\frac{\lambda / N_0}{1+\lambda / N_0}} \tag{11.93}$$

In this case, we can theoretically evaluate the BEP.

Single Multiuser Detection Method

In this section, we propose a simple multiuser detection method. If the preamble contains information on the spreading codes used by the simultaneous users, any user can know it easily (although providing multiuser information in the downlink channel is questionable from the viewpoint of security). In this method, the user first estimates information for simultaneous other *J-1* users using the orthogonality restoring single-user detection method. After removing the interference component from the received signal, the user detects its own information using the frequency domain maximum ratio combining method. If the decisions for the other users are correct, this detection method can minimize the BEP. In this case, we need to resort to computer simulation to analyze the BEP.

11.6.4 Numerical Results

We assume the following system parameters to demonstrate BEP performance:

- $R = 3.0$ Mbits/sec;
- $K_{DS} = 31$ (Gold codes);
- $K_{MC} = 32$ (Hadamard Walsh codes).

BEP Performance of the DS-CDMA System

Figures 11.43, 11.44, and 11.45 show the BEP performance of the DS-CDMA system in frequency selective slow Rayleigh fading channels with 2-path uniform, 7-path uniform, and 7-path exponential multipath delay profiles, respectively. All the delay profiles have the same RMS delay spread τ_{RMS}=20ns.

Compared with the full-finger Rake receiver, the performance of a non-Rake receiver is poor. In particular, for the delay profiles where several delayed paths have the same average power as the first path, there is a large difference in the attainable BEP between the non-Rake and the full-finger Rake receivers. Also, as the number of users increases, the performance gradually degrades. For the 7-path uniform and 7-path exponential multipath delay profiles, the full-finger Rake receiver means a 7-finger Rake receiver. From the practical point of view, its realization could be difficult.

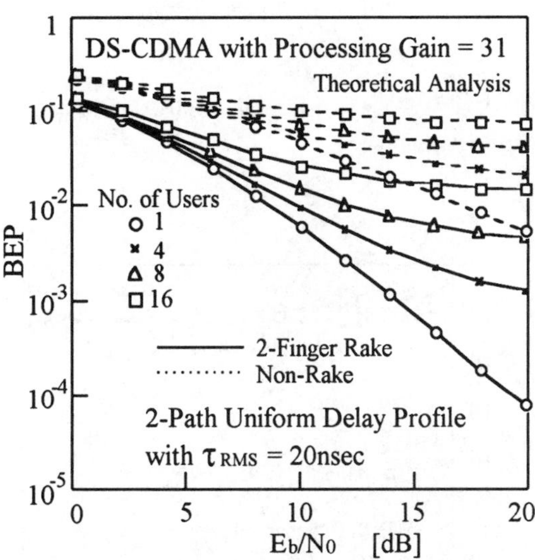

Figure 11.43 BEP of DS-CDMA system in frequency selective slow Rayleigh fading channel with 2-path uniform delay profile.

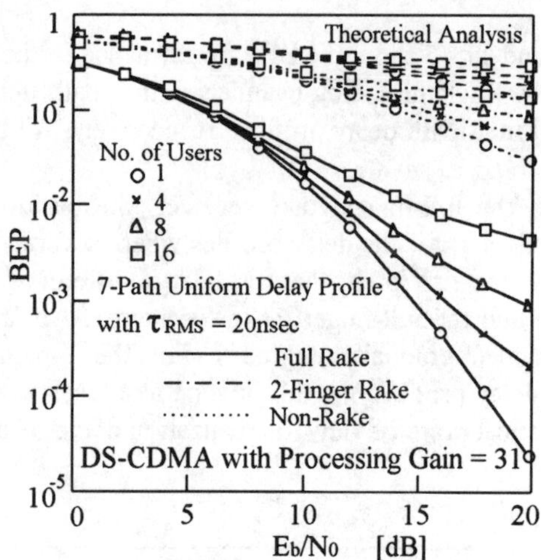

Figure 11.44 BEP of DS-CDMA system in frequency selective slow Rayleigh fading channel with 7-path uniform delay profile.

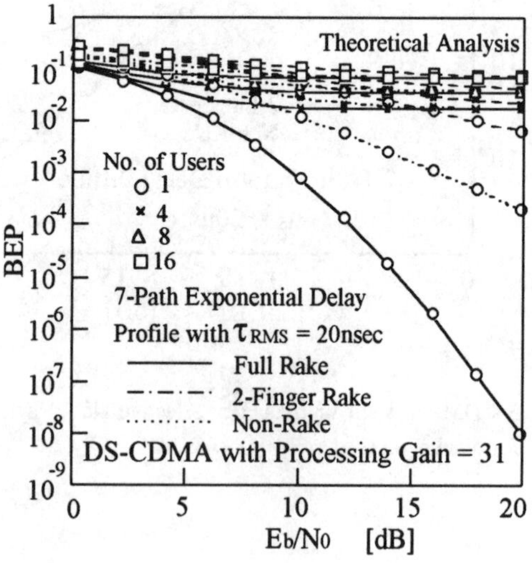

Figure 11.45 BEP of DS-CDMA system in frequency selective slow Rayleigh fading channel with 7-path exponential delay profile.

Design of MC-CDMA System

As the number of subcarriers (N) increases, the transmission performance becomes more sensitive to the time selectivity because the wider symbol duration is less robust to random FM noise. On the other hand, as N decreases, it becomes poor because the wider power spectrum of each subcarrier is less robust to the frequency selectivity. Therefore, there exists an optimum value in N to minimize the BEP [68].

Also, as the guard duration (Δ) increases, the transmission performance becomes poor because the signal transmission in the guard duration introduces the power loss. On the other hand, as Δ decreases, it becomes more sensitive to the frequency selectivity because the shorter guard duration is less robust to the delay spread. Therefore, there exists an optimum value in Δ to minimize the BEP [68].

In [68], it is shown that when the product of the maximum Doppler frequency f_D and the RMS delay spread τ_{RMS} introduced in the channel satisfies the following condition:

$$\tau_{RMS} \cdot f_D < 1.0 \times 10^{-6} \tag{11.94}$$

the multicarrier modulation scheme can achieve almost the same BEP performance as a single-carrier modulation scheme with equalization.

To determine the number of subcarriers (N) and guard duration (Δ), we assume the following channel parameters:

$f_D = 10$ Hz
$\tau_{RMS} = 20$ ns.

Figure 11.46 shows the optimum values of the number of subcarriers and guard period. The MC-CDMA system with N = 1024 and Δ = 100 ns can minimize the BEP. Consequently, the original information sequence is first converted into 32 parallel sequences (P=32), and then each sequence is mapped onto 32 subcarriers. The power loss associated with guard period insertion is negligible (the normalized guard duration is about 1%), because the above channel parameters satisfy (11.94).

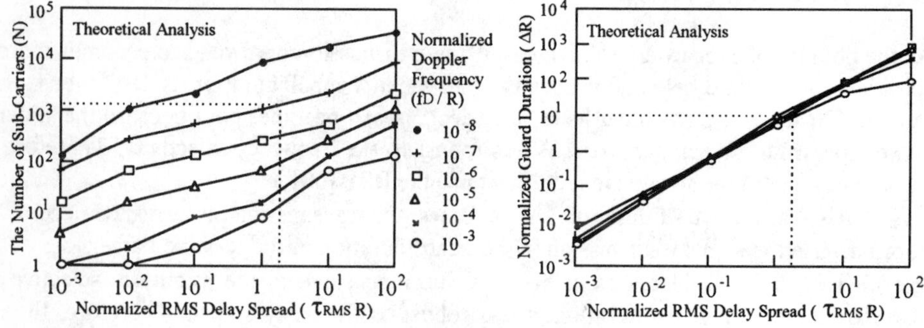

Figure 11.46 Design of the number of subcarriers and guard period.

BEP Performance of the MC-CDMA System

Figure 11.47 shows the BEP performance of the MC-CDMA system where the delay profiles are the same as those used in the analysis of the DS-CDMA-system. The frequency domain Rake receiver with no simultaneous other user can easily achieve the best performance, because it can effectively combine the energy of a received signal scattered in the frequency domain using a lot of subcarriers.

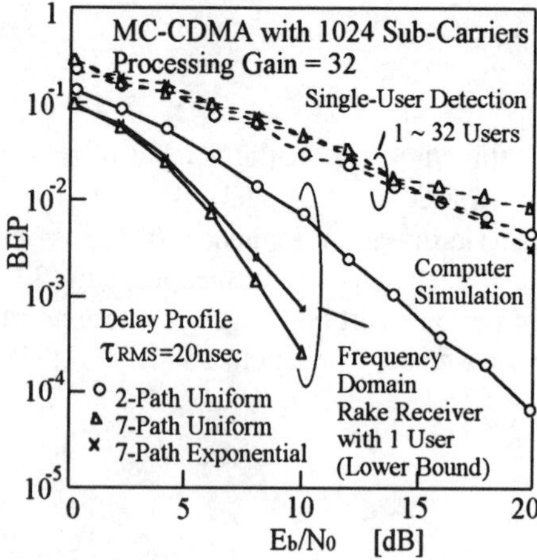

Figure 11.47 BEP of the MC-CDMA system for single-user detection and frequency domain Rake methods in frequency selective slow Rayleigh fading channels.

On the other hand, the performance of the orthogonality restoring single-user detection method is poor, although it is insensitive to the number of users. Among three delay profiles, the performance in the 2-path uniform delay profile is slightly better, because there is less distortion in the frequency domain.

Figure 11.48 shows the BEP performance of the proposed multiuser detection method for the 2-path uniform delay profile.

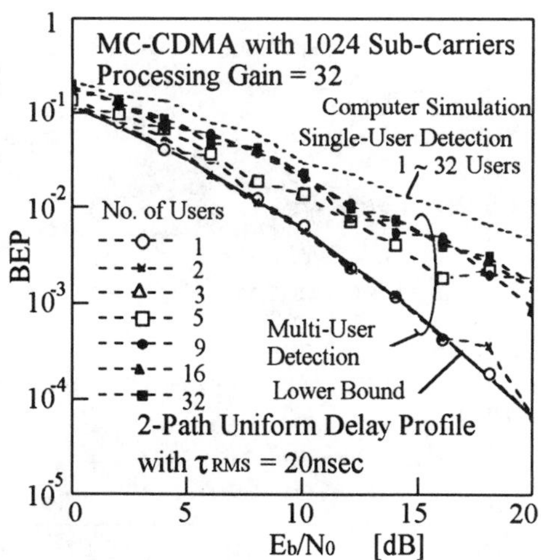

Figure 11.48 BEP of the MC-CDMA system for multiuser detection method in frequency selective slow Rayleigh fading channel with 2-path uniform delay profile.

The proposed multiuser detection method is simple and can improve the BEP as compared with the single-user detection method. However, the performance gradually degrades as the number of users increases. If more sophisticated (but more complicated) multiuser detection methods such as Wiener filtering detection [69], maximum-likelihood detection [7], and the decorrelating interference canceler [70] are employed, performance can be improved.

BEP Comparison of the DS-CDMA and MC-CDMA Systems

Figure 11.49 shows the BEP comparison for the DS-CDMA and MC-CDMA systems. The best performance of the MC-CDMA system agrees well with that of the DS-CDMA system. This implies that the frequency domain Rake receiver is equivalent to the time domain Rake receiver for the case of no simultaneous other user. The next section theoretically proves the equivalence of the frequency domain and time domain Rake receivers.

For the performance with 16 users, the MC-CDMA-system with the single-user detection method does not work well (but it is insensitive to the number of users). Also, the performance of the DS-CDMA system with a full-finger Rake receiver is not very good (it becomes worse as the number of users increases).

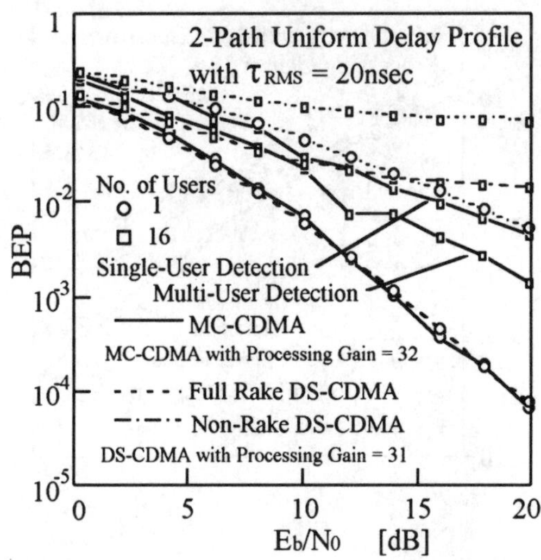

Figure 11.49 BEP comparison of the DS-CDMA and MC-CDMA systems in frequency selective slow Rayleigh fading channel with 2-path uniform delay profile.

The performance of the DS-CDMA system with a non-Rake receiver is much worse than that of the MC-CDMA system. Therefore, a Rake receiver could be necessary for the DS-CDMA system.

11.6.5 Equivalence of Time Domain DS-CDMA Rake Receiver and Frequency Domain MC-CDMA Rake Receiver

As shown in Section 11.6.3, the BEP of the Rake receiver is determined by the eigenvalues of the channel covariance matrix. Therefore, for a multipath delay profile, if the frequency domain covariance matrix given by (11.88) has all the same eigenvalues as the time domain covariance matrix given by (11.78), the BEP of a frequency domain Rake receiver is the same as that of a time domain Rake receiver.

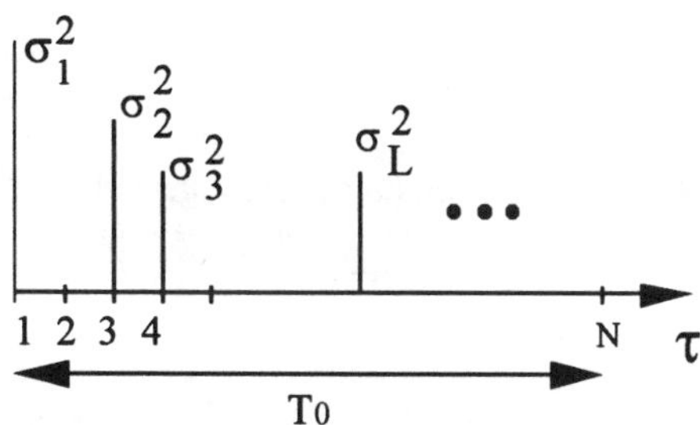

Figure 11.50 Multipath delay profile.

In the FFT, the frequency resolution is determined by the observation period. Therefore, for multicarrier modulation with N subcarriers, an N-point discrete Fourier transform (DFT) is required in the symbol duration T_O. Assume the following $N \times N$ time domain covariance matrix with time resolution of T_O/N, for example, for the multipath delay profile shown in Figure 11.50:

$$\mathbf{M}_t^{'} = \begin{bmatrix} \sigma_1^2 & 0 & \cdots & \cdots & \cdots & 0 \\ 0 & 0 & & & & \vdots \\ \vdots & & \sigma_1^2 & & & \vdots \\ \vdots & & & \sigma_3^2 & & \vdots \\ \vdots & & & & \ddots & 0 \\ 0 & \cdots & \cdots & \cdots & 0 & \ddots \end{bmatrix} \tag{11.95}$$

where the nonzero eigenvalues of $\mathbf{M'}_t$ are $\sigma_1^2, \sigma_2^2, \ldots, \sigma_L^2$.

Using $\mathbf{M'}_t$, the frequency domain covariance matrix is written by

$$\mathbf{M}_t^{'} = \mathbf{W}\mathbf{M}_t^{'}\mathbf{W}^{*} \tag{11.96}$$

where $\mathbf{W}$ is the $N \times N$ DFT matrix given by

$$\mathbf{W} = \{w^{i,j}\}$$
$$w^{i,j} = e^{j2\pi\frac{ij}{N}} \tag{11.97}$$

We define $\mathbf{r}_l$ as the eigenvector corresponding to the eigenvalue σ_l^2:

$$M_t' r_l = \sigma_l^2 r_l \qquad (l = 1,2,\cdots,L). \tag{11.98}$$

Also, we define $\mathbf{z}_l$ as

$$\mathbf{z}_l = \mathbf{W}\mathbf{r}_l \qquad (l = 1,2,\cdots,L). \tag{11.99}$$

Now, we can theoretically prove that the frequency domain covariance matrix has all the same eigenvalues as the time domain covariance matrix as follows:

$$\begin{aligned} \mathbf{M}_t' &= \mathbf{W}\mathbf{M}_t'\mathbf{W}^*\mathbf{W}\mathbf{r}_l \\ &= \mathbf{W}\mathbf{M}_t'\mathbf{W}\mathbf{r}_l \\ &= \mathbf{W}\sigma_l^2\mathbf{r}_l \\ &= \sigma_l^2\mathbf{W}\mathbf{r}_l \\ &= \sigma_l^2\mathbf{z}_l. \end{aligned} \tag{11.100}$$

The above equation clearly shows that the eigenvalues of $\mathbf{M'}_f$ are σ_1^2, σ_2^2, ... , σ_L^2.

Also, we can see that the assumption of independent fading characteristic at each subcarrier implies a frequency selective fading at each subcarrier as long as we employ OFDM signaling, because it requires N paths uniformly scattered in the symbol duration.

11.6.6 Conclusions

This section discussed the advantages and disadvantages of DS-CDMA and MC-CDMA systems, and analyzed the BEP performance in given frequency selective Rayleigh fading channels.

The MC-CDMA system can accommodate more users than the DS-CDMA system for a given frequency bandwidth. On the other hand, it has to make an extra effort in terms of FFT window synchronization at receiver and linear amplification at the transmitter.

For the DS-CDMA system, a Rake receiver is necessary to improve the BEP in frequency selective fading channels. However, from the hardware limitation, it is difficult to use all the received signal energy scattered in the time domain.

The MC-CDMA system can easily combine the signal energy scattered in the frequency domain. However, a multiuser detection method is also necessary.

In the best BEP comparison, there is no difference between the DS-CDMA and MC-CDMA systems. This chapter also theoretically showed the equivalence of a time domain DS-CDMA Rake receiver and a frequency domain MC-CDMA Rake receiver for the case of no simultaneous other user. Therefore, if the DS-CDMA system does not work well for a given channel condition and a given system condition, in other words, a given frequency selectivity and a given processing gain, the MC-CDMA system can be attractive, even at the sacrifice of cost, for subcarrier synchronization, because it can effectively combine the signal energy scattered in the frequency domain for any frequency selectivity and any processing gain using a number of subcarriers.

REFERENCES

[1] R. Kohno, P.B. Rapajic and B.S. Vucetic, "An overview of adaptive techniques for interference minimization in CDMA-systems," *Wireless Personal Comm.*, Vol. 1, pp. 3-22, January/February 1994.

[2] R.K. Sukdeo, H.S. Misser and R. Prasad, "A CDMA-system with multiuser interference cancellation for indoor wireless communications," *Proceedings IEEE ISSSTA'96,* Mainz, Germany, 22-25 September 1996.

[3] F. van der Wijk, G.J.M. Janssen and R. Prasad, "Groupwise successive interference cancellation in a DS/CDMA-systems," *Proceedings PIMRC'95*, Toronto, Canada, pp. 742-746, 27-29 September 1995.

[4] P. Jung, J. Blanz, M. Nabhan and P.W. Baier, "Simulation of the uplink of JD-CDMA mobiel radio systems with coherent receiver antenna diversity," *Wireless Personal Comm.*, Vol. 1, pp. 61-89, January/February 1994.

[5] M. Nabhan and P. Jung, "New results on the application of antenna diversity and turbo-codes in a JD-CDMA mobile radio system," *Proceedings PIMRC'94*, The Hague, The Netherlands, pp. 524-528, September 1994.

[6] A. Klein, B. Steiner and A. Steil, "Known and novel diversity approaches in a JD-CDMA-systems concept developed within COST 231," *Proceedings PIMRC'95*, Toronto, Canada, 22-29 September 1995.

[7] H. van Roosmalen, J. Nijhof and R. Prasad, "Performance analysis of a hybrid CDMA/ISMA protocol for indoor wireless computer communications," *IEEE J. Selected Areas Comm.*, Vol. 12, pp. 909-916, June 1994.

[8] R. Prasad, R.G.A. Rooimans, L. Vandendorpe and F. Çakmak, "Packet switched hybrid DS/SFH CDMA network with B-, Q- and DPSK modulation in an indoor Rician fading environment," *Proceedings GLOBECOM'94*, San Francisco, USA, pp. 69-73, November/December 1994.

[9] R. Prasad, "An overview of TU Delft Research on CDMA," *Proceedings IEEE ISSSTA '94*, Oulu, Finland, pp. 319-324, June 1994.

[10] R. Prasad and J. Nijhof, "Multiple access protocols for terrestrial cellular and satellite systems," *Proceedings 2nd Joint COST 227/231 Workshop on Mobile and Personal Communications*, Firenze, Italy, 20-21 April 1995.

[11] P.W. Baier, "Flexible hybrid multiple access schemes for 3rd generation mobile radio systems," *Proceedings 2nd Joint COST 227/231 Workshop on Mobile and Personal Communications*, Firenze, Italy, 20-21 April 1995.

[12] J.K. Holnes, *Coherent Spread-Spectrum Systems*, John Wiley & Sons, New York, 1982.

[13] F.M. Gardner, *Phaselock Techniques*, 2nd ed., John Wiley & Sons, New York, 1979.

[14] W.J. Weber, "Performance of phase-locked loops in the presence of fading communication channels," *IEEE Trans. Comm.*, Vol. 24, pp. 487-499, May 1976.

[15] B.W. 't Hart, R.D.J. van Nee and R. Prasad, "Bit error probability degradation due to code tracking errors in spread-spectrum communication systems," *Proceedings PIMRC'95*, Toronto, Canada, pp. 1016-1020, 27-29 September 1995.

[16] R.D.J. van Nee, "Spread-spectrum code and carrier synchronization errors caused by multipath and interference," *IEEE Trans. Aerospace Electron. Syst.*, Vol. 29, No. 4, pp. 1359-1365, October 1993.

[17] W.J. Weber, "Performance of phase-locked loops in the presence of fading communication channels," *IEEE Trans. Comm.*, Vol. 24, pp. 487-499, May 1976.

[18] P. Patel and J. Holzman, "Analysis of a simple successive interference cancellation scheme in a DS/CDMA-system," *IEEE J. Selected Areas Comm.*, Vol. 12, No. 5, pp. 796-807, June 1994.

[19] R. Kohno, H. Imai, M. Hatori and S. Pasupathi, "Combination of an adaptive array antenna and a canceller of interference for direct-sequence spread-spectrum multiple access system," *IEEE J. Selected Areas Comm.*, Vol. 8, No. 4, pp. 675-682, May 1990.

[20] Y.C. Yoon, R. Kohno and H. Imai, "Spread-spectrum multi-access system with co-channel interference cancellation for multipath fading channels," *IEEE J. Selected Areas Comm.*, Vol. 11, No. 7, pp. 1067-1075, May 1992.

[21] J.S. Lehnert and M.B. Pursley, "Multipath diversity reception of spread-spectrum multiple-access communications," *IEEE Trans. Comm.*, Vol. COM-35, No. 11, pp. 1189-1198, November 1987.

[22] H.A. David, *Order Statistics*, Wiley, New York, 1981.

[23] M.B.K. Widjaja, H.S. Misser and R. Prasad, "Performance analysis of an overlay CDMA-system with multiple BPSK interferers," *Proceedings IEEE ISSSTA '94*, Oulu, Finland, pp. 510-514, July 1994.

[24] R.A. Iltis and L.B. Milstein, "Performance analysis of narrowband interference rejection techniques in DS spread-spectrum systems," *IEEE Trans. Comm.*, Vol. COM-32, No. 11, pp. 1169-1177, November 1984.

[25] E. Masry and L.B. Milstein, "Performance of DS spread-spectrum receiver employing interference-suppression filters under a worst-case jamming condition," *IEEE Trans. Comm.*, Vol. COM-34, No. 1, pp. 13-21, January 1986.

[26] R.L. Pickholz, L.B. Milstein and D.L. Schilling, "Spread-spectrum for mobile communications," *IEEE Trans. Vehicular Technol.*, Vol. 40, No. 2, pp. 313-322, May 1991.

[27] J. Wang and L.B. Milstein, "Applications of suppression filters for CDMA overlay situations," *Proceedings ICC'92 (IEEE),* Chicago, USA, pp. 0245-0249, June 1992.

[28] L.B. Milstein, S. Davidovici and D. Schilling, "The effect of multi-tone interfering signals on a direct-sequence spread-spectrum communication system," *IEEE Trans. Comm.*, Vol. COM-30, No. 3, pp. 436-446, March 1982.

[29] D.L. Schilling, L.B. Milstein, R.L. Pickholz and R.W. Brown, "Optimization of the processing gain of an M-ary direct sequence spread-spectrum communication system," *IEEE Trans. Comm.*, Vol. COM-28, pp. 1389-1398, August 1980.

[30] D. Torrieri, *Principles of Secure Communication Systems*, 2nd. ed., Artech House, Norwood Mass., 1992.

[31] L.B. Milstein, "Interference rejection techniques in spread-spectrum communications," *Proceedings of the IEEE*, Vol. 76, No. 6, pp. 657-671, June 1988.

[32] H. Yueming, S.F. Lei, P. Das and G.J. Saulnier, "Suppression of narrowband jammers in a DS spread-spectrum receiver using modified adaptive filtering technique," *IEEE Global Telecommunications Conference & Exhibition*, Florida, pp. 0540-0545, November-December 1988.

[33] L. Sadiq and A.H. Aghvami, "Performance of a coded hybrid spread-spectrum communiction system in the presence of partial band noise and multiple access interference," *IEEE Military Communications Conference (MILCOM)*, San Diego, California, pp. 0817-0821, October 1988.

[34] J.G. Proakis, *Digital Communications*, 2nd edition, McGraw-Hill, New York, 1989.

[35] A. Carlson, *Communication Systems*, McGraw-Hill, New York, 1986.

[36] A. Baier, "Multi-rate DS-CDMA: a promising access technique for third-generation mobile radio systems," *Proceedings PIMRC'93*, Yokohama, Japan, pp. 114-118, September 1993.

[37] D.L. Schilling and L.B. Milstein, "Broadband CDMA for indoor and outdoor personal communications," *Proceedings PIMRC'93*, Yokohama, Japan, pp. 104-105, September 1993.

[38] N. Yee and J.P.M.G. Linnartz, "Wiener filtering of multicarrier CDMA in Rayleigh fading channel," *Proceedings PIMRC'94 and WCN,* The Hague, The Netherlands, pp. 253-257, September 1994.

[39] O. van de Wiel and L. Vandendorpe, "Adaptive equalization structures for multitone CDMA-systems," *Proceedings PIMRC'94 and WCN*, The Hague, The Netherlands, pp. 253-257, September 1994.

[40] Special issue on Code Division Multiple Access Networks I, *IEEE J. Selected Areas Comm.*, Vol. 12, May 1994.

[41] Special issue on Code Division Multiple Access Networks II, *IEEE J. Selected Areas Comm.*, Vol. 12, June 1994.

[42] *Proceedings IEEE ISSSTA'94*, Volume 1 and 2, Oulu, Finland, June 1994.

[43] Special issue on Multicarrier Communications, *Wireless Personal Comm.*, Vol. 2, 1995.

[44] Special issue on Signal Separation and Interference Cancellation for PIMRC, *Wireless Personal Comm.*, Vol. 2, 1995.

[45] A. Klein and. P.W. Baier, "Simultaneous cancellation of cross interference and ISI in CDMA mobile radio communications," *Proceedings of the IEEE International Symposium on Personal, Indoor and Mobile Radio Communications (PIMRC'92)*, Boston, pp. 118-122, 1992.

[46] A. Klein and P.W. Baier, "Linear unbiased data estimation in mobile radio systems applying CDMA," *IEEE J. Selected Areas Comm.*, Vol. 11, pp. 1058-1066, 1993.

[47] P. Jung, J. Blanz and P.W. Baier, "Coherent receiver antenna diversity for CDMA mobile radio systems using joint detection," *Proceedings of the IEEE International Symposium on Personal, Indoor and Mobile Radio Communications (PIMRC'93*), Yokohama, Japan, pp. 488-492, 1993.

[48] P. Jung, J. Blanz, M. Naßhan and P.W. Baier, "Simulation of the uplink of JD-CDMA mobile radio systems with coherent receiver antenna diversity," *Wireless Personal Comm.*, Vol. 1, pp. 61-89, 1994.

[49] P. Jung and J. Blanz, "Joint detection with coherent receiver antenna diversity in CDMA mobile radio systems," *IEEE Trans. Vehicular Technol.*, Vol. 44, pp. 76-88, 1995.

[50] B. Steiner and P. Jung, "Optimum and suboptimum channel estimation for the uplink of CDMA mobile radio systems with joint detection," *European Trans. Telecomm. Related Technol.*, Vol. 5, pp. 39-50, 1994.

[51] B. Steiner and P.W. Baier, "Low cost channel estimation in the uplink receiver of CDMA mobile radio systems," *Frequenz,* Vol. 47, pp. 292-298, 1993.

[52] A. Klein, G.K. Kaleh and P.W. Baier, "Equalizers for multiuser detection in code division multiple access mobile radio systems," *Proceedings of the IEEE International Conference on Vehicular Technology (VTC'94),* Stockholm, Sweden, pp. 762-766, 1994.

[53] P. Jung, M.M. Naßhan and J. Blanz, "Application of Turbo-codes to a CDMA mobile radio system using joint detection and antenna diversity," *Proceedings of the IEEE International Conference on Vehicular Technology (VTC'94)* Stockholm, Sweden, pp. 770-774, 1994.

[54] M Naßhan and P. Jung, "New results on the application of antenna diversity and Turbo-codes in a JD-CDMA mobile radio system," *Proceedings of the IEEE International Symposium on Personal, Indoor and Mobile Radio Communications (PIMRC'94),* The Hague, The Netherlands, pp. 524-528, 1994.

[55] W. van Etten, "Maximum likelihood receiver for multiple channel transmission systems," *IEEE Trans. Comm.*, Vol. 24, pp. 276-283, 1976.

[56] S. Verdú, "Minimum probability of error for asynchronous Gaussian multiple-access channels," *IEEE Trans. Info. Theory*, Vol. 32, pp. 85-96, 1986.

[57] R. Lupas and S. Verdú, "Near-far resistance of multiuser detectors in asynchronous channels," *IEEE Trans. Comm.*, Vol. 38, pp. 496-508, 1990.

[58] Z. Zvonar, "Multiuser detection in asynchronous CDMA frequency-selective fading channels," accepted for publication in *Wireless Personal Comm.*

[59] R.S. Mowbray, R.D. Pringle and P.M. Grant, "Increased CDMA-system capacity through adaptive cochannel interference regeneration and cancellation," *IEE Proceedings-I*, Vol. 139, pp. 515-524, 1992.

[60] R. Kohno, P.B. Rapajic and B.S. Vucetic, "An overview of adaptive techniques for interference minimization in CDMA-systems," *Wireless Personal Comm.*, Vol. 1, pp. 3-21, 1994.

[61] M. Naßhan, P. Jung, A. Steil and P.W. Baier, "On the effects of quantization, nonlinear amplification and band-limitation in CDMA mobile radio systems using joint detection," *Proceedings of the 5th Annual International Conference on Wireless Communications (Wireless'93),* Calgary, Calgary, pp. 173-186, 1993.

[62] N. Yee, J-P. Linnartz and G. Fettweis, "Multicarrier CDMA in indoor wireless radio networks", *Proceedings of IEEE PIMRC'93*, Yokohama, Japan, pp. 109-113, 1993.

[63] K. Fazel and L. Papke, "On the performance of convolutionally-coded CDMA/OFDM for mobile communication systems," *Proceedings of IEEE PIMRC'93*, Yokohama, Japan, pp. 468-472, 1993.

[64] H. Sari and I. Jeanclaude, "An analysis of orthogonal frequency-division multiplexing for mobile radio applications," *Proceedings of 1994 IEEE VTC'94*, Stockholm, Sweden, pp. 1635-1639, 1994.

[65] P. Monsen, "Digital transmission performance on fading dispersive diversity channels," *IEEE Trans. Commun.*, Vol. COM-21, pp. 33-39, 1973.

[66] K. Fazel, "Performance of CDMA/OFDM for mobile communication system," *Proceedings of IEEE ICUPC'93*, Ottawa, Canada, pp. 975-979, 1993.

[67] N. Yee and J-P. Linnartz, "Controlled equalization of multicarrier CDMA in an indoor Rician fading channel," *Proceedings of IEEE VTC'94*, Stockholm, Sweden, pp. 1665-1669, 1994.

[68] S. Hara, M. Mouri, M. Okada and N. Morinaga, "Transmission performance analysis of multicarrier modulation in frequency selective fast Rayleigh fading channel," accepted for publication in *Wireless Personal Comm.*

[69] N. Yee and J-P. Linnartz, "Wiener filtering of multicarrier CDMA in a Rayleigh fading channel," *Proceedings of IEEE PIMRC'94*, The Hague, The Netherlands, pp. 1344-1347, 1994.

[70] Y. Bar-Ness, J-P. Linnartz and X. Lin, "Synchronous multiuser multicarrier CDMA communication system with decorrelating interference canceler," *Proceedings of IEEE PIMRC'94*, The Hague, The Netherlands, pp. 184-188, 1994.

[71] S. Hara, T.H. Lee and R. Prasad, "BER comparison of DS CDMA and MC CDMA for frequency selective fading channels," *Proceedings of 7th Tyrrhenian Workshop,* Viareggio, Italy, 11-14 September 1995.

[72] R. Prasad and S. Hara, "An overview of multicarrier CDMA," *Proceedings ISSSTA '96,* Mainz, Germany, 22-25 September 1996.

About the Author

Ramjee Prasad was born in Babhnaur (Gaya), Bihar, India, on July 1, 1946. He received a B.Sc. (Eng.) from Bihar Institute of Technology, Sindri, India, and a M.Sc. (Eng.) and Ph.D. degrees from Birla Institute of Technology (BIT), Ranchi, India, in 1968, 1970 and 1979, respectively.

He joined BIT as a senior research fellow in 1970 and became associate professor in 1980. While he was with BIT, he supervised many research projects in the area of microwave and plasma engineering. During 1983-1988 he was with the University of Dar es Salaam (UDSM), Tanzania, where he became a professor of telecommunications at the Department of Electrical Engineering in 1986. At UDSM he was responsible for the collaborative project "Satellite Communications for Rural Zones" with Eindhoven University of Technology, The Netherlands. Since February 1988, he has been with the Telecommunications and Traffic-Control Systems Group, Delft University of Technology (DUT), The Netherlands, where he is actively involved in the area of personal indoor and mobile radio communications (PIMRC). He has published over 200 technical papers. His current research interest lies in wireless networks, packet communications, multiple access protocols, adaptive equalizers, spread-spectrum CDMA systems, and multimedia communications. He is currently involved in the European ACTS project FRAMES (Future Radio Wideband Multiple Access System) as a project leader of DUT.

He has served as a member of advisory and program committees for several IEEE international conferences. He has also presented keynote speeches, invited papers, and tutorials on PIMRC at various universities, technical institutions, and IEEE conferences. He is also a member of a working group of European cooperation in the field of scientific and technical research (COST-231) for a project dealing with "Evolution of Land Mobile Radio (Including Personal) Communications" as an expert for The Netherlands.

He is listed in the U.S. *Who's Who in the World*. He was organizer and interim chairman of the IEEE Vehicular Technology/Communications Society Joint Chapter, Benelux Section. He is now the elected chairman of the joint chapter. He is also founder of the IEEE Symposium on Communications and Vehicular Technology (SCVT) in Benelux and he was the Symposium Chairman of SCVT'93 and SCVT'95. He is the coordinating editor and one of the editors-in-chief of a Kluwer international journal on *Wireless Personal Communications* and also a member of the editorial board of other international journals including *IEEE Communications Magazine* and *IEE Electronics and Communications Engineering Journal*. He was the technical program chairman of the PIMRC '94 International Symposium, held in The Hague, The Netherlands, during September 19-23, 1994, and also of the Third Communication Theory Mini-Conference held in conjunction with GLOBECOM '94, San Francisco, California, November 27-30, 1994. He is a fellow of IEE, a fellow of the Institution of Electronics & Telecommunication Engineers, a senior member of IEEE, and a member of the New York Academy of Sciences and of NERG (The Netherlands Electronics and Radio Society).

Index

The Artech House Telecommunications Library

Vinton G. Cerf, Series Editor

Access Networks: Technology and V5 Interfacing, Alex Gillespie

Advanced High-Frequency Radio Communications, Eric E. Johnson, Robert I. Desourdis, Jr., et al.

Advanced Technology for Road Transport: IVHS and ATT, Ian Catling, editor

Advances in Computer Systems Security, Vol. 3, Rein Turn, editor

Advances in Telecommunications Networks, William S. Lee and Derrick C. Brown

Advances in Transport Network Technologies: Photonics Networks, ATM, and SDH, Ken-ichi Sato

An Introduction to International Telecommunications Law, Charles H. Kennedy and M. Veronica Pastor

Asynchronous Transfer Mode Networks: Performance Issues, Second Edition, Raif O. Onvural

ATM Switches, Edwin R. Coover

ATM Switching Systems, Thomas M. Chen and Stephen S. Liu

Broadband: Business Services, Technologies, and Strategic Impact, David Wright

Broadband Network Analysis and Design, Daniel Minoli

Broadband Telecommunications Technology, Byeong Lee, Minho Kang and Jonghee Lee

Cellular Mobile Systems Engineering, Saleh Faruque

Cellular Radio: Analog and Digital Systems, Asha Mehrotra

Cellular Radio: Performance Engineering, Asha Mehrotra

Cellular Radio Systems, D. M. Balston and R. C. V. Macario, editors

CDMA for Wireless Personal Communications, Ramjee Prasad

Client/Server Computing: Architecture, Applications, and Distributed Systems Management, Bruce Elbert and Bobby Martyna

Communication and Computing for Distributed Multimedia Systems, Guojun Lu

Community Networks: Lessons from Blacksburg, Virginia, Andrew Cohill and Andrea Kavanaugh, editors

Computer Networks: Architecture, Protocols, and Software, John Y. Hsu

Computer Mediated Communications: Multimedia Applications, Rob Walters

Computer Telephone Integration, Rob Walters

Convolutional Coding: Fundamentals and Applications, Charles Lee

Corporate Networks: The Strategic Use of Telecommunications, Thomas Valovic

Digital Beamforming in Wireless Communications, John Litva and Titus Kwok-Yeung Lo

Digital Cellular Radio, George Calhoun

Digital Hardware Testing: Transistor-Level Fault Modeling and Testing, Rochit Rajsuman, editor

Digital Switching Control Architectures, Giuseppe Fantauzzi

Digital Video Communications, Martyn J. Riley and Iain E. G. Richardson

Distributed Multimedia Through Broadband Communications Services, Daniel Minoli and Robert Keinath

Distance Learning Technology and Applications, Daniel Minoli

EDI Security, Control, and Audit, Albert J. Marcella and Sally Chen

Electronic Mail, Jacob Palme

Enterprise Networking: Fractional T1 to SONET, Frame Relay to BISDN, Daniel Minoli

Expert Systems Applications in Integrated Network Management, E. C. Ericson, L. T. Ericson and D. Minoli, editors

FAX: Digital Facsimile Technology and Applications, Second Edition, Dennis Bodson, Kenneth McConnell and Richard Schaphorst

FDDI and FDDI-II: Architecture, Protocols, and Performance, Bernhard Albert and Anura P. Jayasumana

Fiber Network Service Survivability, Tsong-Ho Wu

Future Codes: Essays in Advanced Computer Technology and the Law, Curtis E. A. Karnow

Guide to Telecommunications Transmission Systems, Anton A. Huurdeman

A Guide to the TCP/IP Protocol Suite, Floyd Wilder

Implementing EDI, Mike Hendry

Implementing X.400 and X.500: The PP and QUIPU Systems, Steve Kille

Inbound Call Centers: Design, Implementation, and Management, Robert A. Gable

Information Superhighways Revisited: The Economics of Multimedia, Bruce Egan

Integrated Broadband Networks, Amit Bhargava

International Telecommunications Management, Bruce R. Elbert

International Telecommunication Standards Organizations, Andrew Macpherson

Internetworking LANs: Operation, Design, and Management, Robert Davidson and Nathan Muller

Introduction to Document Image Processing Techniques, Ronald G. Matteson

Introduction to Error-Correcting Codes, Michael Purser

An Introduction to GSM, Siegmund Redl, Matthias K. Weber and Malcom W. Oliphant

Introduction to Radio Propagation for Fixed and Mobile Communications, John Doble

Introduction to Satellite Communication, Bruce R. Elbert

Introduction to T1/T3 Networking, Regis J. (Bud) Bates

Introduction to Telephones and Telephone Systems, Second Edition, A. Michael Noll

Introduction to X.400, Cemil Betanov

LAN, ATM, and LAN Emulation Technologies, Daniel Minoli and Anthony Alles

Land-Mobile Radio System Engineering, Garry C. Hess

LAN/WAN Optimization Techniques, Harrell Van Norman

LANs to WANs: Network Management in the 1990s, Nathan J. Muller and Robert P. Davidson

Minimum Risk Strategy for Acquiring Communications Equipment and Services, Nathan J. Muller

Mobile Antenna Systems Handbook, Kyohei Fujimoto and J.R. James, editors

Mobile Communications in the U.S. and Europe: Regulation, Technology, and Markets, Michael Paetsch

Mobile Data Communications Systems, Peter Wong and David Britland

Mobile Information Systems, John Walker

Networking Strategies for Information Technology, Bruce Elbert

Packet Switching Evolution from Narrowband to Broadband ISDN, M. Smouts

Packet Video: Modeling and Signal Processing, Naohisa Ohta

Personal Communication Networks: Practical Implementation, Alan Hadden

Personal Communication Systems and Technologies, John Gardiner and Barry West, editors

Practical Computer Network Security, Mike Hendry

Principles of Secure Communication Systems, Second Edition, Don J. Torrieri

Principles of Signaling for Cell Relay and Frame Relay, Daniel Minoli and George Dobrowski

Principles of Signals and Systems: Deterministic Signals, B. Picinbono

Private Telecommunication Networks, Bruce Elbert

Radio-Relay Systems, Anton A. Huurdeman

RF and Microwave Circuit Design for Wireless Communications, Lawrence E. Larson

The Satellite Communication Applications Handbook, Bruce R. Elbert

Secure Data Networking, Michael Purser

Service Management in Computing and Telecommunications, Richard Hallows

Smart Cards, José Manuel Otón and José Luis Zoreda

Smart Highways, Smart Cars, Richard Whelan

Successful Business Strategies Using Telecommunications Services, Martin F. Bartholomew

Super-High-Definition Images: Beyond HDTV, Naohisa Ohta, Sadayasu Ono and Tomonori Aoyama

Television Technology: Fundamentals and Future Prospects, A. Michael Noll

Telecommunications Technology Handbook, Daniel Minoli

Telecommuting, Osman Eldib and Daniel Minoli

Telemetry Systems Design, Frank Carden

Teletraffic Technologies in ATM Networks, Hiroshi Saito

Toll-Free Services: A Complete Guide to Design, Implementation, and Management, Robert A. Gable

Transmission Networking: SONET and the SDH, Mike Sexton and Andy Reid

Troposcatter Radio Links, G. Roda

Understanding Emerging Network Services, Pricing, and Regulation, Leo A. Wrobel and Eddie M. Pope

Understanding GPS: Principles and Applications, Elliot D. Kaplan, editor

Understanding Networking Technology: Concepts, Terms and Trends, Mark Norris

UNIX Internetworking, Second Edition, Uday O. Pabrai

Videoconferencing and Videotelephony: Technology and Standards, Richard Schaphorst

Wireless Access and the Local Telephone Network, George Calhoun

Wireless Communications in Developing Countries: Cellular and Satellite Systems, Rachael E. Schwartz

Wireless Communications for Intelligent Transportation Systems, Scott D. Elliot and Daniel J. Dailey

Wireless Data Networking, Nathan J. Muller

Wireless LAN Systems, A. Santamaría and F. J. López-Hernández

Wireless: The Revolution in Personal Telecommunications, Ira Brodsky

Writing Disaster Recovery Plans for Telecommunications Networks and LANs, Leo A. Wrobel

X Window System User's Guide, Uday O. Pabrai

For further information on these and other Artech House titles, contact:

Artech House
685 Canton Street
Norwood, MA 02062
617-769-9750
Fax: 617-769-6334
Telex: 951-659
email: artech@artech-house.com

Artech House
Portland House, Stag Place
London SW1E 5XA England
+44 (0) 171-973-8077
Fax: +44 (0) 171-630-0166
Telex: 951-659
email: artech-uk@artech-house.com

WWW: http://www.artech-house.com